# Topics in Applied Physics  Volume 12

# Topics in Applied Physics   Founded by Helmut K. V. Lotsch

# Turbulence

Edited by P. Bradshaw

With Contributions by
P. Bradshaw   T. Cebeci   H.-H. Fernholz
J. P. Johnston   B. E. Launder   J. L. Lumley
W. C. Reynolds   J. D. Woods

Second Corrected and Updated Edition

With 47 Figures

Springer-Verlag Berlin Heidelberg GmbH 1978

PETER BRADSHAW

Department of Aeronautics, Imperial College of Science and Technology,
University of London, London SW7 2BY, Great Britain

ISBN 978-3-540-08864-6      ISBN 978-3-540-35805-3 (eBook)
DOI 10.1007/978-3-540-35805-3

Library of Congress Cataloging in Publication Data. Main entry under title: Turbulence. (Topics in applied physics; v. 12) Bibliography: p. Includes index. 1. Turbulence. I. Bradshaw, Peter. TA357.T87   1978   620.1'064   78-18822

© Springer-Verlag Berlin Heidelberg 1976 and 1978
Originally published by Springer-Verlag Berlin Heidelberg New York in 1978

Typesetting, Brühlsche Universitätsdruckerei, Lahn-Gießen
2153/3130-543210

# Preface to the Second Edition

The first edition has sold out so quickly that a corrected and updated version, rather than a completely revised second edition, seemed appropriate—even in so rapidly developing a field as turbulence. By reprinting in paperback we have been able to produce the book at a price within the reach of students' pockets. The minor errors and misprints in the first edition have been corrected, a few extra explanations have been added, and the list of additional references (p. 325) has been updated to early 1978, with cross-references to pages in the main text.

The comments of reviewers, as well as the volume of sales, indicate that our aim of producing an integrated treatment (as distinct from a collection of noninteracting review articles) has been achieved. One or two reviewers remarked, quite correctly, that the book does not cover all the branches of science or engineering in which turbulence is important. Our more modest intention, as outlined in the original preface, was to review those branches that have *contributed* to our general knowledge of turbulence via research on their own special problems. Branches of engineering such as acoustics or architectural aerodynamics indeed have their own turbulence problems, but because these problems appear in the complicated contexts of sound generation, strong viscous-inviscid interaction and so on, in situ study is inherently unlikely to produce generally applicable results. Therefore these subjects and their turbulence problems have not been treated explicitly. Basic experiments on turbulence done with a view to application in acoustics, hydraulics, and so on, are, of course, frequently quoted in the book.

London, April 1978                                      PETER BRADSHAW

# Contents

# Contributors

BRADSHAW, PETER
Department of Aeronautics, Imperial College of Science and Technology, London SW7 2BY, Great Britain

CEBECI, TUNCER
Douglas Aircraft Co., Long Beach, CA 90846, USA

FERNHOLZ, HANS-HERMANN
Herman-Föttinger Institut für Thermo-Fluiddynamik, Technische Universität, D-1000 Berlin 12, Fed. Rep. of Germany

JOHNSTON, JAMES PAUL
Thermosciences Division, Department of Mechanical Engineering, Stanford University, Stanford, CA 94305, USA

LAUNDER, BRIAN EDWARD
Department of Mechanical Engineering, University of California, Davis, CA 95616, USA

LUMLEY, JOHN LEASK
School of Mechanical and Aerospace Engineering, Cornell University, Ithaca, NY 14853, USA

REYNOLDS, WILLIAM CRAIG
Department of Mechanical Engineering, Stanford University, Stanford, CA 90305, USA

WOODS, JOHN DAVID
Institut für Meereskunde, Universität Kiel, D-2300 Kiel 1, Fed. Rep. of Germany

# 1. Introduction

P. Bradshaw

With 6 Figures

This chapter is a description of the physical processes that govern turbulence and the mathematical equations that in turn govern them. It is self-contained, but the treatment of the mathematics, already available in many other textbooks, has been abbreviated in favor of discussions of the physical consequences of the equations. The chapter is intended to contain all the main results assumed without proof in later chapters. In some cases, a topic is outlined in Chapter 1 and developed in one or more later chapters; in these cases forward references are given.

## 1.1 Equations of Motion

The one uncontroversial fact about turbulence is that it is the most complicated kind of fluid motion. It is generally accepted that turbulence in simple liquids and gases is described by the Navier-Stokes equations, which express the principle of conservation of momentum for a continuum fluid with viscous stress directly proportional to rate of strain. Although the principle and the stress law are the simplest that can be imagined, some of the possible solutions of the equations, even for simple flow geometries, are too complicated to be comprehended by the human mind.

The Navier-Stokes momentum-transport equations are the second-order Chapman-Enskog approximation to the Boltzmann equation for molecular motion. For a masterly, if slightly inaccessible, review see GOLDSTEIN [1.1]. The first-order approximation, leading to the Euler equations, neglects viscosity altogether, while the more complicated molecular-stress terms yielded by higher-order approximations are not important in common gases at temperatures and pressures of the order of atmospheric. It is easy to show that the smallest turbulent eddies have a wavelength many times the mean free path unless the Mach number (velocity divided by speed of sound) is exceptionally high; the continuum approximation is a good one. Similarly, the constitutive equations of common liquids are close to the linear Newtonian viscous-stress law. Therefore, we shall use the Navier-Stokes equations throughout this book except for the discussion of "non-Newtonian" fluids in Chapter 7.

Since the equations are needed in their most general three-dimensional form, we shall use Cartesian tensor notation for compactness, ignoring the distinction between covariant and contravariant tensors and using the repeated-suffix summation convention so that $a_i b_i = a_1 b_1 + a_2 b_2 + a_3 b_3$ (but $a_i + b_i = a_1 + b_1$ or $a_2 + b_2$ or $a_3 + b_3$). Sometimes $\partial f/\partial x_i$ – say – is denoted by $f_{,i}$; this form is used for brevity in the extensive mathematics of Chapters 5 and 7. In special cases, $x, y, z$ notation will be used. Unless otherwise stated, $x$ or $x_1$ is the general direction of flow and $y$ or $x_2$ is normal to the plane of a shear layer. Occasionally vector notation will appear; in particular, $u$ will be used for the velocity vector, whose components are $u_1, u_2, u_3$ or $u, v, w$. In general, capital $U$ is used to denote a mean velocity (time-average or ensemble average) and small $u$ denotes a fluctuation about that mean. However, in most of the discussion of Sections 1.1 to 1.5, the presence or absence of a mean velocity is immaterial, so the symbol $u$ is used for simplicity; if desired it can be interpreted as the instantaneous (mean plus fluctuating) velocity, denoted by $U + u$ in later sections. The conventional division into mean and fluctuating components exists for the convenience of technologists. It is not as arbitrary as is sometimes claimed, because it leads to self-consistent equations with useful physical interpretations, but it is artificial because the motion at a given point and time receives no information about mean values, which necessarily depend on averages over large distances or long times (Sect. 1.3). Flows in which the *mean* velocity vector is everywhere parallel to a plane (usually taken as the $xy$ plane) are called "two dimensional"; an example is the flow over a very long cylindrical body such as an unswept, untapered wing, normal to the oncoming stream. Note that this definition does not require the fluctuations to be parallel to the plane; in this book at least, motion which is two dimensional at every instant is not regarded as turbulence.

In words, the principle of conservation of momentum (Newton's second law of motion), as applied to a fluid subjected to any kind of molecular forces, is

acceleration (following the motion of the fluid)

= molecular force per unit mass

+ body force per unit mass . (1.1)

Molecular force is a surface force and reduces to a sum of stress gradients divided by the density, while body force is a volume force and is usually

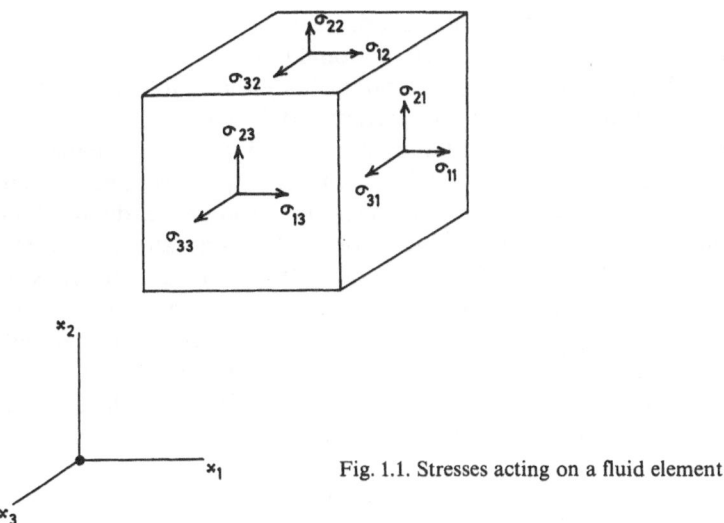

Fig. 1.1. Stresses acting on a fluid element

specified directly as a force per unit mass or equivalent acceleration. The molecular stress is conventionally divided into a scalar pressure $p$ equal to $(-1/3)$ times the sum of the three normal (tensile) stress components, and a stress due to deformation and bulk dilatation which is a second-order tensor with components $\sigma_{il}$. For compatibility with the definition of $p$, the sum of the normal components of the stress due to deformation must be zero, leading to the term in (1.3), below, containing $(2/3)\mu$. However, the individual normal-stress components are *not* zero. According to these definitions the total molecular stress acting in the $x_i$-direction on a plane normal to the $x_l$-direction (Fig. 1.1) is $-p\delta_{il}+\sigma_{il}$, where $\delta_{il}=1$ if $i=l$ and $\delta_{il}=0$ if $i\neq l$. Equation (1.1) can now be written for the $x_i$-component of velocity, $u_i$ (where $i=1, 2$ or $3$) as

$$\frac{\partial u_i}{\partial t}+u_l\frac{\partial u_i}{\partial x_l}=\frac{1}{\varrho}\frac{\partial}{\partial x_l}(-p\delta_{il}+\sigma_{il})+f_i\equiv-\frac{1}{\varrho}\frac{\partial p}{\partial x_i}+\frac{1}{\varrho}\frac{\partial \sigma_{il}}{\partial x_l}+f_i$$

or

$$u_{i,t}+u_lu_{i,l}=(1/\varrho)\,(-p\delta_{il}+\sigma_{il})_{,l}+f_i=-(1/\varrho)p_{,i}+(1/\varrho)\sigma_{il,l}+f_i \quad (1.2)$$

in the compact suffix notation for differentiation. Here $f_i$ is the $x_i$-component of body force per unit mass and $\varrho$ is the (instantaneous) density. In a gravitational field, $f_i=g_i$ where $g_i$ is a component of the gravitational acceleration. Unless density fluctuations or a free surface is present, this simply leads to an extra pressure gradient, which balances $g_i$ so that both can be forgotten. Gravitational body forces are considered in Chapters 4 and 6, and Coriolis apparent body forces in Chapter 3.

These are "Eulerian" equations, expressed in terms of the velocity components at a fixed point. Corresponding (Lagrangian) equations can be derived in terms of the velocity of a marked particle. The Lagrangian equations are much less convenient for studying ordinary fluid motion and will not be needed in this book, though Lagrangian concepts are used in the discussion of particle-laden flows in Chapter 7. Note that (1.2) applies to any fluid, whatever the constitutive law for $\sigma_{il}$. It even applies to the mean velocity in turbulent flow if all symbols denote mean quantities and $\sigma_{il}$ is understood to include apparent turbulent (Reynolds) stresses (Section 1.3) as well as the molecular stresses. For a Newtonian viscous fluid, the instantaneous stress due to deformation is

$$\sigma_{il} = \mu \left( \frac{\partial u_i}{\partial x_l} + \frac{\partial u_l}{\partial x_i} \right) + (\beta - \tfrac{2}{3}\mu)\delta_{il} \frac{\partial u_m}{\partial x_m} \tag{1.3}$$

where the last term enforces $\sigma_{ii} = 0$, and $\beta$ is the bulk viscosity, of the same order as $\mu$ [1.1]. In the most general case $\partial\sigma_{il}/\partial x_l$ is quite complicated and forbidding; for a discussion see Howarth [Ref. 1.2, pp. 49–51] or Schlichting [Ref. 1.3, Chapter 3], who neglect the bulk viscosity. Clearly $\sigma_{il} = \sigma_{li}$ in Newtonian fluids, so $\sigma_{il}$ is a "diagonally symmetric" tensor.

Any fluid obeys the law of conservation of mass, obtainable from simple control-volume analysis as

$$\frac{\partial \varrho}{\partial t} + \frac{\partial}{\partial x_l}(\varrho u_l) = 0 . \tag{1.4}$$

If the density is constant, (1.2) is unaltered but (1.4) reduces to

$$\frac{\partial u_l}{\partial x_l} = 0 \tag{1.5}$$

in steady or unsteady flow. Thus the last term of (1.3) vanishes in constant-density flow and the remainder of (1.3) implies that the viscous stress due to deformation is

$$\sigma_{il} = 2\mu e_{il} \tag{1.6}$$

where

$$e_{il} \equiv \frac{1}{2}\left(\frac{\partial u_i}{\partial x_l} + \frac{\partial u_l}{\partial x_i}\right) \tag{1.7}$$

which defines the rate of strain $e_{il}$; the factor $1/2$ is inserted for compatibility with the usual definition of strain in solid mechanics, but some fluids textbooks omit this and the factor 2 in (1.6). In a pure rotation about an axis normal to the $x_i x_l$-plane $\partial u_i/\partial x_l = -\partial u_l/\partial x_i$, both being numerically equal to the angular velocity, and $e_{il}$ and $\sigma_{il}$ are zero.

It is sometimes useful to add $u_i/\varrho$ times the "continuity" equation (1.4) to (1.2); the left-hand side of the resulting equation is

$$\frac{1}{\varrho}\left[\frac{\partial \varrho u_i}{\partial t} + \frac{\partial}{\partial x_l}(\varrho u_i u_l)\right]$$

called the "divergence" form, as opposed to the "acceleration" form of (1.2). In this form, the equation shows that the rate of accumulation of $x_i$-component momentum in a unit control volume, plus the rate at which $x_i$-component momentum leaves the control volume, equals the force applied to the fluid instantaneously in the control volume by molecular and body forces. The addition of multiples of the continuity equation often helps to simplify or clarify equations.

In constant-property flow (constant density and constant viscosity[1]) the viscous-stress gradients can be simplified by neglect of viscosity gradients and further use of the continuity equation (1.5); then (1.2) becomes

$$\frac{\partial u_i}{\partial t} + u_l\frac{\partial u_i}{\partial x_l} = -\frac{1}{\varrho}\frac{\partial p}{\partial x_i} + v\frac{\partial^2 u_i}{\partial x_l^2} + f_i \tag{1.8}$$

where $v$ is the kinematic viscosity $\mu/\varrho$. Note that the three elements of $\partial^2 u_i/\partial x_l^2$ do not individually equal the three elements of $\partial \sigma_{il}/\partial x_l$ for constant-property flow, because part of each term has been removed by using the continuity equation.

Very fortunately, viscosity does not usually affect the larger-scale eddies which are chiefly responsible for turbulent mixing (in fact, as we shall see in Section 7.4, turbulence processes are usually the same in the

---

1 In the study of gas flows the word "incompressible" is used instead of "constant property"; this usage is misleading in liquid flows, which can often be assumed incompressible in the strict sense of constant density but whose viscosity varies very rapidly with temperature. A "constant pressure" flow is one in which $\partial p/\partial x_i$ is negligible in (1.2).

simpler types of non-Newtonian fluids). Equally fortunately, the effects of density fluctuations on turbulence are small if, as is usually the case (Section 2.5), the density fluctuations are small compared to the mean density. Part of the discussion below will be concerned with the two major exceptions to these statements: the effect of viscosity on turbulence in the "viscous sublayer" very close to a solid surface (Sections 1.8, 2.3) and the effect of temporal fluctuations and spatial gradients of density in a gravitational field (Chapters 4 and 6); elsewhere, we will usually neglect the direct effect of viscosity and compressibility on turbulence. Several different definitions of the Reynolds number (velocity scale) × (length scale)/$v$ will be used below. They fall into two main classes: "bulk" Reynolds numbers in which the scales are those of the mean flow, and "turbulent" or "local" Reynolds numbers, in which the scales are those of the turbulence or even of part of the turbulence.

A fluctuating velocity field will cause fluctuations to develop in an initially smooth spatial variation of a scalar such as enthalpy, or concentration in a two-component flow. It is important to note that the fluctuating velocity field drives the fluctuating scalar field while the effect of the latter on the former, applied via mean gradients and fluctuations of density, is usually weak or even negligible. The conservation equation for a scalar $c$, equating the rate of change of $c$ with time (following the motion of the fluid) to the sum of molecular diffusion and sources within the fluid, is

$$\frac{\partial c}{\partial t} + u_l \frac{\partial c}{\partial x_l} = \frac{1}{\varrho} \frac{\partial}{\partial x_l} \left( \varrho \gamma \frac{\partial c}{\partial x_l} \right) + S \qquad (1.9)$$

where $\gamma$ is the molecular diffusivity of $c$ (having the same dimensions as the kinematic viscosity $v$) and $S$ is the rate of generation of $c$ per unit volume (by chemical reaction, say) at the point considered. Compare (1.2) and (1.9): equations like these, whose left-hand sides contain the "transport operator" $\partial/\partial t + u_l \partial/\partial x_l$, are called "transport equations"; sometimes this name is reserved for equations containing the time- or ensemble-averaged version of this operator. An important case of (1.9) in engineering is when $c$ represents enthalpy and $\gamma$ represents the thermal diffusivity $k/\varrho c_p$; the dimensionless group of fluid properties $\mu c_p/k \equiv v/(k/\varrho c_p) \equiv \sigma$ is the Prandtl number, and, in compressible flow, $S$ is the sum of compression work and viscous dissipation of kinetic energy into heat. If $c$ represents mass concentration, $v/\gamma$ is the Schmidt number, $Sc$. If $\gamma$ is constant, (1.9) becomes

$$\frac{\partial c}{\partial t} + u_l \frac{\partial c}{\partial x_l} = \gamma \frac{\partial^2 c}{\partial x_l^2} + S. \qquad (1.10)$$

The obvious similarities between (1.8), without pressure gradient or body force, and (1.10), without the source term $S$, are the bases for analogies between momentum transfer and the transfer of heat and other scalars. The similarities between the general equations (1.2) and (1.9) are less obvious. Bearing in mind the relative unimportance of viscosity in turbulent mixing at all but the lowest Reynolds numbers, we can deduce the relative unimportance of molecular diffusivity (constant or otherwise) at least if $v/\gamma$ is not much smaller than unity. The presence of pressure gradients in (1.8) and their absence from (1.10) prevent the analogies from being exact in turbulent flow, even if $v/\gamma = 1$, because pressure fluctuations always accompany velocity fluctuations. To see this, take the divergence of the Navier-Stokes equations [i.e., differentiate (1.8) with respect to $x_i$] neglecting density variations and body forces; after rearranging and using (1.5) we get the Poisson equation

$$\frac{1}{\varrho} \frac{\partial^2 p}{\partial x_i^2} = -\frac{\partial u_i}{\partial x_l} \frac{\partial u_l}{\partial x_i} = -\frac{\partial^2 u_i u_l}{\partial x_i \partial x_l}. \tag{1.11}$$

However, the analogies between momentum transfer and heat or mass transfer are sufficiently accurate for this introduction to be confined to momentum transfer, leaving the corresponding qualitative results for heat or mass transfer to be inferred by the reader. We return to the heat-transfer equations in Chapter 6: see also Subsection 2.3.9.

Detailed discussions of the equations of motion, some with specific application to turbulence, are given in [1.2–7].

## 1.2 Shear-Layer Instability and the Development of Turbulence

In steady "inviscid" flow[2] without body forces, the Navier-Stokes equations (1.2) reduces to the requirement that the total pressure

$$P \equiv \varrho \left( \frac{1}{2} |u|^2 + \int \frac{dp}{\varrho} \right), \tag{1.12}$$

where $u$ is the velocity vector, shall be constant along a streamline (the envelope of the velocity vector); here the integral is evaluated along a

---

[2] Meaning a flow with negligible viscous stresses, the result of negligible rate of strain rather than negligible viscosity.

streamline starting from the point where $u=0$. In constant-density flow,

$$P=p+\tfrac{1}{2}\varrho|u|^2 .\qquad(1.13)$$

$P$ may vary normal to the streamlines because of the previous influence of viscosity or body forces, and in flow with significant viscous stresses (or turbulent stresses) it will in general vary along and normal to the stream-lines. A flow with a total-pressure gradient normal to the streamlines (a working definition of a "shear layer") may be unstable to infinitesimal, or small but finite, time-dependent disturbances. Other kinds of in-stability may occur, but if turbulence develops it is almost always via the stage of shear layer instability. Naturally, the shear layer is most unstable to the type of disturbance which travels downstream with the fluid, that is, a "traveling-wave" disturbance. Unstable shear flows give rise to complicated flow patterns [1.8, 9] and complicated mathe-matics [1.10], both of great beauty. Our present concern is with the further development of amplified unstable disturbances, which in the simplest cases can be two-dimensional sinusoidal fluctuations, into the three-dimensional continuous-spectrum fluctuations of turbulence.

The key phenomenon in both the further development of instabilities and the maintenance of fully developed turbulence is the intensification of vorticity in three-dimensional flows [Ref. 1.8, p. 266]. Vorticity is defined as

$$\omega \equiv \operatorname{curl} u \equiv \nabla_{\wedge} u \qquad(1.14a)$$

in vector notation or

$$\omega_i \equiv \varepsilon_{ijk}\frac{\partial u_k}{\partial x_j}\qquad(1.14b)$$

in tensor notation. Here the unit alternating tensor $\varepsilon_{ijk}$ is defined to be unity if $i, j, k$ are in cyclic order (1 2 3 1 2 3...), $-1$ if $i, k, j$ are in cyclic order, and zero otherwise (i.e., if two indices are equal); it exists simply to provide a tensor representation of a vector (cross) product. The vorticity of an element of fluid in unstrained ("solid body") rotation about the $x_i$-axis with angular velocity $\Omega$ is $\omega \equiv \omega_i \equiv 2\Omega$. The "transport" equation for vorticity in a Newtonian viscous fluid is obtained by taking the curl of the Navier-Stokes equations. For incompressible, constant-property flow without body forces, or with a body-force vector $f$ that satisfies $\operatorname{curl} f=0$, as is the case for many simple body forces, including

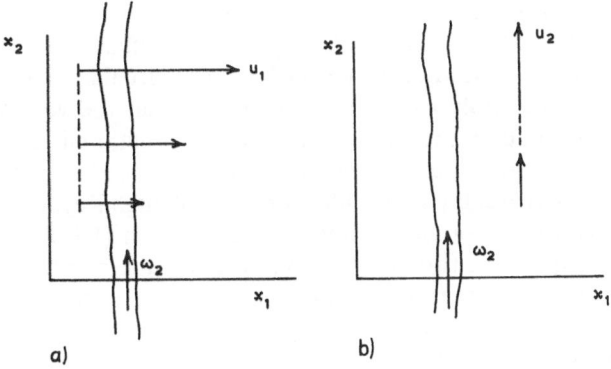

Fig. 1.2a and b. Effect of a velocity gradient on a vortex line. (a) Tilting: vorticity $\omega_2$, velocity gradient $\partial u_1/\partial x_2$. (b) Stretching: vorticity $\omega_2$, velocity gradient $\partial u_2/\partial x_2$

gravity, (1.8), with some use of (1.5), gives

$$\frac{\partial \omega_i}{\partial t} + u_l \frac{\partial \omega_i}{\partial x_l} = \omega_l \frac{\partial u_i}{\partial x_l} + v \frac{\partial^2 \omega_i}{\partial x_l^2}.$$
(1.15)

Note that the pressure term has disappeared but that a new term appears on the right, in addition to the viscous diffusion term. The vorticity/velocity-gradient interaction term $\omega_l \partial u_i/\partial x_l$, which is a nonlinear term because $\omega$ depends on $u$, has two main effects. It is convenient to discuss these with reference to a slender element of fluid, rotating about its axis (Fig. 1.2). We shall call such elements "vortex lines", the length of the line being nominally infinite and the distribution of vorticity over the cross section being immaterial for present purposes. Except for the effects of viscous diffusion, which tends to increase their cross section, vortex lines move with the fluid, a consequence of Kelvin's circulation theorem [Ref. 1.8, p. 273]. A vortex sheet is an envelope of vortex lines, and a finite body of fluid with vorticity can be regarded as a continuous distribution of vortex lines. Vortex lines induce an irrotational (zero-vorticity) velocity field at other points in space according to the Biot-Savart law, which also governs the magnetic field due to a current-carrying conductor.

The first effect of the term $\omega_l \partial u_i/\partial x_l$ in (1.15) is that if $i \neq l$ (say, $i=1$, $l=2$, see Fig. 1.2a) it represents an exchange of vorticity between components, because a velocity gradient $\partial u_1/\partial x_2$ (say) tilts a vortex line which was initially in the $x_2$-direction so that it acquires a component in the $x_1$-direction. Secondly, if $i=l$ ($=2$, say, see Fig. 1.2b) the vortex line is stretched by the rate of tensile strain along its axis, without any change of the direction of that axis. Neglecting viscous diffusion (and, strictly, requiring the cross section of the vortex line to be circular so that pressure gradients cannot apply a torque to it) we see that the vortex line will conserve its angular momentum as its cross-sectional area decreases under the influence of axial stretching; therefore, its vorticity (angular velocity) will increase. If viscous diffusion is small (high Reynolds number) the vorticity/velocity-gradient interaction terms in (1.15) can change an initially simple (but three-dimensional) flow pattern into an unimaginably complicated distribution of vorticity and velocity—turbulence.

Three dimensionality is essential to the genesis and maintenance of turbulence [1.11]: in an instantaneously two-dimensional flow, by definition, the velocity vector would be everywhere parallel to a plane, the vorticity vector would be normal to that plane, and $\omega_l \partial u_i / \partial x_l$ would be zero. It appears that although the most unstable infinitesimal disturbance in a steady two-dimensional shear flow is a two-dimensional traveling wave, amplified disturbances of sufficient amplitude (which can be regarded as packets of vortex lines with spanwise axes) are themselves unstable to infinitesimal three-dimensional perturbations. A small kink in an otherwise straight vortex line is distorted and enlarged by the induced velocity field of the vortex itself, and if viscous diffusion is small enough (i.e., if the vortex Reynolds number is large enough), the distortion will continue indefinitely. Therefore, once the primary unstable disturbances have reached a sufficient amplitude they rapidly become more complicated and unsteady, because of the stretching and tilting by the induced velocity field of the vortex lines themselves, as well as by the basic shear flow. The simplest way of explaining how non-periodic unsteadiness arises is to note that in real life the wavelength of the primary disturbance is bound to be slightly unsteady. The percentage unsteadiness in the wavelength of the first harmonic disturbance will be roughly twice as large, and so on for higher harmonics. Sum-and-difference wave numbers[3] appear because of the nonlinearity of the interaction of different packets of vortex lines via their induced velocity fields, and the wave-number spectrum eventually becomes continuous.

As the motion becomes increasingly complicated the effects are felt of a theorem in random processes, known as the theorem of the random walk or "Drunkard's Walk", which states that a particle subjected to random impulses will, on the average, increase its distance from its

---

[3] Wave number $= 2\pi/$(wavelength): it is a vector with the same direction as the wavelength (which is not necessarily the direction of propagation of the wave). Note that small scale = small wavelength = large wave number.

starting point. The phenomenon is known to, and regretted by, cab drivers. An obvious corollary states that the distance between two randomly perturbed particles will, on the average, increase. If those two particles are situated at the ends of a given element of a vortex line, then, in a flow field approximating to random disturbances, the length of the element will on the average increase, and its vorticity will be increased by this stretching. Moreover, the typical length scales of the region of high vorticity—the diameter of the vortex line in our simple model—will also decrease. This is the key mechanism of fully developed turbulence: interaction of tangled vortex lines maintains random fluctuations of the vorticity and velocity, while the random-walk mechanism transfers vorticity to smaller and smaller length scales. It remains to provide the transfer process with a beginning and an end. If there is a mean rate of strain it deforms the fluctuating vorticity field and intensifies vortex lines whose axes are, at any given instant, near the axis of the largest positive principal strain rate. Because of the nonlinearity of the process, this intensification usually predominates over the weakening of those vortex lines whose axes are near that of the negative principal strain rate. Thus the mean strain rate helps to maintain the level of vorticity fluctuation. A more rigorous analysis shows that the main effect of the mean strain rate is on the larger-scale motions, which then distort motions of smaller scale and so on. A limit to the decrease of vortex-line diameter by stretching is set when viscous stress gradients diffuse vorticity away from the axis as fast as stretching reduces the diameter.

If we now consider the kinetic energy of the fluctuating motion, we can see that vortex stretching increases the rotational kinetic energy of the vortex line. Angular momentum is proportional to $\omega r^2$ while energy is proportional to $\omega^2 r^2$, so if the former is conserved while $r$ decreases, the latter increases. The kinetic energy comes from the velocity field that does the stretching, so that kinetic energy passes from the mean flow (if a mean strain rate is present) down through vortex motions of smaller and smaller length scale until it is converted into thermal internal energy via work done against viscous stresses. If there is no mean strain field to do work on the fluctuating motion the latter gradually decays. Now this process of energy transfer to smaller scales, aptly called the "energy cascade", is independent of viscosity except in the final stages, as can be seen from the description above. It therefore follows that the rate of energy transfer to the smallest, viscous-dependent motions that dissipate it into "heat" is *independent* of viscosity. Viscosity causes dissipation but does not control its rate; the intensity and length scale of the small-scale motion adjust themselves so as to dissipate all the energy transferred from larger scales, and the smaller the viscosity the

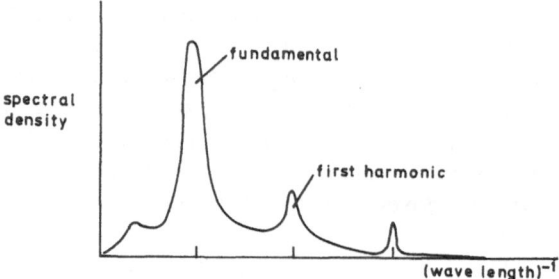

Fig. 1.3. Spectral distribution of velocity fluctuations in a late stage of transition to turbulence

smaller the motions that can survive. The rate of viscous dissipation of energy per unit volume by turbulent fluctuations is the mean product of the fluctuating rate of strain and the fluctuating viscous stress (clearly this is the mean rate at which the turbulence does work against viscous stresses). The viscous stress is equal to $2\mu$ times the rate of strain, so the dissipation is proportional to the mean square of the rate of strain and is therefore non-negative as required by the second law of thermodynamics.

In summary, the stages in the development of turbulence from an initial unstable shear layer (or from other unstable situations like that of a fluid whose density decreases in the direction of a body-force vector) are:
1) The growth of disturbances with *periodic* fluctuations of vorticity.
2) Their secondary instability to three-dimensional (infinitesimal) disturbances if the primary fluctuations are two dimensional.
3) The growth of three dimensionality and higher harmonics of the disturbance, leading to spectral broadening by vortex-line inter-action.
4) The onset of the random-walk mechanism when the vorticity field becomes sufficiently complicated, leading to a general transfer of energy across the spectrum to smaller and smaller scales.

It is not useful to agonize about the exact point at which the motion can properly be called "turbulence". Flows in the later stages of transition from laminar to turbulent (at the spectral state shown in Fig. 1.3, say) are even more difficult to understand and calculate than turbulence in an undoubted state of full development. Therefore, equating "turbulence" and "incalculability"—the unconscious basis of many definitions—is to be deprecated. So is the use of the word "turbulence" in plasma physics to describe miscellaneous instabilities in current-carrying fluids, and in meteorology to describe the synoptic-scale motion (whose horizontal length scales are many times the depth of the atmosphere and which is therefore close to instantaneous two dimensionality). The essential characteristic of turbulence is the transfer of energy to smaller spatial scales across a continuous wave-number spectrum; this is a three-dimensional, nonlinear phenomenon.

Note that we have discussed the main mechanisms of turbulence without any mathematics other than the qualitative use of the vorticity equation. The vortex-line model used above is clearly artificial, but the spatial distribution of vorticity in a real turbulence field is almost discontinuous, the ratio of the width of typical high-vorticity regions to the width of the flow decreasing with increasing Reynolds number. There is some controversy [1.12] over whether the high-vorticity regions are best approximated by rods, strips or sheets (known to some as the "pasta problem"). An alternative concept is that of an "eddy". In qualitative discussion an eddy can be thought of as a typical turbulent flow pattern, covering a moderate range of wavelengths so that large eddies and small eddies can coexist in the same volume of fluid. Flow-visualization experiments (to be heartily recommended to all who work with turbulence and especially those who seek to calculate it) show the usefulness of the concept and the difficulty of a precise definition [1.13]. In quantitative work one uses statistical-average equations based on Fourier analysis of the velocity patterns (Section 1.4), and since sinusoidal modes have no relation to the actual modes (eddies, velocity patterns, high-vorticity regions...) the physical processes are obscured. Note however that an optimum mode for spectral decomposition can be found only *after* the problem has been solved!

## 1.3 Statistical Averages

The reason for working with statistical averages [1.6, 11, 14, 15] is that one is generally not interested in complete details of the behavior of the three velocity components and the pressure as functions of three space coordinates and time. In most cases, indeed, only very simple statistics, such as the mean rates of transfer of mass or momentum, are required for engineering or geophysical purposes.

The simplest form of statistical average is the mean, with respect to time, at a single point. This is useful only if the mean is independent of the time at which the averaging process is started; in symbols

$$\bar{f} \equiv \operatorname*{Lt}_{T \to \infty} \frac{1}{T} \int_{t_0}^{t_0 + T} f \, dt \tag{1.16}$$

must be convergent and independent of $t_0$. This is true of, say, flow through a pipe with constant pressure difference maintained between its ends, but not of, say, decaying motion in a closed container. In cases like the latter, *ensemble averages* must be used:

$$\bar{f}(t) = \underset{N \to \infty}{\mathrm{Lt}} \frac{1}{N} \Sigma_1^N f(t) \qquad (1.17)$$

where the summation is over a set of samples, each taken at a time $t$ after the beginning of one of a set of $N$ repeats of the experiment.

It is plausible to suppose that for a stationary process (one in which time averages are independent of $t_0$) time and ensemble averages will be identical. This is one version of the "ergodic hypothesis", whose proof is a complicated matter but which has never been shown to be incorrect. Therefore, while theoretical work should strictly be based on ensemble averages, experimenters almost always use time averages in statistically stationary flows; below we shall simply refer to "mean values", denoted by an overbar.

The simplest statistical properties of a turbulent flow [Ref. 1.11, Chapt. 2] are the mean squares and second-order mean products of the velocity fluctuation components at a fixed point. The velocity fluctuation is the instantaneous velocity less the mean velocity, and turbulence statistics are almost always based on the fluctuating parts of the variables; in the definitions below, therefore, small $u$ denotes the fluctuation only. The sum of the mean squares of the three velocity fluctuation components, $\overline{u_i^2}$ in tensor notation or $\overline{q^2}$ in $x, y, z$ notation, is independent of the coordinate axes (i.e., it is a scalar) and can be regarded as twice the kinetic energy of the fluctuating motion, per unit mass of fluid. The quantities $\varrho \overline{u_i u_j}$ are mean rates of momentum transfer, per unit area, by the fluctuating motion: $\varrho u_i$ is the fluctuating part of the rate of mass flow through a unit area normal to the $x_i$-direction, and $u_j$ is the fluctuating part of the $x_j$-component momentum per unit mass, so that $\varrho \overline{u_i u_j}$ is the mean rate of transfer of $x_j$-component momentum through a unit area normal to the $x_i$-direction (Fig. 1.1). Newton's second law of motion implies that, *from the viewpoint of an observer who sees only the mean flow and not the turbulence*, the momentum transfer per unit area $\varrho \overline{u_i u_j}$ is equivalent to a stress $\sigma_{ij} = -\varrho \overline{u_i u_j}$, acting in the $x_j$-direction on a plane normal to the $x_i$-direction (of course $\sigma_{ij} = \sigma_{ji}$ so only six of the nine $\sigma_{ij}$ are different). These are *apparent* stresses, just as viscous stresses are the apparent result of molecular motion as seen by an observer who cannot distinguish the molecules. It is not meaningful to call $\varrho u_i u_j$ an "instantaneous stress"; the only instantaneous stresses are the viscous stresses and the pressure. The $\sigma_{ij}$ are called the

Reynolds stresses; we do not usually assimilate $-(1/3)\sigma_{ii}$ into the pressure as was done with the viscous stresses. Similar arguments applied to a scalar quantity whose fluctuating part is $c$ show that the mean rate of transfer of the scalar in the $x_i$-direction by the turbulence is $\overline{cu_i}$. Because turbulence is essentially three dimensional the three normal stresses $(i=j)$ are almost always of the same order, and the shear stresses $(i \neq j)$ are of the same order as the normal stresses unless they are zero for reasons of symmetry. Mean viscous stresses act in addition to Reynolds stresses but are usually much smaller. Equation (1.8) refers to instantaneous quantities. The equivalent for mean quantities in constant-property flow can be derived from (1.8) by replacing $u_i$ by $U_i + u_i$ (mean plus fluctuation) and ensemble averaging, using the continuity equation to simplify the Reynolds-stress terms. It is

$$\frac{\partial U_i}{\partial t} + U_l \frac{\partial U_i}{\partial x_l} = -\frac{1}{\varrho}\frac{\partial \overline{p}}{\partial x_i} + v \frac{\partial^2 U_i}{\partial x_l^2} - \frac{\partial \overline{u_i u_l}}{\partial x_l} + F_i$$

$$\equiv -\frac{1}{\varrho}\frac{\partial \overline{p}}{\partial x_i} + \frac{\partial}{\partial x_l}\left(v \frac{\partial U_i}{\partial x_l} - \overline{u_i u_l}\right) + F_i. \tag{1.18}$$

Probability distributions play a limited role in the study of turbulence. The joint probability distribution for the velocity at all points in space does completely specify the statistical properties [Ref. 1.14, Chapt. 2] but is too complicated to use. The single-point or simpler joint-probability distributions are usually nearly Gaussian but the departures from a Gaussian distribution are a manifestation and measure of the essential nonlinear behavior of the turbulence. Briefly, the "central limit theorem" states that the sum of a very large number of *independent* processes is Gaussian; although turbulence has a very large number of degrees of freedom, they interact.

## 1.4 Two-Point Statistics and Spectral Representation

The Navier-Stokes equations refer to motion at a single point. To investigate the statistics of motions of different spatial scales (i.e., different eddy sizes), we need information at two points (at least), and the most basic piece of information is the coefficient of correlation between the velocity fluctuations at two points. Let $u_i$ be the $x_i$-component of velocity at a point $P$ with position vector $x$ (Fig. 1.4) and $u_j'$ the $x_j$-component at a point $P'$, distant $r$ from $P$ so that its position vector is $x' \equiv x + r$. As usual in defining correlation coefficients we subtract the means from both quantities, and therefore continue to suppose that $u_i$ and $u_j$ are fluctuations with zero mean, giving

$$\text{correlation coefficient} \equiv \overline{u_i'u_j'}/\sqrt{(\overline{u_i^2}\,\overline{u_j'^2})} \tag{1.19}$$

where we do *not* sum over repeated suffixes. In theoretical studies it is usually simpler to work with the "covariance" $\overline{u_i u_j}$ alone, while experimental results are often presented as $\overline{u_i'u_j'}/\overline{u_i u_j}$ (again unsummed); both quantities are denoted by $R_{ij}(r)$, and authors usually make it clear which they mean.

Multiply the $x_i$-component Navier-Stokes equation, (1.8), written for the point P, by $u_j'$; add $u_i$ times the $x_j$-component Navier-Stokes equation written for the point P', and take an ensemble average (denoted by an overbar). We get, neglecting body forces,

$$\frac{\partial \overline{u_i u_j'}}{\partial t} + \left(\overline{u_j'u_l \frac{\partial u_i}{\partial x_l}} + \overline{u_i u_l' \frac{\partial u_j'}{\partial x_l'}}\right)$$

$$= -\frac{1}{\varrho}\left(\overline{u_j' \frac{\partial p}{\partial x_i}} + \overline{u_i \frac{\partial p'}{\partial x_j'}}\right) + v\left(\overline{u_j' \frac{\partial^2 u_i}{\partial x_l^2}} + \overline{u_i \frac{\partial^2 u_j'}{\partial x_l'^2}}\right) \tag{1.20}$$

where the use of $l$ as the repeated suffix at P' as well as P does not imply any loss of generality. Now (1.8), whose first term is $\partial u_i/\partial t$, represents the rate of change of $u_i$ with respect to *time*, following the motion of the fluid. In the same way (1.20) represents the rate of change of any of the nine components of $\overline{u_i u_j'}$ with respect to time (note that $\overline{u_i u_j'} \neq \overline{u_j u_i'}$ in general) with P following the ensemble-average motion of the fluid and $r$ constant. If there is in fact a mean motion as well as fluctuations (in which case the $u$-symbols must be allowed to revert to their original definitions as mean-plus-fluctuating velocity), part of the second group of terms can be rearranged to represent the spatial part of the rate of change of $\overline{u_i u_j'}$, analogous to the second term of (1.2). In this generality the equation becomes forbiddingly complicated, but there are some special cases in which it can be simplified mathematically without losing physical meaning. Before discussing these we note the generality of the technique used to obtain this equation for $\overline{u_i u_j'}$. To obtain an equation for the rate of change of a covariance $\overline{q_A q_B}$, where $q_A$ and $q_B$ are any fluctuating quantities for which transport equations like (1.8) or (1.9) can be written, we add $q_B$ times the transport equation for $q_A$ to $q_A$ times the transport equation for $q_B$, and average. In particular we note that if the primes are dropped from the $x_j$-component quantities in (1.20) it becomes a transport equation for the one-point second-order product $\overline{u_i u_j}$ whose relation to

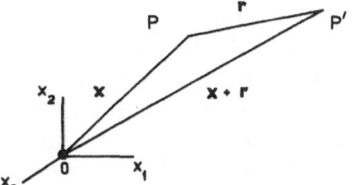

Fig. 1.4. Notation for spatial covariances

an apparent stress was discussed above. This equation will be treated in Section 1.6.

A special case in which (1.20) simplifies greatly is "homogeneous turbulence", in which statistical quantities are independent of position in space (for example, $\overline{u_i u_j'}$ is independent of $x$ although it still depends on $r$). It is usual to make the further restriction that the mean velocity be constant also, so that it can be taken as zero without loss of generality. However there has been much recent interest in homogeneous turbulence with a mean rate of strain; in this book "homogeneous turbulence" may have a mean rate of strain (necessarily constant throughout space) while "unstrained homogeneous turbulence" does not. In homogeneous turbulence, the spatial part of the rate of change of $\overline{u_i u_j}$ is zero. The practical case of unstrained homogeneous turbulence is that of fluid in a large container, randomly stirred and left for the turbulent motion to decay. Among the more obvious simplifications that result is

$$\frac{\partial}{\partial x_i'} = \frac{\partial}{\partial r_l}, \qquad \frac{\partial}{\partial x_l} = -\frac{\partial}{\partial r_l} \tag{1.21}$$

because a positive change in the vector $x$ implies an equal negative change in $r$ if $x' \equiv x + r$ remains unchanged (see Fig. 1.4). With these and other simplifications, the equation for the decay of the two-point covariance $\overline{u_i u_j'}$ in unstrained homogeneous turbulence becomes

$$\frac{\partial \overline{u_i u_j'}}{\partial t} = \frac{\partial}{\partial r_l}(\overline{u_i u_j' u_l} - \overline{u_i u_j' u_l}) + \frac{1}{\varrho}\left(\frac{\partial \overline{p u_j'}}{\partial r_i} - \frac{\partial \overline{p' u_i}}{\partial r_j}\right) + 2v\frac{\partial^2 \overline{u_i u_j'}}{\partial r_l^2} \tag{1.22}$$

where the terms on the right represent, respectively, the transfer between motions of different spatial scales by the nonlinear initial terms, the redistribution over different values of $i$ and $j$ by the action of pressure fluctuations (with no effect on the typical length scales of the motion as a whole), and the change produced by viscous stress fluctuations (principally a reduction at small spatial scale). Now $\overline{u_i u_j'}$ for given $r$ can be thought of as the contribution to the one-point mean, $\overline{u_i u_j}$, from all motions with spatial scales *larger* than $|r|$ (since motions of smaller scale cannot contribute simultaneously to $u_i$ and $u'_j$ or to the covariance $\overline{u_i u'_j}$). Length scales for turbulence can be arbitrarily defined from geometrical properties of $\overline{u_i u'_j}(r)$. The most common are the integral scales, $\int_0^\infty \overline{u_i u'_j}(r_l) dr_l$ (no summation over $l$) of which the most frequently quoted and easily measured is the "longitudinal integral scale" $\int_0^\infty \overline{u_1 u'_1} dr \equiv L_{11}$ or $L_x$.

The covariance $\overline{u_i u'_j}$ is not suitable for direct analysis of the importance of different scales of motion, and in quantitative work it is better to use the three-dimensional Fourier transforms of $\overline{u_i u'_j}$, and of (1.20) as a whole, with respect to $r$. The transformed variable is the three-dimensional wave-number vector $k \equiv (k_1, k_2, k_3)$. We define the wave-number spectral density as

$$\Phi_{ij}(k) \equiv \frac{1}{(2\pi)^3} \int \overline{u_i u'_j} \exp(-ikr) dr$$

$$\equiv \frac{1}{(2\pi)^3} \int \int \int \overline{u_i u'_j} \exp(-ik_1 r_1) \exp(-ik_2 r_2)$$

$$\times \exp(-ik_3 r_3) dr_1 dr_2 dr_3 \tag{1.23}$$

where symbol $i$ represents $\sqrt{-1}$, and where the integrals are over the whole volume of the fluid. As a very simple example, if $\overline{u_i u'_j} = \cos k^* r_1$, $\Phi_{ij}$ is a "delta function"—a spike of infinite height along the $\Phi$ axis— situated at $k = k_1 = k^*$. It can be shown that if $\overline{u_i u'_j}$ has a continuous range of wavelengths $\Phi_{ij}(k)$ has a continuous distribution in wave-number space. We can rigorously regard $\Phi_{ij}(k) dk_1 dk_2 dk_3$ as the contribution of the elementary volume $dk_1 dk_2 dk_3$ (centered at wave number $k$ and therefore representing a wave of length $2\pi/|k|$ in the direction of the vector $k$) to the value of $\overline{u_i u'_j}$ — hence the name "spectral density". This is consistent with the behavior of the inverse transform

$$\overline{u_i u'_j}(r) = \int \Phi_{ij}(k) \exp(ikr) dk \tag{1.24}$$

where the integral is over the whole volume of wave-number space, which for $r = 0$ becomes

$$\overline{u_i u_j} = \int \Phi_{ij}(k) dk \; ; \tag{1.25}$$

thus the integral of $\Phi_{ij}$ over all wave-number space is $\overline{u_i u_j}$. Particularly in experimental work, much use is made of the one-dimensional transform of the covariance $\overline{u_i u'_j}$ or, more frequently, $\overline{u_i u'_j}/\overline{u_i u_j}$, with separation $r$

along one of the coordinate axes. For instance, the one-dimensional wave-number spectrum of $\overline{u_i u_j}$, for a wave-number component in the $x_1$-direction, is

$$\phi_{ij}(k_1) = \frac{1}{2\pi} \int_{-\infty}^{\infty} \overline{u_i u_j}(r_1) \exp(-ik_1 r_1) dr_1 \tag{1.26}$$

whose inverse is

$$\overline{u_i u_j}(r_1) = \int_{-\infty}^{\infty} \phi_{ij}(k_1) \exp(ik_1 r_1) dk_1 . \tag{1.27}$$

Equating (1.27) to (1.24) with $r_2 = r_3 = 0$ shows that

$$\phi_{ij}(k_1) = \int_{-\infty}^{\infty} \int_{-\infty}^{\infty} \Phi_{ij}(k) dk_2 dk_3 \tag{1.28}$$

not $\Phi_{ij}(k_1, 0, 0)$, because $\phi_{ij}(k_1)$ represents the spectral density due to all motions of $x_1$-wave number $k_1$ whatever their values of $k_2$ and $k_3$. Simplicity of definition has to be paid for in uncertainty of interpretation; for instance $\phi_{ij}(0)$ is almost always nonzero although $\Phi_{ij}(0, 0, 0)$ is always zero (there are no waves of infinite length but there are some whose wavelengths are perpendicular to the $x_1$-axis). In experimental work $\phi_{ij}(k_1)$ is often related to $\overline{u_i u_j}(r_1)$ by (1.26) and (1.27) with the integrals taken from 0 to $\infty$ only (and the factor $1/(2\pi)$ in (1.26) consequently replaced by $2/\pi$) or by $1/\overline{u_i u_j}$ times these definitions. See [1.11, 15] for details of this and of the *frequency spectrum*, $\phi_{ij}(\omega)$, defined as the one-dimensional Fourier transform in $(\omega, \Delta t)$ of the *autocovariance* or *autocorrelation* $\overline{u_i(x, t) u_j(x, t + \Delta t)}$. It is also possible to define and transform *Lagrangian* correlations like $\overline{u_i(x, t_0) u_j(x', t_0 + \Delta t)}$, where $x'$ is the position at time $t_0 + \Delta t$ of a marked particle (of fluid or otherwise) that was at $x$ at time $t_0$; the overbar denotes an average over $t_0$. Strictly, the integrals over an infinite range of $r$ required in the definition of $\Phi$ and $\phi$ can be reliably performed only in homogeneous turbulence. In practice the same definitions are used in inhomogeneous turbulence, the imaginary part resulting from the asymmetry of $\overline{u_i u_j}$ as a function of $r$ usually being small.

In addition to making physical interpretation easier, Fourier transforming converts spatial differentiation, e.g., $\partial/\partial r_l$, to simple multiplication, in this case by $ik_l$.

Equation (1.22) for unstrained homogeneous turbulence becomes, on Fourier transforming,

$$\frac{\partial}{\partial t} \Phi_{ij}(k) = \Gamma_{ij}(k) + \Pi_{ij}(k) - 2\nu k_l^2 \Phi_{ij}(k) \tag{1.29}$$

where $\Gamma$ and $\Pi$ are the transforms of the triple-product and pressure terms, respectively [unlike the viscous term they cannot be exactly expressed in terms of $\Phi_{ij}(k)$]. This last parenthesis reminds one of what is fairly obvious from (1.20): that *these are not soluble equations* because the number of unknowns exceeds the number of equations. The instantaneous Navier-Stokes equations, plus the continuity equation, are a closed soluble set, but averaging loses information. (Another fairly obvious fact is that Fourier transforming, like other mathematical operations, neither adds to nor reduces the information but simply rearranges it). The lost information must somehow be replaced *empirically*.

The problem posed by this excess of unknowns over equations, in any finite set of averaged equations derived from the Navier-Stokes equations, is called the "closure problem", and its solutions are called "turbulence models" or "closures". It can be attacked by considering only one-point statistics, in which case (1.20) can be regarded as an equation for the Reynolds stresses—$\overline{u_i u_l}$ appearing in (1.18). When we discuss this "transport-equation closure" approach in Chapter 5 we shall see that information about a length scale of the turbulence is always required, and this implies an explicit or implicit empirical assumption—perhaps a very crude one—about two-point statistics or spectra. Detailed discussion of two-point statistics in inhomogeneous turbulence becomes very complicated, and most of the existing "spectral closure" work therefore relates to homogeneous turbulence.

The first serious attacks on the spectral closure problem for unstrained homogeneous turbulence, in the period 1940–1950 [1.14] sought a direct representation of $\Gamma$ and $\Pi$ in terms of $\Phi(k)$, or of $\overline{u_i u_j' u_l}$ and $\overline{u_j' \partial p / \partial x_i}$ in terms of $\overline{u_i u_j'}$. We can derive a transport equation for the third-order product $\overline{u_i u_j' u_l}$ from the Navier-Stokes equation by an extension to three quantities of the rule for the $q_A q_B$ equation explained above, but it contains *fourth*-order products. Mathematically simple relations between these and the lower-order products, especially those based on assumptions valid only for Gaussian probability distributions, are liable to give physically implausible results. There is no guarantee that the importance of a product will decrease with increasing order so that a rough truncation at high enough order will give acceptable predictions of $\Phi_{ij}(k)$. Several proposals have been made for truncations which do not correspond exactly to the higher-order truncations outlined here but which have an easier physical interpretation. Some of the most impressive work in this area has been done by R. H. KRAICHNAN, collected and interpreted by LESLIE [1.16]. While the calculations needed to obtain a solution for $\Phi$ are not as lengthy as the solution of the time-dependent Navier-Stokes equations for an equivalent case, the equations are very much more complex. The extension from unstrained homogeneous turbulence to shear flow, also discussed in [1.16], creates further difficulties, which can be overcome only by approximations and empiricisms suspiciously like those made at a much lower theoretical level in engineering calculation methods. Several workers previously interested in closure of the equations for correlations or spectra have now transferred their attention to "computer simulations" of turbulence, solutions of the time-dependent Navier-Stokes equations (Sect. 5.1) which may supplement or replace spectral closure work as the highest-level attack on turbulence. Therefore, no detailed account of current spectral closure work will be given here. Work up to about 1950 is described by BATCHELOR [1.14], an excellent review of the problems at 1970 was given by ORSZAG [1.17], and Leslie's book is very thorough up-to-date guide with special reference to Kraichnan's work. An extensive review of turbulence and statistical concepts is given by MONIN and YAGLOM [1.18].

## 1.5 Local Isotropy

The vortex-stretching "cascade" process, being essentially random and three dimensional, tends not to transmit directional preferences from large-scale motions to small scales; therefore, motions at scales many times smaller than those of the main energy-containing motions tend to be statistically isotropic. If the whole of the turbulence were statistically isotropic, we would have $\overline{u_1^2} = \overline{u_2^2} = \overline{u_3^2}$, etc.; this small-scale "local" isotropy leads to simple relations between the various $\Phi_{ij}$ for sufficiently high $k$. Unfortunately, some information about the large-scale structure does pass down to the small eddies, complicating their analysis. It is obvious that in a shear layer, where the rate of energy transfer to the smallest, energy-dissipating, eddies varies across the width of the flow, the intensity of the small-scale motion will vary likewise. It is less obvious that, even in statistically homogeneous turbulence, the instantaneous rate of energy transfer to the small eddies will fluctuate in space and time. It is even less obvious that, as a consequence, the intensity of the small-scale motion, *obtained as an average over a time much longer than the typical time-scale[4] of the small motions but much shorter than the typical time-scale of the large motions*, will also fluctuate in space and time, at a rate depending on the large-scale motion. Therefore, assumptions of

---

[4] Note that this time scale is (eddy length scale)/(eddy velocity scale) whereas the period of fluctuations seen by a fixed observer is, according to Taylor's hypothesis [1.11], closely related to (eddy length scale)/(*mean* velocity).

the statistical independence of widely separated wave numbers are not quite correct [1.19]. This is extremely unfortunate: simple and useful results can be obtained from the Kolmogorov "universal equilibrium" assumption [1.14] that the statistics of the small-scale motion are determined uniquely by the rate of transfer of turbulent kinetic energy per unit mass down the "cascade" to the dissipating eddies, $\varepsilon$, by the wave-number magnitude $k \equiv |\mathbf{k}|$, and by the kinematic viscosity $v$. We will now derive some of these results, whose inaccuracy turns out to be no more than the likely error in experiments to check them.

The typical length and velocity scales of the dissipating eddies (the Kolmogorov scales) are, according to the universal equilibrium assumption, simply $\eta \equiv (v^3/\varepsilon)^{1/4}$ and $v_\varepsilon \equiv (v\varepsilon)^{1/4}$. The fact that the Reynolds number based on these scales is exactly unity is largely an accident but, since the smallest motions are by definition those for which inertial effects (i.e., gain of energy by spectral transfer) and viscous effects (i.e., loss of energy by dissipation) are equal on the average, we expect the Reynolds number, representing as always the ratio of inertial effects to viscous effects, to be of the order of unity. Another result, obtainable from simple dimensional analysis, is

$$\Phi_{ii}(\mathbf{k}) = v_\varepsilon^2 \eta^3 f_1(k\eta) \tag{1.30}$$

where $\Phi_{ii} \equiv \Phi_{11} + \Phi_{22} + \Phi_{33}$ is a spherically symmetric function of $\mathbf{k}$ in the locally isotropic region of wave-number space. Also $\Phi_{ij}$ and $\phi_{ij}$ are zero for $i \neq j$ in this region. The one-dimensional spectra for $i = j$ obey relations like

$$\phi_{11}(k_1) = v_\varepsilon^2 \eta f_2(k\eta) . \tag{1.31}$$

Equations (1.30) and (1.31) are valid—to the universal-equilibrium approximation—for all wave numbers much higher than the typical wave numbers of the energy-containing range (a factor between 10 and 100 in wave number is required). Another requirement of the above scaling on $\varepsilon$ and $v$ is that the mean-square vorticity $\overline{\omega_i^2}$, which resides almost entirely in the smallest eddies, should be proportional to $(v_\varepsilon/\eta)^2 \equiv \varepsilon/v$: in fact, the constant of proportionality is unity.

Instead of $\phi_{ij}(k_1)$ some workers prefer to use the "energy spectrum function" or "wave-number-magnitude spectrum", $E(k)$, defined as $(1/2)\int\Phi_{ii}(\mathbf{k})dA(k)$, where $dA(k)$ is an element of area of the surface of a sphere of radius $k \equiv |\mathbf{k}|$. In the locally isotropic region of $k$-space it is just $2\pi k^2 \Phi_{ii}(\mathbf{k})$, or in terms of the one-dimensional spectrum of the longitudinal component of velocity fluctuation it is $k_1^2 d^2 \phi_{11}(k_1)/dk_1^2 - k_1 d\phi_{11}(k_1)/dk_1$. Both these results rely on the presence of local isotropy and not on universal equilibrium as such (see [1.14, pp. 49–50]).

Now if the lowest wave number at which (1.30) is valid is smaller than the lowest wave number at which viscous effects are significant (about $0.1/\eta$), then, according to the universal equilibrium assumptions, (1.30) should reduce to a viscosity-independent form in the intervening range, which is therefore called the "inertial subrange": $k > 0.1/\eta$ defines the "viscous subrange". The only dimensionally correct form of (1.30) in the inertial subrange is

$$\Phi_{ii}(k) = \text{constant} \times \varepsilon^{2/3} k^{-11/3} \tag{1.32}$$

while (1.31) becomes

$$\phi_{11}(k_1) = \text{constant} \times \varepsilon^{2/3} k^{-5/3}, \tag{1.33}$$

the constant in the latter case being about 0.5. The analogous result for the scalar-fluctuation spectrum in the inertial subrange, valid for $v/\gamma \ll 1$, is

$$\phi_{cc}(k_1) = \text{constant} \times 2\varepsilon_c \varepsilon^{-1/3} k_1^{-5/3}. \tag{1.34}$$

Here $\varepsilon_c$ is the rate of destruction of $\overline{c^2}/2$, where $\overline{c^2}$ is the mean-square concentration fluctuation, by the molecular diffusivity $\gamma$, so that $\partial \overline{c^2}/\partial t = -2\varepsilon_c$ in the absence of other terms in the transport equation for $\overline{c^2}$. Sometimes $2\varepsilon_c$ is given the symbol $\chi$; here it is convenient to use $\varepsilon_c$ because of its analogy with $\varepsilon$, but usage is not uniform in the literature. The constant in (1.34) is poorly documented but seems to be about 0.6 [1.20]. In practice, $-5/3$ power ranges are frequently found at wave numbers so low that cross spectra like $\phi_{12}$ are not zero as required by local isotropy; this is more likely to be coincidence than failure of the universal-equilibrium concept.

Various attempts have been made to measure, or allow for, departures from the simple results of universal-equilibrium theory. One of the most conclusive tests is a measurement of the skewnesses of the spatial derivatives of velocity components, the simplest of these being $S \equiv \overline{(\partial u_1/\partial x_1)^3}/[\overline{(\partial u_1/\partial x_1)^2}]^{3/2}$. According to the theory this skewness should be a universal constant; in fact it varies slowly with the Reynolds number of the energy-containing eddies (an example of the "turbulent Reynolds number" mentioned on p. 6). It is usual to define this Reynolds number as $\overline{(u_1^2)}^{1/2} \lambda/v$, where $\lambda$ is the Taylor "microscale" defined by $\lambda^2 = \overline{u_1^2}/\overline{(\partial u_1/\partial x_1)^2}$. Unfortunately $\lambda$ is a hybrid scale (the denominator in its definition being determined by the *small-scale* motion) but one of the results of isotropy of the dissipating (smallest) eddies is that the dissipation rate per unit mass, $\varepsilon$, equal in general to $1/2v\overline{(\partial u_i/\partial x_j + \partial u_j/\partial x_i)^2}$ according to the definition on p. 12, simplifies to $15v\overline{(\partial u_1/\partial x_1)^2} \equiv 15v\overline{u_1^2}/\lambda^2$.

The square of the microscale Reynolds number is therefore proportional to $\overline{(u_1^2)}^2/(\varepsilon\nu) \equiv \overline{u_1^2}/v_\varepsilon^4$. This is $\overline{(u_1^2)}^{1/2}L_\varepsilon/\nu$, where $L_\varepsilon \equiv \overline{(u_1^2)}^{3/2}/\varepsilon$ is a typical length scale of the energy-containing eddies, which determine the energy-transfer or dissipation rate $\varepsilon$. A second approximation to the universal-equilibrium theory, based on the assumption that the short-time-average rate of viscous dissipation in the small eddies has a log-normal probability distribution in $x$-space (that is, its logarithm has a three-dimensional normal or Gaussian distribution), predicts that the skewness $S$ should increase proportional to some power of the micro-scale Reynolds number. This has been confirmed by experiment [1.21], with $S \sim \mathrm{Re}^{3/8}$ approximately. The exponent is related to the standard deviation of the log-normal distribution. The corresponding result for the wave-number spectrum $\phi_{11}$ is, approximately, an increase in the numerical value of the exponent in (1.33) from 5/3(1.67) to 1.71, and the consequent changes in values of $\varepsilon$ inferred from values of $\phi_{11}$—the main use of (1.33) in experimental work—is small. However the moral is obvious: simple physical ideas should be carefully examined, particularly if they relate to so complex a phenomenon as turbulence.

The above ideas have been developed for homogeneous turbulence but the universal-equilibrium concept should be no less accurate in inhomogeneous flows, where universality or near universality of the small-scale motion is a great experimental and theoretical simplification. For further details of the basic theory of homogeneous turbulence, see the standard text [1.14]. Spectral theory will not often appear explicitly in the rest of this book, but the concepts of the last two sections underlie much of the discussion below.

## 1.6 Inhomogeneous Turbulence

In homogeneous turbulence the transport equations for $\overline{u_i u_j}$ (i.e., (1.20) with $r=0$) degenerate because there is no spatial transport. In inhomogeneous flows, Reynolds-stress gradients affect the mean motion, as can be seen from (1.18), and the Reynolds-stress transport equations become general: for both reasons the latter equations have received close attention [1.22]. They are sometimes called "conservation" equations although the only quantities which are conserved in the strict thermodynamic sense are mass, momentum and energy (including turbulent kinetic energy, $(1/2)\overline{u_i^2}$). The terms in the equations are most easily explained as contributions to the maintenance of the transported quantity in a unit control volume (Fig. 1.5).

Following the rule for obtaining transport equations for products set out in Section 1.4 and considering constant-property flow for simplicity,

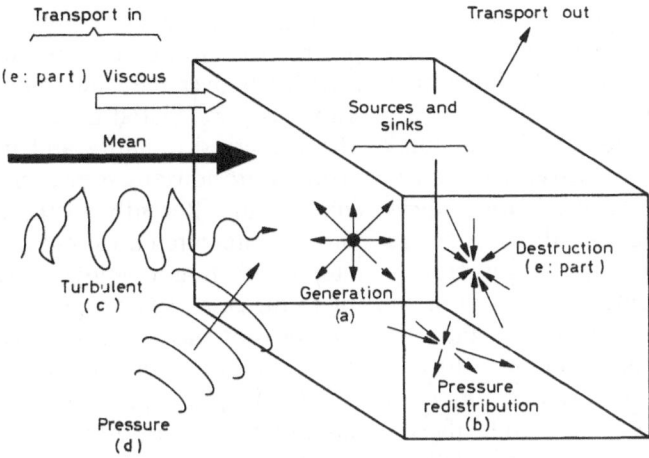

Transport in

Transport out

(e: part ) Viscous

Mean

Sources and sinks

Turbulent (c)

Generation (a)

Destruction (e: part )

Pressure (d)

Pressure redistribution (b)

Fig. 1.5. Control volume for transport equations in turbulent flow

we take the $x_i$-component Navier-Stokes equation for the instantaneous velocity ($U_i + u_i$ in our notation for flows with nonzero mean velocity) and multiply by the fluctuation $u_j$. We then add $u_i$ times the equations for $U_j + u_j$, and average. The form usually quoted is

$$\frac{\partial \overline{u_i u_j}}{\partial t} + U_l \frac{\partial \overline{u_i u_j}}{\partial x_l} = -\left( \overline{u_i u_l} \frac{\partial U_j}{\partial x_l} + \overline{u_j u_l} \frac{\partial U_i}{\partial x_l} \right) \tag{a}$$

$$+ \frac{\overline{p'}}{\varrho} \left( \frac{\partial u_i}{\partial x_j} + \frac{\partial u_j}{\partial x_i} \right) \tag{b}$$

$$- \frac{\partial}{\partial x_l} (\overline{u_i u_j u_l}) - \frac{1}{\varrho} \left( \frac{\partial}{\partial x_i} (\overline{p' u_j}) + \frac{\partial}{\partial x_j} (\overline{p' u_i}) \right) \tag{c, d}$$

$$+ v \left( u_i \frac{\overline{\partial^2 u_j}}{\partial x_l^2} + u_j \frac{\overline{\partial^2 u_i}}{\partial x_l^2} \right) \tag{e}$$

$$+ (\overline{f_i' u_j} + \overline{f_j' u_i}) . \tag{f} \quad (1.35)$$

Choosing different values of $i$ and $j$ gives us six equations for the six independent components of the stress tensor $-\varrho \overline{u_i u_j}$. Here $f'$ is the body-force fluctuation, $p'$ is the pressure fluctuation, and we have rearranged some terms into one part expressible as spatial gradients of statistical averages and another part not so expressible. Note that the spatial part of the mean transport term, $U_l \partial(\overline{u_i u_j})/\partial x_l$, can be rearranged as a pure spatial gradient, $\partial(U_l \overline{u_i u_j})/\partial x_l$, by using the continuity equation. The reason for making the distinction is that spatial-gradient terms always represent transport of the conserved quantity from one place to another, as can be seen by noting that the integrals of such terms over the whole flow volume must be zero. Terms not expressible as pure spatial gradients represent generation or destruction of the conserved quantity and in general their integrals over the whole flow volume are nonzero; sometimes they are called "source" (or "source/sink") terms. Recently Lumley [1.23] has pointed out that the separation of the pressure terms into the source/sink term (b) and the transport term (d) is not unique. His alternative separation involves generalizing to the case $i \neq j$ an argument which strictly applies only to the case $i = j$, but is at least as reasonable as the usual separation into (b) and (d). Lumley's suggestion is to replace (b) by $-(\overline{u_i \partial p'/\partial x_j + u_j \partial p'/\partial x_i})/\varrho + (2/3)\delta_{ij}\overline{\partial p' u_l}/\partial x_l/\varrho$ and (d) by $-(2/3)\delta_{ij}\overline{\partial p' u_l}/\partial x_l/\varrho$. Thus there is no pressure transport term if $i = j$. The available evidence is that the pressure transport is in any case small even if $i \neq j$.

The source terms are as follows [letters refer to (1.35) or Fig. 1.5]:

a) Generation by interaction of the turbulent motion with the mean rate-of-strain field (actually part of the term represents exchange between the particular $\overline{u_i u_j}$ under consideration and the others, analogous to the "tilting"-terms in the vorticity equation). The generation term is not necessarily of the same sign as $\overline{u_i u_j}$, but clearly $\overline{u_i u_j}$ tends to decay rather rapidly in cases where it is not.

b) Generation, or destruction, or redistribution between components, by means of pressure fluctuations: given symbol $\phi_{ij}$ in this book, and called the "pressure-strain" term because it is the mean product of the fluctuating pressure and fluctuating strain rate. Generally, the pressure fluctuations tend to make the turbulence more nearly isotropic, increasing the weaker normal stresses at the expense of the stronger and reducing the magnitude of shear stresses.

e (part) Destruction, or generation, by means of viscous-stress fluctuations. The viscous sink terms in the normal-stress equations $(i = j)$ are negative definite and provide the dissipation of turbulent energy, but generation of Reynolds stress by viscous action can occur during transition from laminar to turbulent flow. The viscous terms in the shear-stress equations $(i \neq j)$ are negligibly small if the small-scale motions are locally isotropic, an assumption sometimes too glibly made.

f) Generation, or destruction, by means of body-force fluctuations. The most common example is buoyancy-force fluctuation in a variable density fluid in a gravitational field in the negative $x_m$-direction (say): then $f_i' = -g\delta_{im}\varrho'/\bar{\varrho}$ and the complete body-force term is $-g(\delta_{im}\overline{\varrho' u_j} + \delta_{jm}\overline{\varrho' u_i})/\bar{\varrho}$. For the shear stress $\overline{u_1 u_2}$ in the case $m = 2$, this becomes $-g\overline{\varrho' u_1}/\bar{\varrho}$ and for the turbulent energy $(1/2)\overline{u^2}$ it is $-g\overline{\varrho' u_2}/\bar{\varrho}$.

The transport terms are:

Transport by the mean flow — the left-hand side of the equation. In statistically stationary ("steady") flows, $\partial \overline{u_i u_j}/\partial t$ is zero: unsteady flows present no extra problems in turbulence modelling as long as the total transport term is no larger than in corresponding steady flows.

c)   Transport by velocity fluctuations: $\overline{u_i u_j u_l}$ represents a transport of $\overline{u_i u_j}$ in the $x_l$ direction, for the reasons already given to explain the momentum transfer rate $\overline{u_i u_j}$, and so the contribution to the rate of change of $\overline{u_i u_j}$ at a given point is (minus) the $x$-derivative (transport out of a unit control volume less transport in).

d)   Transport by pressure fluctuations, the physical interpretation of the terms being rather difficult.

e (part)  Transport by viscous stress fluctuations, which is generally small except when spatial gradients of Reynolds stress are extremely large.

The terms can be interpreted more rigorously in the case of the turbulent energy equation, which is half the sum of the three normal-stress equations,

$$\frac{\partial}{\partial t}(\tfrac{1}{2}\overline{u_i^2}) + U_l \frac{\partial}{\partial x_l}(\tfrac{1}{2}\overline{u_i^2}) = -\overline{u_i u_l}\frac{\partial U_i}{\partial x_l} - \frac{\partial}{\partial x_l}(\overline{p'u_l}/\varrho + \tfrac{1}{2}\overline{u_i^2 u_l}) - \varepsilon$$

$$- \text{viscous transport} + \overline{f_i' u_i}. \qquad (1.36)$$

Note that the pressure "source" term $\overline{p'(\partial u_1/\partial x_1 + \partial u_2/\partial x_2 + \partial u_3/\partial x_3)}$ vanishes by continuity, being only an exchange between the three components of $\overline{u_i^2}$. The mean-flow generation term, called "production" is strictly $-\overline{u_i u_l}(1/2)(\partial U_i/\partial x_l + \partial U_l/\partial x_i)$, which is the product of each Reynolds stress with the corresponding mean rate of strain and is therefore the rate at which the mean flow does work on the turbulence. The "dissipation" term can be interpreted as the mean rate at which the turbulence does work against viscous stresses. The turbulent transport "diffusion" terms are just a special case of the corresponding terms in the $\overline{u_i u_j}$ equation. The left-hand side of the turbulent energy equation is called the "advection". The names advection, production, diffusion and dissipation are used in this book *only* for the terms in the turbulent energy equation and not for terms in the Reynolds-stress transport equations in general. Sometimes the terms on the left-hand side of any transport equation are called the "convective" terms: we avoid this usage here as it can lend to confusion with thermal convection (Chapt. 4).

Transport equations for triple products and higher-order structural parameters will be mentioned briefly in Chapt. 5.

## 1.7 Turbulent Flows in Practice: Shear Layers

This completes the outline of the physical processes of turbulence and
the equations that govern them. It remains to discuss the properties of the
solutions of these equations for the boundary conditions of practical
interest. This book is intended to serve people working on a wide range
of problems and a correspondingly wide range of boundary conditions.
For this reason the foregoing analysis has been kept general and the
details of specialist problems have been deferred until later chapters.
However, some sets of boundary conditions or flow geometries occur so
frequently that a general discussion will be given here. For simplicity
we consider only statistically stationary flows.

We start by asking what flow geometries result in Reynolds stress
gradients which are significant enough, compared to pressure gradients,
body forces, etc., to have an appreciable effect on the mean acceleration.
Analogous questions can be asked about the gradients of the scalar-
concentration flux $\overline{cu_i}$ (p. 15): the answers are the same. Rapid changes
of Reynolds stress with respect to time, following the mean motion of
the fluid (i.e., large gradients along a mean streamline) are rare, because
the mean transport terms in the Reynolds-stress transport equations
are never very large. Therefore, large gradients are usually in a direction
normal to the streamline. Since Reynolds stresses are rarely more than,
say, one percent of the mean dynamic pressure $(1/2)\varrho U^2$, while pressure
gradients are in general of the order of $(1/2)\varrho U^2/l$ where $l$ is a typical
streamwise length of the flow, it follows that Reynolds stresses must
change by a large fraction of their typical value in a cross-stream distance
of order one percent of $l$ if they are to have a significant effect on the mean
flow. Now the Reynolds stress whose cross-stream gradient affects the
streamwise acceleration is the shear stress, referred to coordinates along
and normal to the stream. The equation for the mean motion along a
mean streamline in two-dimensional constant-density flow, neglecting
viscous stresses and streamwise gradients of Reynolds stress for simplicity
and taking the $x$ axis locally tangential to the streamline, is

$$U \frac{\partial U}{\partial x} = -\frac{1}{\varrho} \frac{\partial p}{\partial x} - \frac{\partial \overline{uv}}{\partial y} \qquad (1.37a)$$

or

$$\frac{\partial}{\partial x}(p + \tfrac{1}{2}\varrho U^2) = -\frac{\partial \overline{uv}}{\partial y} \qquad (1.37b)$$

so that the streamwise spatial gradient of total pressure $p + (1/2)\varrho U^2$
equals the cross-stream shear-stress gradient. A varying, a cross-stream

gradient of shear stress leads to a cross-stream gradient of total pressure, and thus of streamwise velocity. Indeed, it can be seen from (1.35) that a large velocity gradient in *some* direction is necessary for Reynolds stresses to be maintained, at least in the absence of body forces. The part of a flow in which shear stress interacts with a large cross-stream gradient of streamwise velocity is called a *shear layer* [1.24]: the thickness $\delta$ of a shear layer is always fairly small compared to the distance from its origin $l$. Sometimes, but by no means always, the thickness is so small that streamwise gradients of $U$, or of other quantities, are very small compared with cross-stream gradients of the same quantity, which permits approximations, such as the neglect of some stress gradients, to be made in the equations of motion [Ref. 1.3, Chapt. 7; Ref. 1.24, Chapt. 3]. Because it was first developed for shear layers on a plane solid boundary this simplification is called the "boundary layer" approximation. Since it applies to other flows, such as plane jets and wakes, a better name is "thin-shear-layer" approximation. There is also a slightly less general "slender-shear-layer" approximation for flows which are thin in both cross-stream directions, while slender axisymmetric flows obey a simple version of this.

The thin-shear-layer approximations were first developed for laminar flows, in which they are generally more accurate because laminar shear layers are usually very thin[5]. The approximations are sometimes used rather uncritically in turbulent flow, to the extent that "thin shear layers" and "flows with significant Reynolds stress gradients" are occasionally implied to be synonymous. While the thin-shear-layer simplifications need not be dismissed as applicable only to flows over streamlined bodies, workers in other fields should beware of the—for them—excessive emphasis on thin shear layers in aeronautically oriented textbooks and research papers. The addition of other rate-of-strain components to the "simple shear" (cross-stream gradient of streamwise velocity) can have a large effect on turbulence. For instance curvature of the streamlines in the plane of the shear (Sect. 3.1) can greatly augment or decrease turbulent mixing in much the same way as buoyancy forces (Chapt. 4). The effects can be significant even if the extra strain rate is small enough for the thin-shear-layer approximation to apply; it is convenient to call a flow which is free of these effects a "simple shear layer".

The mean-motion equations and continuity equation for a thin shear layer confined to a region near the $x, z$ plane, but without the very

---

[5] And also because the neglected stress gradients are of order $(\delta/l)^2$ times the retained ones, while in turbulent flow the ratio is only $\delta/l$.

special choice of coordinates used in deriving (1.37), are

$$\varrho \left( U \frac{\partial U}{\partial x} + \hat{V} \frac{\partial U}{\partial y} + W \frac{\partial U}{\partial z} \right) = - \frac{\partial p}{\partial x} + \frac{\partial}{\partial y} \left( \mu \frac{\partial U}{\partial y} - \varrho \overline{uv} \right) + \varrho F_x ,$$

(1.38)

$$0 = - \frac{\partial p}{\partial y} + \varrho F_y ,$$    (1.39)

$$\varrho \left( U \frac{\partial W}{\partial x} + \hat{V} \frac{\partial W}{\partial y} + W \frac{\partial W}{\partial z} \right) = - \frac{\partial p}{\partial z} + \frac{\partial}{\partial y} \left( \mu \frac{\partial W}{\partial y} - \varrho \overline{vw} \right) + \varrho F_z ,$$

(1.40)

and

$$\frac{\partial \varrho U}{\partial x} + \frac{\partial \varrho \hat{V}}{\partial y} + \frac{\partial \varrho W}{\partial z} = 0 .$$    (1.41)

Here $F_x$, $F_y$, and $F_z$ are the mean components of the body force. The equations in this form also apply to compressible flow, except for the lack of some small terms containing the density fluctuation $\varrho'$, if $\hat{V}$ is taken as $V + \overline{\varrho' v}/\varrho$; in constant-density flow, $\hat{V} = V$. The enthalpy equation [a special case of (1.9)] is

$$\varrho \left( U \frac{\partial h}{\partial x} + \hat{V} \frac{\partial h}{\partial y} + W \frac{\partial h}{\partial z} \right)$$

$$= U \frac{\partial p}{\partial x} + W \frac{\partial p}{\partial z} + \frac{\partial}{\partial y} \left( \frac{k}{c_{\mathrm{p}}} \frac{\partial h}{\partial y} - \varrho \overline{h' v} \right) + \Phi$$    (1.42)

where $h = c_{\mathrm{p}} T$ for perfect gases, $h'$ is the enthalpy fluctuation, and $\Phi$ is the rate of viscous dissipation of mean and turbulent kinetic energy into heat, per unit volume. For details of the order-of-magnitude arguments used to derive the thin-shear-layer equations from the Navier-Stokes equations see SCHLICHTING [1.3] or CEBECI and BRADSHAW [1.24]. In a two-dimensional mean flow, gradients with respect to $z$ are zero and $\overline{vw}$ is zero by symmetry; in axisymmetric flow analysed in $x$, $r$, $\theta$ coordinates, mean gradients with respect to $\theta$ are zero.

The most remarkable simplification in the thin-shear-layer equations is that of the second equation, (1.39), which in the absence of body forces allows $p$ to be equated to its value just outside the shear layer and thus changes $p$ from a dependent variable to a boundary condition. This changes the type of the above system of differential equations from elliptic (in which a disturbance at one point affects the whole flow field) to parabolic (in which a disturbance affects only the flow further downstream, so that a "marching" solution, progressing steadily downstream from given initial values, becomes possible). For an introduction to numerical methods for fluid flows see [1.25]. In Chapter 2 we discuss the momentum integral equation and other equations obtained by integrating the thin-shear-layer equations through the thickness of the layer. These equations embody no new physical principles but are frequently used in simple calculation methods.

There is a significant class of shear layers which, although fairly thin and dominated by shear-stress gradients, does not obey the thin-shear-layer equations with acceptable accuracy— in particular, the normal pressure gradient may be appreciable because the streamlines are curved. However the terms neglected in the derivation of the thin-shear-layer equations will usually be *small*, and a more important property of this class is that extra strain rates (additional to the simple shear) may be large enough to affect the turbulence structure greatly. For a fuller discussion of types of shear layers see Section 2.2.

For the record, we write down the transport equations for turbulent energy and shear stress in $x$, $y$, $z$ notation for the special case of a steady two-dimensional constant-property thin shear layer (without body forces) whose dominant mean rate of strain is $\partial U/\partial y$. In this case $\overline{uw}$ and $\overline{vw}$, like all mean quantities containing $W$ or odd powers of $w$, are zero (by symmetry) and gradients with respect to $x$ are small compared to gradients with respect to $y$ (by definition of a thin shear layer). We use the common symbol $q^2$ for $u_i^2$, assume that the local Reynolds number is high enough for *all* viscous terms except the dissipation (denoted by $\varepsilon$) to be negligible, and get

$$\left(U\frac{\partial}{\partial x}+V\frac{\partial}{\partial y}\right)\tfrac{1}{2}\overline{q^2}=-\overline{uv}\frac{\partial U}{\partial y}-\frac{\partial}{\partial y}(\overline{p'v}+\tfrac{1}{2}\overline{q^2 v})-\varepsilon, \tag{1.43}$$

$$\left(U\frac{\partial}{\partial x}+V\frac{\partial}{\partial y}\right)(-\overline{uv})=\overline{v^2}\frac{\partial U}{\partial y}-\overline{p'\left(\frac{\partial u}{\partial y}+\frac{\partial v}{\partial x}\right)}+\frac{\partial}{\partial x}(\overline{p'u}+\overline{uv^2}). \tag{1.44}$$

The corresponding mean-flow equations, again neglecting the viscous terms, are

$$U\frac{\partial U}{\partial x}+V\frac{\partial U}{\partial y}=-\frac{1}{\varrho}\frac{dp}{dx}-\frac{\partial \overline{uv}}{\partial y}, \tag{1.45}$$

$$\frac{\partial U}{\partial x}+\frac{\partial V}{\partial y}=0 \tag{1.46}$$

where an ordinary derivative is used to remind us that to the thin-shear-layer approximation $p$ is a function of $x$ only. We shall see that even if the "bulk" Reynolds number is large the "local" or "turbulent" Reynolds number becomes small very near a solid surface so that viscous terms appear in (1.43)–(1.45). The symbol $\tau$ is commonly used for the total shear stress $-\varrho\overline{uv}+\mu\partial U/\partial y$.

In shear layers, whether strictly thin or not, the largest eddies almost invariably extend across the thickness of the layer. If a shear layer were generated artificially so that the eddies were initially all small compared to the thickness, the rate of (streamwise) growth of the typical eddy size would exceed the rate of increase of the layer thickness until the largest eddies had grown to fill the flow. This can be thought of, rather crudely, as a consequence of the random-walk process induced by vortex-line interaction. It is a great misfortune, because assumptions of weak inhomogeneity, which would be valid if the eddy length scale were small compared to the shear-layer thickness, cannot be used with confidence.

A classical example of the largeness of eddy length scales is that very roughly half the turbulent energy on the centerline of a pipe resides in eddies whose longitudinal wavelength exceeds the pipe diameter $d$ (that is, half the area under the one-dimensional energy spectrum $\phi_{ii}(k_1)$ is at a wave number less than about $2\pi/d$) and the Reynolds shear stress is concentrated even more in the large eddies. The smallest eddies near the centerline of a pipe have wave numbers of the order of $(v^3/\varepsilon)^{-1/4}$, as always, and measurements of $\varepsilon$ show that their typical wavelength is roughly $30d(Ud/v)^{-3/4}$, or 0.005 times the diameter at a Reynolds number $Ud/v$ of $10^5$.

From the point of view of the Reynolds-stress transport equations discussed in Section 1.6, the consequence of the largeness of the largest eddies is that the transport terms in (1.35) and (1.36) are not usually negligible. In particular, large eddies play an important part in the turbulent transport across the layer, while their large size implies a long streamwise lifetime so that spatial transport of Reynolds stress by the mean flow is significant compared with, say, viscous dissipation. The extreme cases are that in the fully turbulent part of the inner layer of a turbulent wall flow (Sect. 1.8) both mean and turbulent transport are negligible compared with the source/sink terms, while near the free-stream edge of almost any shear layer mean and turbulent transport are nearly equal and opposite and much larger than source/sink terms. Elsewhere in a wall flow, mean and turbulent transport are small but not negligible, while turbulent transport is important in all parts of a free shear layer. Turbulent energy balances for the common types of shear layer are given by TOWNSEND [1.22] and, although in some cases superseded by more accurate results (referenced below), give a useful qualitative idea of the flow of energy.

When all transport terms are negligible, the turbulence is said to be in local (energy) equilibrium. Local equilibrium in $x$-space should be distinguished from the universal equilibrium in $k$-space discussed in Section 1.5, and from the generalized profile similarity called "self-preservation" by TOWNSEND but "equilibrium" by some other authors. In self-preserving flows (Chapt. 2) transport terms are not negligible but the terms in the Reynolds-stress transport equations obey simple scaling laws; instead of generation and destruction terms being equal, as in local-equilibrium regions, their ratio is a function of $y/\delta$ only. Equation (1.36) shows that equality of, or simple relations between, the generation and destruction terms implies simple relations between turbulence quantities and the mean rate of strain (which appears in the generation terms). It follows that in local-equilibrium or self-preserving flows fairly simple empirical relations can be found between the Reynolds-stress components and the components of the mean rate-of-strain tensor.

It is important to realize that this simplicity is a consequence of simple scaling laws rather than unusual smallness of the turbulent eddies, and that it is restricted to these special classes of flow. However it has encouraged a search for a generally valid relation between local Reynolds stress and local mean rate of strain, like the viscous stress relation (1.6) but with an "eddy viscosity" replacing the molecular viscosity (and with the addition of a term $(1/3)\delta_{ij}\sigma_{ii}$ to the right-hand side to account for the fact that $\partial U_i/\partial x_i$ is zero in constant-density flow whereas $\sigma_{ii} = -\varrho\overline{u_i^2}$ is not). As a definition of a fourth-order tensor quantity having the dimensions of viscosity but determined by the properties of the flow considered this is unexceptionable! The fallacy lies in expecting the eddy viscosity to be a scalar, simply related *in general* to the mean-flow scales, or even the turbulence scales, near the point considered; it is a hybrid quantity, depending on both the turbulence and the mean flow. As will be seen in Chapter 5, empirical eddy-viscosity relations are quite widely used in calculation methods, not only for correlating empirical information about Reynolds stress (turbulent transport of momentum) but also, under the general name of "eddy diffusivity" or "gradient transport", for representing turbulent transport of Reynolds stress or turbulent energy in (1.35) or (1.36). How far one should trust correlations based on a physically erroneous concept is a matter of opinion.

Any method of predicting Reynolds stresses is an explicit or implicit "closure" (p. 20) of (1.35). Shear-layer calculation methods are discussed in Chapt. 5. Here we simply note that the two main classes are eddy-viscosity methods (or less specific algebraic relations between Reynolds stress and mean rate of strain) and explicit closures of one or more of the component equations of (1.35) including the transport terms. Similar comments apply to transport equations and calculation methods for heat or pollutant transfer (Chapt. 6).

## 1.8 Turbulent Flow Near a Solid Surface (Fig. 1.6)

The turbulent flow at a height $y$ above a smooth solid surface necessarily depends on $y$, on the shear stress at the surface, $\tau_w$, and on fluid properties $\varrho$ and $v$. It may also depend on conditions at smaller $x$, or larger $y$ or different $z$, and on the pressure gradient. However, if $y$ is small compared both to the thickness of the shear layer and to the distances in the $x$ or $z$ directions over which significant changes of (say) $\tau_w$ occur, and if the pressure gradient in the $x$ or $z$ directions is small compared with (say) $\tau_w/y$, then it is plausible to suppose that the flow is statistically determined by $y$, $\tau_w$, $\varrho$ and $v$ only. Conditions elsewhere in the flow appear only via their effect on $\tau_w$, so that the shear-layer thickness is immaterial. The acceleration is small—in local terms—so that the total shear-stress gradient $\partial\tau/\partial y$ is small also. The wall layer is sometimes called the "constant-stress" layer. Typically it occupies 10 to 20 percent of the shear-layer thickness. Note that the above arguments are independent of the applicability of the thin-shear-layer approximation (Sect. 1.7) to the flow as a whole, although they imply its applicability at distances from the surface of order $y$ and less. Dimensional analysis gives

$$\frac{U}{u_\tau} = f_1\left(\frac{u_\tau y}{v}\right) \qquad (1.47a)$$

$$\frac{\partial U}{\partial y} = \frac{u_\tau}{y} f_2\left(\frac{u_\tau y}{v}\right) \qquad (1.47b)$$

$$-\frac{\overline{uv}}{u_\tau^2} = f_3\left(\frac{u_\tau y}{v}\right) \qquad (1.47c)$$

where $u_\tau \equiv \sqrt{(\tau_w/\varrho)}$ is called the friction velocity, and $f_1(0)=f_3(0)=0$ because all velocity components are zero at the surface. Often, $y^+$ is used to denote $u_\tau y/v$, and $U^+$ to denote $U/u_\tau$.

Now the largest eddies at height $y$ cannot have wavelengths much larger than $y$ in the $y$ direction, and if the largest wavelengths in the $x$ or $z$ direction were much larger than $y$ the large-scale motion would be instantaneously two dimensional, like the synoptic-scale motion in the

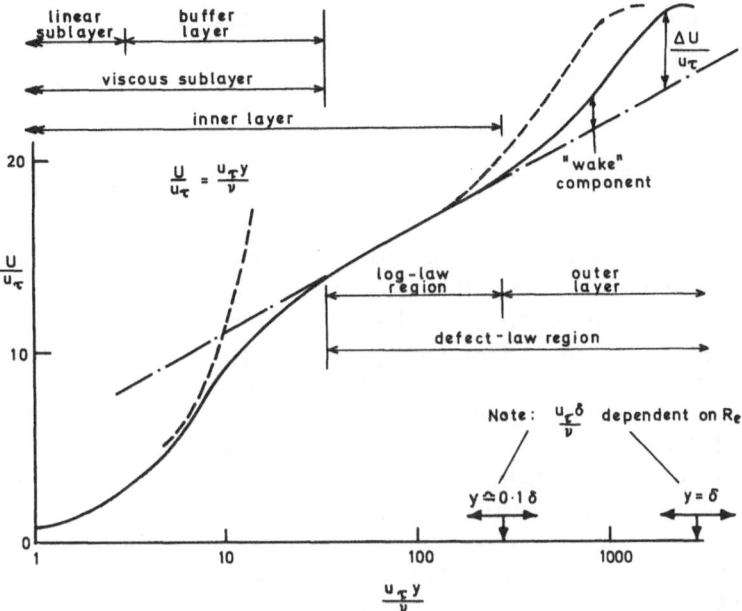

Fig. 1.6. Wall-layer nomenclature, on inner-layer scales. Thickness of inner layer is roughly constant fraction of shear-layer thickness: for details of outer layer see Section 2.3

atmosphere, and therefore not truly turbulent. The remarks in Sect. 1.6 about turbulent eddies tending to fill the flow imply, in this context, that the largest wavelengths will not be an order smaller than $y$. Therefore, the largest wavelengths of the genuinely turbulent motion will be of order $y$, and we may use $y$ as a typical turbulent length scale. Equation (1.47c) shows that $u_\tau$ may be used as a typical velocity scale of the energy-containing turbulence, since $(-\overline{uv})^{1/2}$ is undoubtedly such a scale. Therefore, $u_\tau y/v$ is a typical Reynolds number of the energy-containing turbulence, another example of a "turbulence Reynolds number" (p. 6). We expect the viscous shear stress, and direct viscous effects on the turbulent eddies and their interaction with the mean flow, to be small at large $u_\tau y/v$ so that $-\overline{uv}=u_\tau^2$, i.e., $f_3(\infty)=1$, and $f_2(\infty)=$ constant $=1/\kappa$, say. That is, we expect

$$\frac{\partial U}{\partial y} = \frac{u_\tau}{\kappa y} \tag{1.48}$$

at large $u_\tau y/v$. Experimentally we find $\kappa \approx 0.41$ and (1.48) is acceptably accurate for $u_\tau y/v > 40$ approx. (Values between 30 and 50, or even higher, are often quoted, and of course the value depends on the accuracy required of (1.48); in this book we choose $u_\tau y/v = 40$, where $-\overline{uv}/u_\tau^2 \approx 0.94$). Integrating (1.48) and requiring compatibility with (1.47a), we get

$$\frac{U}{u_\tau} = \frac{1}{\kappa} \ln \frac{u_\tau y}{v} + C \qquad (1.49)$$

where, according to the above analysis, $C$ is an absolute constant for a Newtonian flow over a smooth surface and is found experimentally to be 5.0–5.2. The "logarithmic law" (1.49) is a special case of the "inner law" or "law of the wall" (1.47a). There are some types of non-Newtonian flows (Chapt. 7) in which molecular stress effects are still confined to the smaller eddies and in these cases (1.48) and (1.49) become valid at large enough distances from the surface for the non-Newtonian equivalent of $u_\tau y/v$ to be large. In such a case the constant of integration depends on the non-Newtonian constitutive law and no general information can be given.

There is some experimental support for the empirical generalization of (1.48) to the case where the total shear stress $\tau$ varies with $y$,

$$\frac{\partial U}{\partial y} = \frac{(\tau/\varrho)^{1/2}}{\kappa y} = \frac{(-\overline{uv})^{1/2}}{\kappa y} \qquad (1.50)$$

(the formula being, at best, valid only outside the "viscous sublayer" (i.e., $u_\tau y/v > 40$) so that $\tau = -\varrho\overline{uv}$). This involves the assumptions that the velocity scale for the turbulent eddies at height $y$ depends only on the local shear stress, while the eddies retain sufficient awareness of their surroundings for their length scale to be proportional to the distance from the surface. These are conflicting assumptions; even if we accept that the qualitative arguments leading to (1.48) are still valid if $\tau$ varies with $y$, we should expect $\partial U/\partial y$ to depend on the behavior of $\tau$ all the way from the surface to height $y$ or rather further. The point is discussed in Subsect. 2.3.1 where various integrals of (1.50) are presented (see also [1.26]). If $\partial\tau/\partial x$ is large, no local-equilibrium formula can apply [1.27].

There is no reliable model for the behavior of (1.47a) in the viscous sublayer $u_\tau y/v < 40$; turbulence in the presence of viscous effects is even more complicated than turbulence at high local Reynolds number. It is convenient, though not essential, to recast (1.47b), using (1.47c), in the equivalent form

$$\frac{\partial U}{\partial y} = \frac{(-\overline{uv})^{1/2}}{\kappa y f[(\tau/\varrho)^{1/2} y/v]} . \qquad (1.51)$$

Empirical forms for the "damping function" $f$ in (1.51) are discussed in Chapt. 5.

Equation (1.51) can be derived [1.26] by applying the local-equilibrium assumption (p. 33) to the turbulent energy equation, leading to the equality of production and dissipation. To the thin-shear-layer approximation, we have

$$\text{production} = -\overline{uv}\,\frac{\partial U}{\partial y} = \text{dissipation} = \varepsilon \equiv (-\overline{uv})^{3/2}/L \qquad (1.52)$$

where the last element of the equation defines the dissipation length parameter $L$, and outside the viscous sublayer $-\overline{uv} = u_\tau^2$. If the dissipation (i.e., the rate of energy transfer from the energy-containing eddies) is determined by $u_\tau$ and $y$ only (which implies most of the assumptions leading to (1.47)) then $\varepsilon = u_\tau^3/\kappa y$ and (1.48) follows. This derivation, like that above, is invalid in the viscous sublayer. Because of the very large gradients of turbulence properties normal to the wall, significant turbulent transport of energy towards the wall occurs, so that the flow is not in local equilibrium.

The above inner-layer formulae and the assumptions needed to derive them have been discussed in great detail because one or other of them is used in almost all calculation methods for turbulent wall flows. Inner-layer scaling is also the basis of several methods of measuring surface shear stress (one of which is to fit measured velocity profiles to (1.49) or its extensions and deduce $u_\tau$). Equation (1.49) is indeed one of the cornerstones of fluid dynamics; (1.33) is another, upon which doubts have been cast. Doubts have also been cast on (1.49) [1.15, 27, 28]; the assumptions on which it is based are to be regarded as engineering approximations rather than exact principles, but there is at present no general agreement on their limits of validity. It is perhaps a sound principle to suspect *any* simple formula that applies to turbulence, because the derivation of simple results from complicated equations generally implies crudity of approximation. However the main reason for the scatter in quoted values of $\kappa$ and $C$ is undoubtedly the difficulty of measuring the slope of a line through data points; most of the values proposed give closely coincident values of $U/u_\tau$ near the middle of the logarithmic range (say, $u_\tau y/\nu = 100$). The heat-transfer analogues of (1.49) and its consequences are discussed in Chapt. 6.

Another derivation of (1.50) uses the "mixing length" hypothesis. The physical model on which this hypothesis is based is one in which eddies interact by intermittent collisions, in much the same way as molecules. The analysis yields

$$\frac{\partial U}{\partial y} = \frac{(-\overline{uv})^{1/2}}{l} \tag{1.53}$$

where $l$ is the "mixing length", analogous to the mean free path between molecular collisions. It is plausible to assume $l = \kappa y$ in the inner layer of a wall flow but outside the region of significant viscous effects, and (1.50) follows. However, the physical model is now known to be completely erroneous; turbulent eddies (however one defines them) interact continuously, and are not small compared to the width of the mean flow. Unfortunately the collision model is still presented in some engineering textbooks, and leads to confusion when the student encounters the realities of turbulence structure. In [1.29] it is speculated that the model resulted from observations of free-surface patterns in turbulent open-channel flow. The requirement that the vorticity vector shall meet the free surface at right angles leads to motion similar to the synoptic-scale motion in the atmosphere which also has the vorticity vector normal to the surface; the "mixing length" collision model is a good description of the patterns seen. The success of (1.53), with $l = \kappa y$ and $\kappa$ found empirically, in predicting the mean velocity profile in a wall layer led to its use in other flows. However (1.53) is merely the definition of a length $l$, which will be related to the true length scale of the motion only if both mean flow and turbulence properties have the *same* length scale. This is the case only in the local-equilibrium wall layer [1.26] and in certain self-preserving shear layers (Chapt. 2), for purely dimensional reasons, just as in the case of eddy viscosity (Sect. 1.7).

Note that when the mixing length is well-behaved for dimensional reasons, the kinematic eddy viscosity $\nu_T$ defined by

$$\nu_T = -\overline{uv}(\partial U/\partial y) \tag{1.54}$$

also behaves simply. In a constant-stress wall layer, for instance

$$\nu_T = \kappa u_\tau y. \tag{1.55}$$

Since the mixing length and eddy viscosity concepts are both (erroneously) based on an analogy with the gradient-transport or "gradient-diffusion" mechanism of the kinetic theory of gases, and since simple assumptions about either lead to accurate results only in local-equilibrium regions and self-preserving flows, they are usually lumped together in general discussions and referred to as the "eddy viscosity" or "eddy diffusivity" or "gradient transport" concept (see Sect. 1.7).

However, even in more general flows the behaviour of the eddy viscosity may be simpler to correlate empirically than that of the shear stress itself, and it is in this spirit that it is used in simple calculation methods (Sect. 5.2.3).

## 1.9  Conditions Near a Turbulent/Non-Turbulent Interface

It can be deduced from (1.15) that, in general, vorticity can be transferred to initially irrotational fluid only by viscous stresses (the exceptions are the Bjerknes mechanism, involving pressure gradients in the presence of density gradients, and possibly some special kinds of body force with a nonzero curl). The interface between a turbulent region and the non-turbulent region outside it therefore consists of a viscous "superlayer" [1.30] whose thickness is of the order of the wavelength of the smallest eddies near the inner side of the interface—that is, the Kolmogorov length scale $\eta$ defined on p. 22. In fact, the superlayer is just the envelope of the vortex lines which make up the smallest eddies (which contain most of the mean-square vorticity and in which viscous diffusion and vortex stretching just balance on the average). Just as the dissipation is effected by the smallest eddies but determined by energy transfer from the larger ones, so the propagation of the interface into the irrotational fluid is effected by viscous stresses but at a rate set by the spatial transport of turbulent energy from below by the large eddies. As can easily be seen from flow-visualization pictures, the interface is highly irregular. It can be shown that, at least in simple cases, the mean velocity of propagation normal to the mean position of the interface—in the $x_2$ direction, say—is $(\overline{p'u_2}/\varrho + (1/2)\overline{u_i^2 u_2})/((1/2)\overline{u_i^2})$ where the spatial derivative of the numerator is the turbulent transport ("diffusion") term in the turbulent energy equation for $(1/2)\overline{u_i^2}$.

Outside the interface, vorticity fluctuations are zero but velocity fluctuations are not, being driven by pressure fluctuations generated within the turbulence. The mean-square intensity of these "irrotational" fluctuations decreases rapidly with distance outside the interface, the theory of PHILLIPS [1.31] predicting an asymptotic variation as $y^{-4}$. The velocity fluctuations within the turbulence also contain an irrotational, pressure-driven mode. Both in theory and in experiment it is difficult to distinguish the rotational and irrotational contributions. The external irrotational motion [1.32] is of interest in certain special cases; in stably stratified flows it takes the form of gravity waves which may propagate independently of the turbulence, while the pressure fluctuations which drive the irrotational field produce fluctuating loads on nearby solid bodies.

The interface thickness is negligibly small at high Reynolds numbers, though in a wall flow it is probably rather thicker than the sublayer (whose thickness is about 15 times the value of $\eta$ at its edge) because $\varepsilon$ is smaller, and $\eta$ therefore larger, near the edge of a shear layer than near a solid surface. At lower Reynolds numbers both viscous layers thicken, and, because the interface is highly contorted, the volume of the viscous superlayer per unit plan area may be many times the actual thickness measured normal to the interface. The measurements of PAIZIS [1.33] in a boundary layer suggest a factor of 7 at a low laboratory Reynolds number; this is almost certainly a gross underestimate because Paizis' technique distinguished only the major corrugations of the interface. This behavior of the superlayer is probably responsible for the apparent effects of viscosity on the intermittent region of low-Reynolds-number flows. For $(\tau/\varrho)_{max}^{1/2}\delta/\nu < 2000$, approximately, the velocity defect law in a boundary layer (Subsect. 2.3.1) becomes dependent on Reynolds number while the law of the wall appears to be unaffected. The Reynolds number quoted here corresponds to $U_e\theta/\nu \approx 5000$ in a constant-pressure boundary layer, to which all the existing information relates; it is intended as a typical Reynolds number of the large eddies, and should apply roughly to other boundary layers and even to other flows.

The effect of free-stream turbulence on the development of a shear layer depends, in principle, on all the statistical properties of the free-stream turbulence, as well as those of the shear layer. It is discussed in Chapt. 2.

The effect of a finite value of $\partial U/\partial y$ in the external flow has been investigated mainly for two special cases, the internal boundary layer downstream of a change in surface roughness and the interaction between the two halves of a duct or jet flow. In the former case the velocity gradient in the "external" flow is the same as that of the shear layer while in the latter case it is of the opposite sign; in both cases, of course, the "external" flow is turbulent. A finite (positive) velocity gradient without significant turbulence occurs in the supersonic flow over a blunt-nosed body, where the total-pressure loss through the curved bow shock wave decreases with increasing distance from the axis. Unfortunately a large density gradient occurs as well and complicates the interpretation of the results. The flow in the internal boundary layer behind large changes in roughness is discussed in Subsect. 2.3.7, and the interaction of two shear layers of opposite sign is discussed in Chapt. 3.

## 1.10 The Eddy Structure of Shear Layers

The word "structure" is used to describe dimensionless properties of the turbulence, such as spectrum or correlation shapes or ratios of Reynolds stress components. We now have a good deal of information about the structure of the common shear layers, which will be reviewed in detail below (see especially Sect. 2.3). It is necessary to appreciate that there are no simple rules that govern eddy behavior, except of course for the Navier-Stokes equations. The dimensional analysis that produces simple formulae for restricted types of flow or flow regions, such as the logarithmic formula of Sect. 1.8, is *based* on assumptions about eddy behavior and does not imply any particularly simple kind of eddy structure. Just as the small-scale eddies (Sect. 1.5) are controlled by the larger-scale eddies that bear most of the Reynolds stresses, so the larger eddies are controlled by the mean rate of strain, which appears in the Reynolds-stress transport equations of Sect. 1.6. The great difference is that while the small-eddy structure is nearly universal and does not depend on the details of the large eddies, the large eddies do depend strongly on the rate-of-strain field and its history, and have a significantly different structure in different kinds of shear layer.

In 1956 TOWNSEND [1.22], writing at a time when methods for pre-dicting shear-layer behavior were at best at the conceptual level of the "eddy viscosity" gradient-transport models mentioned in Sect. 1.7, put forward the hypothesis that the large eddies controlled the flow behavior. The "large eddies" in this context are those whose transverse wavelength is of the order of the shear-layer thickness and whose longitudinal wavelength is a few shear-layer thicknesses; they are the largest eddies that can be fully three dimensional. The details of Townsend's analysis have been superseded (for instance he postulated that the large eddies controlled the smaller, Reynolds-stress-bearing motion, rather than themselves bearing most of the Reynolds stress which we now know to be the case). However the attention he drew to the large eddies, and in particular his argument that they should have a fairly simple preferred shape characteristic of the shear-layer considered, inspired a generation of research work, beginning with that by Townsend's students GRANT [1.34] and TRITTON [1.35]. The preferred shape of the large eddies can be inferred from the behavior of spatial correlations at large separation. The predominant feature, of course, is an eruption from regions of high turbulent intensity but low/high mean velocity towards regions of low turbulent intensity but high/low mean velocity, leading to a large contribution to Reynolds shear stress. The standard deviation about the preferred shape is very large, and the correlation coefficient between two points half a large-eddy wavelength apart is only $-0.1$ to $-0.2$ instead of the value of $-1$ indicating a perfectly repeated pattern. The irregularity of the large eddies is also clear in photographs of smoke-filled shear layers (e.g., Fig. 2.4); the large eddies determine the shape of the interface (Sect. 1.9) but flair or faith is needed to distinguish a pattern. However the large-eddy concept is of great qualitative or semi-quantitative use in developing realistic models of turbulence structure.

In recent years there have been two completely contradictory developments in our attitude to large eddies. On the one hand, many developers of turbulence models for machine computation have used gradient-transport concepts to describe turbulent transport of Reynolds stress [term 3 in (1.35)] or even returned to the eddy-viscosity concept of gradient transport of momentum. Such concepts would be valid, as distinct from pragmatically justifiable, only if the turbulent eddies were all small compared to the flow width. On the other hand, much recent experimental work is related to a search for what has been called "orderly structure" in turbulence. The strategy of calculation methods will be discussed in Chapt. 5. Here a few comments on "orderly structure" may be useful although it is to be hoped that much of the current confusion is transient.

In this author's view, confusion has arisen because of failure to distinguish properly between five different possible causes of large correlation coefficients, high spectrum peaks or other manifestations of organization in turbulence:

a) Large eddies with the properties outlined above
b) Exceptionally well-organized large eddies in flows not previously studied
c) Persistence of flow patterns arising in the laminar-turbulent transition region, or dependent on the initial conditions
d) Irrotational fluctuations just outside the turbulence proper, which usually have a sharp spectral cutoff above the large-eddy wavenumber range for reasons explained in [1.31]
e) Resonances such as edge tones or vortex streets.

Of course these five causes can occur in combination. For instance, the flow-visualization work of Crow and Champagne [1.36] in circular jets revealed an axisymmetric pulsation, with a frequency proportional to (jet speed)/(diameter) and an amplitude depending on Reynolds number. The fact that the frequency was independent of streamwise distance instead of being related to the thickness of the growing shear layer strongly suggests cause (e) while the dependence on Reynolds number suggests an interaction with (c), although columnar oscillations of circular jets or "flapping" oscillations of plane jets can occur even if the exit boundary layer is turbulent. The very pronounced and apparently two-dimensional large eddies found by BROWN and ROSHKO [1.37] in a plane mixing layer are nominally of type (b) but are thought by some to be of type (a) modified by type (c). For work on irrotational fluctuations in jets see for instance DAVIES *et al.* [1.38], for an older discussion of strong organization in jet turbulence see [1.39], for a recent reminder of the importance of initial conditions see [1.40], and for an example of the wide range of phenomena covered by the term "orderly structure" see [1.41]. The last-mentioned reference reviews several experiments using "conditional sampling". This [1.42] is a technique in which averages are taken, not over the full length of a record of velocity fluctuations, but only over those portions that satisfy some condition (e.g., unusually high fluctuation level) imposed by the experimenter. Since the sampling condition is completely at the whim of the experimenter, this may be thought a fine technique for generating "orderly structure" and random numbers in roughly equal proportions! In careful hands, however, it is a powerful method of investigating eddy structure, supplementing and quantifying information obtained from flow visualization, and should soon lead to resolution of the difficulties outlined in this section.

# References

1.1   S. GOLDSTEIN: J. Math. and Phys. Sci. (Univ. of Madras) 6, 225 (1972)
1.2   L. HOWARTH (ed.): *Modern Developments in Fluid Dynamics. High Speed Flow* (Clarendon Press, Oxford 1953)
1.3   H. SCHLICHTING: *Boundary Layer Theory* (McGraw-Hill, New York 1968) (German edition: G. Braun, Karlsruhe)
1.4   T. CEBECI, A. M. O. SMITH: *Analysis of Turbulent Boundary Layers* (Academic Press, New York 1974)
1.5   J. C. ROTTA: *Turbulente Strömungen* (Teubner, Stuttgart 1972)
1.6   R. ARIS: *Vectors, Tensors and the Basic Equations of Fluid Mechanics* (Prentice-Hall, London 1972)
1.7   J. O. HINZE: *Turbulence* (McGraw-Hill, New York 1959)
1.8   G. K. BATCHELOR: *An Introduction to Fluid Dynamics* (University Press, Cambridge 1967)
1.9   S. GOLDSTEIN (ed.): *Modern Developments in Fluid Dynamics* (Clarendon Press, Oxford 1938)
1.10  C. C. LIN: *The Theory of Hydrodynamic Stability* (University Press, Cambridge 1955)
1.11  P. BRADSHAW: *An Introduction to Turbulence and its Measurement* (Pergamon Press, Oxford 1971)
1.12  A. Y.-S. KUO, S. CORRSIN: J. Fluid Mech. 56, 447 (1972)
1.13  R. W. STEWART: *Turbulence* (Ciné film, 16 mm sound, color, Encyclopedia Britannica Educational Corp., Chicago 1970)
1.14  G. K. BATCHELOR: *The Theory of Homogeneous Turbulence* (University Press, Cambridge 1956)

1.15 H. TENNEKES, J. L. LUMLEY: *A First Course in Turbulence* (MIT Press, Cambridge 1972)

1.16 D. C. LESLIE: *Developments in the Theory of Turbulence* (Clarendon Press, Oxford 1973)

1.17 S. A. ORSZAG: J. Fluid Mech. **4**, 363 (1970)

1.18 A. S. MONIN, A. M. YAGLOM: *Statistical Fluid Mechanics*. 2 vols. (MIT Press, Cambridge 1971, 1975)

1.19 R. H. KRAICHNAN: J. Fluid Mech. **62**, 305 (1974)

1.20 J. P. CLAY: PhD thesis (Engineering Science), University of California, San Diego (1973)

1.21 J. C. WYNGAARD: Statistical models and turbulence, in: *Lecture Notes in Physics*, Vol. 12 (Springer Berlin, Heidelberg, New York 1972)

1.22 A. A. TOWNSEND: *The Structure of Turbulent Shear Flow*. 2nd ed. (University Press, Cambridge 1976)

1.23 J. L. LUMLEY: Phys. Fluids **18**, 750 (1975)

1.24 T. CEBECI, P. BRADSHAW: *Momentum Transfer in Boundary Layers* (McGraw-Hill/Hemisphere Publishing, Washington 1977)

1.25 P. J. ROACHE: *Computational Fluid Dynamics* (Hermosa Publishers, Albuquerque 1972)

1.26 A. A. TOWNSEND: J. Fluid Mech. **11**, 97 (1961)

1.27 R. A. McD. GALBRAITH, M. R. HEAD: Aeronaut. Quart. **26**, 133 (1975)

1.28 H. TENNEKES: AIAA J. **6**, 1735 (1968)

1.29 P. BRADSHAW: Nature **249**, 135 (1974)

1.30 S. CORRSIN, A. L. KISTLER: NACA Rept. 1244 (1955)

1.31 O. M. PHILLIPS: Proc. Camb. Phil. Soc. **51**, 220 (1955)

1.32 L. S. G. KOVASZNAY, V. KIBENS, R. F. BLACKWELDER: J. Fluid Mech. **41**, 283 (1970)

1.33 S. T. PAIZIS, W. H. SCHWARTZ: J. Fluid Mech. **63**, 315 (1974)

1.34 H. L. GRANT: J. Fluid Mech. **4**, 149 (1958)

1.35 D. J. TRITTON: J. Fluid Mech. **28**, 439 (1967)

1.36 S. C. CROW, F. H. CHAMPAGNE: J. Fluid Mech. **48**, 547 (1971)

1.37 G. L. BROWN, A. ROSHKO: J. Fluid Mech. **64**, 775 (1974)

1.38 N. W. M. KO, P. O. A. L. DAVIES: J. Sound Vib. **41**, 347 (1975)

1.39 P. BRADSHAW, D. H. FERRISS, R. F. JOHNSON: J. Fluid Mech. **19**, 591 (1964)

1.40 R. G. BATT: AIAA J. **13**, 245 (1975)

1.41 P. O. A. L. DAVIES, A. J. YULE: J. Fluid Mech. **69**, 513 (1975)

1.42 J. LAUFER: Ann. Rev. Fluid Mech. **7**, 307 (1975)

# 2. External Flows

H.-H. Fernholz

With 7 Figures

"External" turbulent flows are those in which the turbulent region is significantly affected by at most one wall. An example is the boundary layer on an airfoil (Fig. 2.1) and an example without walls is the wake behind the airfoil. Although the interaction between a turbulent shear layer and the non-turbulent "inviscid" flow outside it is often important even in external flows, it is convenient to defer the treatment of interactions until Chapter 3 on internal flow (bounded by at least two walls) in which they are always important. Conversely the treatment of boundary layers in the present chapter is applicable to turbomachine blades and other internal flows. Several topics such as streamline-curvature effects and separation are introduced in the present chapter and illustrated by simple external-flow examples, while the more complicated flow situations are discussed in Chapter 3.

## 2.1 Introduction

We begin with a brief survey of some general turbulent shear flow configurations which, merged or distorted, make up the more complex flow patterns which are often found in practice but which are often not accessible to experiment or calculation. The greater part of this chapter will be concerned with a discussion of steady two- and three-dimensional turbulent boundary layers at low Mach numbers. Boundary layers are the commonest members of the class of "thin shear layers" (Chapt. 1) and provide a convenient framework within which to discuss the effects of boundary conditions on shear layers. If the wall or free stream boundary conditions are changed, the shear layer will feel this change as a more or less strong perturbation. Most shear layers in real life undergo such perturbations, and the difference between this chapter and some of the existing reviews of the subject is that we have tried to give perturbed shear layers a share of space proportional to their importance. Unperturbed free shear layers—the classical wake, jet and mixing layer— are rather rare in practice and the data on growth rate, etc., which are normally needed can be summarized quite briefly (Sect. 5.9). Free shear layers near solid boundaries are usually perturbed and

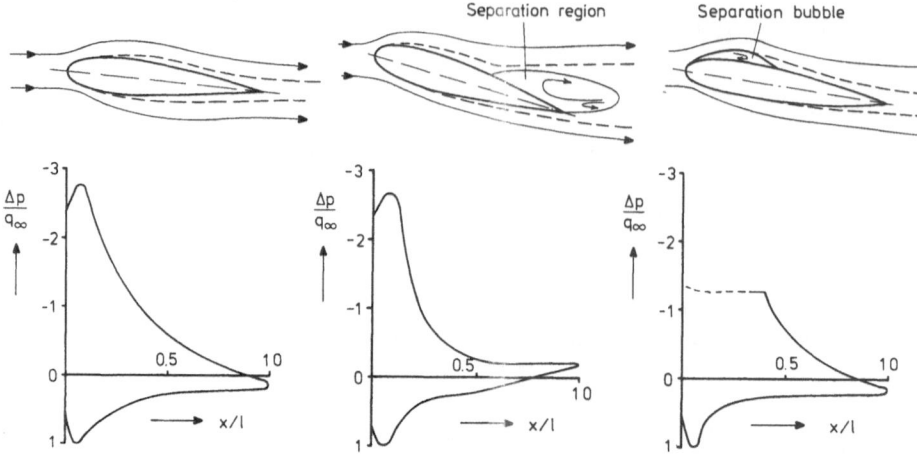

Fig. 2.1. Attached and separating flow (after THWAITES [2.446], Fig. V.13)

often form part of a separation/reattachment flow. Of these flows separation bubbles, in which the free shear layer remains very close to the surface, are discussed in Subsection 2.3.6. The more extensive separated regions, whose turbulence behavior is of most importance in internal flows, are discussed in Section 3.4. Basic references on (unperturbed) free shear layers are given in [1.7, 22, 2.1], and good entries to the current research literature are provided by [2.2, 3]. Ref. [2.2] contains extensive data tabulations. See Section 1.10 for work on orderly structure in jets. A very comprehensive survey of aircraft trailing vortices, with particular reference to turbulence structure, has recently appeared [2.4], and we therefore omit this important subject. Vortices frequently appear as constituent parts of turbulent flow fields, but in that case are better treated as examples of turbulence with superimposed rotation (Sect. 3.1).

For most of the present chapter we concentrate on constant-property ("incompressible") flows for simplicity, but the chapter is concluded by a discussion of compressibility effects on turbulent boundary layers. In all cases discussed here, the boundary conditions can be defined clearly. This has been intentional. It should be mentioned, however, that there exist many external flows, aeronautical or otherwise, where the definition of the conditions at the interface of two merging shear flows is very difficult or where there is no asymptotic transition between the boundary layer and the outer flow if the latter is highly disturbed. However such configurations are even more important in internal flows, and are therefore discussed in Chapter 3; the results of that chapter are at least qualitatively applicable to external flows.

## 2.2 Flow Configurations and Boundary Conditions

Turbulent shear flows can be divided into three groups according to the number of fixed boundaries:
1) Free turbulent shear flows bounded by no wall, such as jets, wakes and mixing layers (Fig. 2.2), and plumes (Chapt. 4).
Jets and wakes differ in principle from each other only in the sign of the momentum creating them. Among them the turbulent jet is the most common flow configuration, ranging in its application from propulsive devices for rockets and aircraft to pneumatic control systems.
2) Turbulent shear flows bounded by one free and one fixed boundary, such as boundary layers or wall-jets, where the wall may be straight or curved, permeable or impermeable.
For both jet and wall-jet the velocity of the ambient flow can have the same or the opposite sign as the shear flow.
3) Turbulent shear flows bounded by two or more fixed boundaries, such as pipe, duct and channel flows which are dealt with in Chapter 3.
In this book a "pipe" is a closed duct of circular cross section, a "duct" has a non-circular cross section and a "channel" has a free surface.

These basic flow phenomena do not always appear in their pure state but often as a combination of two or more configurations, e.g., as a separated and reattaching boundary layer, a jet emanating into a duct flow (ejector problems or a flap configuration with merging shear layers (Fig. 2.3). In this chapter the discussion will be confined almost entirely to simple thin or slender shear layers, having one sign of shear throughout. It is obvious from Fig. 2.2 that it would be rather fortuitous if the shear layers on either side of the velocity maximum in a wall jet flow, say, behaved exactly like an isolated thin shear layer, however well the flow obeyed the thin shear layer approximation; shear-layer inter-actions require further discussion (Chapt. 3). Other cases where the

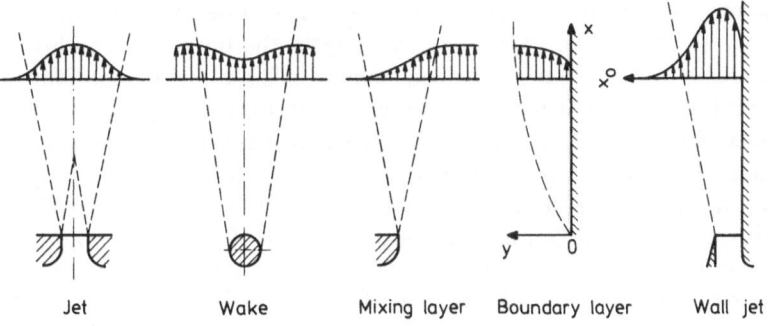

Fig. 2.2. Some basic turbulent shear flow configurations

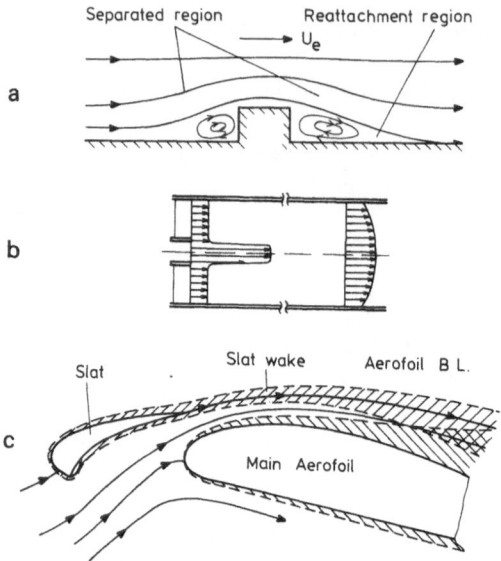

Fig. 2.3a—c. Some complex turbulent shear flow configurations. (a) boundary layer flow over a step, (b) duct flow in an internal ejector, (c) flow over an aerofoil with a front slot (from Horton [2.267])

concept of simple shear layers is violated occur in flow regions in which the thin shear layer is strongly perturbed. Such a perturbation may be caused by a pressure gradient that greatly exceeds the Reynolds stress gradient, as in an impinging jet or in the boundary layer separating in front of a step (Fig. 2.3a). In such regions a quasi-inviscid flow model suffices, few data on turbulence structure are available, and lengthy discussion would be unprofitable.

We shall, however, deal with shear flows, initially in a state of equilibrium[1], which are more or less strongly perturbed by a change of the boundary conditions and which tend to a new state of equilibrium or relax to the original equilibrium according to the boundary conditions downstream from the perturbation. There exists a strong motivation for investigating perturbed shear flows apart from their practical importance. Clauser [2.5] suggested that useful basic information about turbulent flows could be obtained by studying their response to perturbation, after the style of the classical "black box" problem. Many experimenters have taken up this idea. A survey of turbulent boundary layers subject to perturbations in general was provided by Tani [2.6], accompanying the exhaustive review of boundary-layer data by Coles

[1] A concept to be defined in Subsection 2.3.2.

and HIRST [2.7]. In the following the effects of perturbations on simple thin and slender shear layers, in the first instance boundary layers, will be emphasized. These perturbations are caused by changes of the boundary conditions at the outer edge of the boundary layer or at the wall and may affect the outer or inner region of the boundary layer only, or, as in many cases, the boundary layer as a whole.

### Boundary and Initial Conditions

Once the flow configuration, i.e., the type of the shear layer, is established, the (parabolic) partial differential equations describing it (Sect. 1.7) must be supplemented by initial conditions and by boundary conditions at the edges of the shear layer. For many problems in aeronautics these conditions can fortunately be specified without too great difficulties, so that a great number of important flow configurations can be classified as well-posed problems. This has certainly favored the solution of problems in aeronautics, and it is hoped that this experience can be transfered to more complex flows, both in the same field and in other closely related areas such as mechanical and chemical engineering where the occurring problems are often ill defined.

Before we can solve the thin-shear-layer equations (1.38) to (1.41), together with a "turbulence model" to close the equations, the following initial and boundary conditions must be known (for steady flow); for convenience we treat a boundary layer on the solid surface $y=0$.

*Initial conditions* ($x=x_0$ where $x_0$ denotes the starting position of the investigation, see Fig. 2.2)

$$U(y), p(y), T(y) \quad \text{or}^2 \quad e(y).$$

According to the closure model chosen: $\overline{uv}, \overline{vw}, \overline{v\theta}, \overline{q^2}$ etc. (see Chapt. 5). For three-dimensional flow it is necessary to know a further function, being a combination of $\varrho, U, V,$ and $W$ which is determined from a compatibility condition (TING [2.8]).

*Boundary Conditions* ($y=0$ and $y=\delta$) [3]

$y=0$:
    $U=0$ (where $U$ is the relative velocity between the wall and the fluid)
    $V=0$ (impermeable wall) or

---

    [2] $e$ is the internal energy of the flow.
    [3] $\delta$ denotes the free-stream boundary of the boundary layer where the index used is e (for external).

$V<0$ (suction) or $V>0$ (blowing)

$W=0$ (for three-dimensional flow)

$u=v=w=0$ (in the case of a permeable wall $v$ is also assumed to be
zero, which is reasonable if the pressure drop across
the wall is large)

$T=T_w$ (for flow with heat transfer)

$y=\delta$:

$U=U_e$

$W=W_e$   and   $dp_e/\varrho_e+d(U_e^2+W_e^2)2=0$

$T=T_e$

$\overline{uv}|_e=\overline{vw}|_e=\overline{v\theta}|_e=0$ (for high free-stream turbulence these values
must be known).

For complex configurations, for example where there is strong interaction between the viscous and the inviscid flow (see e.g., GREEN [2.9])
or between two shear layers (Fig. 2.3 and Sect. 3.1), iterative matching
of boundary conditions in two regions, or use of the full elliptic equations
of motion, may become necessary. In other cases the boundary conditions
themselves may be given in indirect forms, perhaps by differential
equations as in two-phase flow problems where a vapor and a condensate
boundary layer interact.

It is this multiplicity of boundary conditions which creates such a
variety of thin shear layers with widely differing flow characteristics.
In the following subsections of this chapter we shall deal with some of
these cases by varying the boundary conditions, indicating their effects
on the boundary layer and giving a selection of references for a more
detailed study. Many of the remarks apply to other types of turbulent
wall flow such as wall jets or duct flows, while remarks about free-stream
boundaries apply to free shear layers also.

## 2.3 Two-Dimensional Boundary Layers at Small Mach Numbers

### 2.3.1 The Multi-Layer Model of Turbulent Boundary Layers

As indicated in Chapter 1 (see also Chapt. 5), any solution of the turbulence problem short of a complete solution of the time-dependent
Navier-Stokes equations relies on empirical information. It is therefore
desirable to have as detailed an account of the turbulence structure in the
boundary layer as possible. To put the observed phenomena into some
sort of order it has proven useful to split up the boundary layer into an
inner layer and an outer layer. The elementary dimensional analysis
for the former, which applies to all wall flows, was carried out in Section

Fig. 2.4a and b. Visualization of a boundary layer by means of smoke. Photograph a from [2.282], photograph b by courtesy of P. BRADSHAW. (a) Streamwise view, (b) side view, flow right to left

1.8. We now seek more details of the turbulent motion, and of the characteristic velocity and length scales of the outer layer. Despite the difficulties of the task, obvious from the smoke photographs of Fig. 2.4, some progress can be made, though the outer-layer scales are much more complicated than those of the inner layer.

In turbulent boundary layers one boundary is formed by the wall, where the turbulent shear flow must accommodate itself to the wall through a viscous sublayer and the large velocity gradients associated with it. Very close to the wall the fluctuating velocities and therefore the Reynolds shear stresses tend to zero, while the total shear stress remains constant. Thus the viscous shear stress $\mu \partial U / \partial y$, and the velocity gradient $\partial U / \partial y$, are much larger than elsewhere in the boundary layer. The other boundary is the turbulent/non-turbulent interface where the shear flow with high vorticity accommodates to an essentially irrotational outer flow through the "viscous superlayer" (Sect. 1.9). This latter irregular but very distinct boundary is common to all turbulent thin shear layers with at least one free boundary, but its quantitative properties depend on the characteristic scales and the boundary conditions of the specific shear layer. It has been found useful (see Fig. 1.6) to divide the inner layer at the wall into the linear sublayer and the buffer layer (together comprising the viscous sublayer) and the log-law region. The outer layer which covers about 80% of the boundary layer is bounded

Table 2.1. Multi-layer model of turbulent boundary layer: see text for symbols

| | Inner layer | | Outer layer | |
|---|---|---|---|---|
| | Viscous sublayer | Log-law region | Outer-law region | Viscous superlayer |
| Range within boundary layer | $0 < y < 40\nu/u_\tau$ | $40\nu/u_\tau < y < 0.2\delta$ | $0.2 < y/\delta$ | Inhabits $y > 0.4\delta$ |
| Characteristic length | $(\nu/u_\tau)$, $k_r$ | $(\nu/u_\tau)$, $k_r$ | $\delta$ or $\Delta$ | $(\nu^3/\varepsilon_0)^{\frac{1}{4}}$ or $\nu/V_e$ |
| Characteristic velocity | $u_\tau$ | $u_\tau$ | $u_\tau$, $U_e$ | $V_E \sim K(\nu\varepsilon_0)^{\frac{1}{4}}$ |
| Type of velocity law | $\dfrac{U}{u_\tau} = f\left(\dfrac{yu_\tau}{\nu}, \dfrac{k_r u_\tau}{\nu}\right)$ | $\partial U / \partial y = (-\overline{uv})^{\frac{1}{2}}/Ky$ | $\dfrac{U_e - U}{u_\tau} = g\left(\dfrac{y}{\Delta}, \dfrac{u_\tau}{U_e}, \beta\right)$ | |
| Flow condition | Intermittent turbulent | Fully turbulent | Fully turbulent, but intermittent rotational/irrotational for $y > 0.4\delta$ | Laminar-like |
| Dominating mechanism | Dissipation $\sim$ production of turbulent energy | | Extraction of kinetic mean flow energy by the Reynolds stress gradient | Transfer of vorticity to the irrotational free stream |
| Dependence on viscosity | Dependent on viscosity | Independent of viscosity | | Dependent on viscosity |
| Dominating shear stress term | $-\overline{uv} < \mu \partial U / \partial y$ (for most of the time) | $-\overline{uv} \gg \mu \partial U / \partial y$ | | |

towards the free-stream by the viscous superlayer[4]. Table 2.1 briefly surveys this multi-layer model, specifying the dominating turbulence mechanisms, the characteristic scales and the types of the mean velocity distribution, mainly based on the monographs by TOWNSEND [1.22] and ROTTA [1.5]. A similar survey could be made for other types of turbulent shear flows, but the boundary layer is by far the best documented and most important.

**The Inner Layer** $(0 \leqq y/\delta \leqq 0.1$–$0.2$ roughly; $0 \leqq U/U_e \leqq 0.7$ very roughly)

Initially we consider two-dimensional constant-density flow over an impermeable wall, as outlined in Section 1.8, in a region close enough to the wall for the total shear stress to be nearly equal to the wall value $\tau_w$ and for all derivatives with respect to $x$ to be negligible. The behavior of the mean velocity profile, and of the turbulence, depends on the relative size of the terms on the right-hand side of

$$\tau_w = \mu \partial U/\partial y - \varrho \overline{uv} \tag{2.1}$$

*Linear Sublayer* $(u_\tau y/v < 3$, roughly)

In the innermost part of the sublayer, $\varrho \overline{uv}$ can be neglected compared with $\mu \partial U/\partial y$, and on integrating (2.1) one obtains

$$U/u_\tau = u_\tau y/v \quad \text{or} \quad U^+ = y^+ \tag{2.2}$$

hence the name "linear sublayer". Here $u_\tau$ is the friction velocity $(\tau_w/\varrho_w)^{\frac{1}{2}}$, and we recall from Section 1.8 that $u_\tau y/v = y^+$ is a typical Reynolds number of the energy-containing turbulence. Eq. (2.2) applies for $y^+ < 3$ approximately. Support for the neglect of the Reynolds stresses in the viscous part of the inner layer, based on experiments, has been given recently by ECKELMANN [2.17]. Experimental evidence shows that the pressure gradient has little effect on the sublayer and that (2.2) applies for flows in ducts [2.18] and pipes [2.19] also.

If one measures the instantaneous velocity profile instead of the time-mean profile the sublayer is revealed to have a longitudinally

---

[4] This multi-layer model for turbulent boundary layers has been developed over the past fifty years. Referring to a few papers only one should mention especially: for the sublayer PRANDTL [2.10], KLINE et al. [2.11]; for the logarithmic law PRANDTL [2.10, 12], LUDWIEG and TILLMANN [2.13]; for the outer layer VON KARMAN [2.14], CLAUSER [2.15], COLES [2.16]; for the superlayer CORRSIN and KISTLER [1.30].

streaky, spatial and time-dependent flow structure (Kline and Run-stadler [2.20]), continuously disturbed by small-scale velocity fluctuations of low magnitude and periodically disturbed by fluid elements which penetrate into this region from positions further removed from the wall (Corino and Brodkey [2.21]). For further references the reader is referred to papers by Bakewell [2.19], Kline et al. [2.11], Kim et al. [2.22], Gupta et al. [2.23], Wallace et al. [2.24], Offen and Kline [2.25] discussing the phenomenology and the triggering mechanisms and to Ueda and Hinze [2.26] and Kuo and Corrsin [2.285] for the fine structure of the turbulence. The streaky flow structure seems to have been observed earlier by Hama as pointed out by Corrsin [2.27].

*Buffer Layer* $(3 < u_\tau y/v < 40$ roughly$)$

Since both viscous and Reynolds shear stresses in (2.1) are of the same order of magnitude, no simple relationship for the mean velocity in the buffer layer can be derived. According to the experiments of Laufer [2.28] and Klebanoff [2.29] the production and dissipation of turbulent energy are larger, but not much larger, than the diffusion: this is not a local-equilibrium region. The production of turbulence is apparently strongly connected to the bursts observed in the sublayer (Kim *et al.* [2.22]; the experiments were done at low bulk Reynolds numbers, but this should not matter according to inner-layer similarity arguments). For further details see also Willmarth and Lu [2.30] and Black-welder and Kaplan [2.31] who applied conditional sampling techniques to investigate the link between the bursting phenomenon and the production of shear stress.

*Logarithmic Law Region* $(40v/u_\tau < y < 0.2\delta$ roughly$)$

For large turbulent Reynolds numbers $(u_\tau y/v > 40)$ the viscous term in (2.1) may be neglected compared with the Reynolds shear stress. Then dimensional analysis as in Chapter 1 or in Rotta [2.32] yields the "logarithmic law" or the "law of the wall" for the mean velocity distribution on an impermeable smooth wall with negligible shear stress gradient,

$$U/u_\tau = (1/\kappa)\ln(u_\tau y/v) + C, \tag{2.3}$$

where $\kappa = 0.40$ to $0.41$ and $C = 5.0$ to $5.2$ are determined from experiments[5]. This is one of the most important relationships for turbulent boundary layers since, apart from the floating element, almost all

---

[5] The possibility that $\kappa$ does in fact depend on the bulk Reynolds number is discussed in [1.15, 28].

measuring techniques for skin friction are based on its validity, as well as most semi-empirical relationships for the wall shear stress (Sect. 5.7). For further references see, for example, PRESTON [2.33], CLAUSER [2.5], HEAD and RECHENBERG [2.34] and PATEL and HEAD [2.35]. There is good experimental evidence that the logarithmic law is valid also in wall layers with a streamwise pressure gradient—though the assumptions used for its derivation did not include pressure gradients—but that its range of validity becomes considerably smaller in an adverse pressure gradient compared with that in zero pressure gradient (cf. the Stanford Data Catalogue by COLES and HIRST [2.7]). Deviations from the logarithmic law occur, however, in wall layers with strong favorable pressure gradients (PATEL and HEAD [2.35]). A range of validity for the application of Preston tubes [2.33] to measure skin friction was given by PATEL [2.36].

Analyses of the effect of the pressure gradient on the velocity distribution near the wall were performed by TOWNSEND [1.26], MELLOR [2.37] and McDONALD [2.38] using (1.50) or refinements. For $\tau = \tau_w + \alpha y$, integrating (1.50) and requiring compatibility with (2.3) in the limit $\alpha \to 0$ gives, with $z = \alpha y / \tau_w$,

$$\frac{U}{u_\tau} = \frac{1}{\kappa} \ln \frac{u_\tau y}{v} + C + \frac{1}{\kappa} \left[ 2\ln \left( \frac{2}{(1+z)^{\frac{1}{2}} + 1} \right) + 2((1+z)^{\frac{1}{2}} - 1) \right], \quad (2.4)$$

where $C$ may depend slightly on $\alpha v / (\varrho u_\tau^3)$.

Corresponding to the result $\tau = \tau_w$ for the inner layer in flow over an impermeable wall and negligible acceleration, one obtains, for an inner layer with transpiration, ($V_w$ = constant):

$$\tau = \tau_w + \varrho U V_w, \quad (2.5)$$

where $V_w$ is the normal velocity at the surface. Substituting (2.5) into (1.50), integrating this equation and requiring compatibility with (2.3) yields

$$\frac{2u_\tau}{V_w} \left[ \left( 1 + \frac{U V_w}{u_\tau^2} \right)^{\frac{1}{2}} - 1 \right] = \frac{1}{\kappa} \ln \frac{u_\tau y}{v} + C \quad (2.6)$$

where $C$ is now a function of the dimensionless parameter $V_w / u_\tau$ and, if the surface porosity is not of very small scale, of the properties of that porosity too. For a more detailed discussion see for example COLES [2.39] and BAKER and LAUNDER [2.40]. Other integrals of (1.50), for different shear-stress distributions, can be generated in the same way.

Before turning to the outer region, attention should be drawn to some flow visualization studies in the inner region. As can be seen for instance from the data of KIM et al. [2.22] the flow structure close to the wall is qualitatively very different at periods of oscillatory growth of the "streaky" flow structure from the flow at "non-bursting" periods. The former flow is strongly intermittent and displays features of an organized oscillation or wave packet. This has led to investigations of the turbulence structure where use was made of wave models, e.g., LANDAHL [2.41], BARK [2.42], PHILLIPS [2.43], MOFFATT [2.44], LIGHTHILL [2.45] and HUSSAIN and REYNOLDS [2.46, 47]. So far no decision has been reached whether a random array of deterministic eddies or a model based on travelling waves is more appropriate to describe the structure of turbulence (see also KOVASZNAY [2.48]).

**The Outer Layer** $(0.2 \leq y/\delta \leq 1.0)$

Casual observation of turbulent shear flows with a free-stream boundary made visible by some contaminant (e.g., chimney plumes, clouds) shows that the boundary between contaminated and uncontaminated fluid is highly irregular, unsteady in any coordinate system and moving outwards to entrain previously uncontaminated fluid. In the case of irrotational outer flow, velocity fluctuations, with wavelengths of the order of the largest turbulent eddies but containing negligible vorticity, are found to extend into the irrotational flow region beyond this boundary. Therefore it is advisable to divide the flow into a rotational and an irrotational flow region (see Fig. 2.4). The demarcation line between the two zones gives an intermittent character to the signal of a fixed hot-wire probe (CORRSIN [2.49]), and the fraction of time spent by the probe in the rotational region is a characteristic feature of a turbulent shear layer, called the intermittency factor $\gamma$ (TOWNSEND [2.50]). $\gamma$ equals one near the wall and tends to zero as $y$ goes to infinity. Typical intermittency distributions, which seem to follow a Gaussian (error function) cumulative probability distribution were measured in various shear flows by CORRSIN and KISTLER [1.30] and in boundary layers with variable pressure gradient by FIEDLER and HEAD [2.51] for instance. Recent assessments of discrimination processes for the determination of the rotational/irrotational interface have been given by KAPLAN and LAUFER [2.52], KOVASZNAY et al. [1.32], ANTONIA [2.53], VAN ATTA [2.54], LARUE [2.55], ANTONIA and ATKINSON [2.56] and BRADSHAW and MURLIS [2.57], the latter paper revealing some of the difficulties one may encounter. The scales of the largest surface indentations are comparable with the total width of the shear layer, and the average velocity difference between the rotational and irrotational flow is about 5% of the maximum velocity difference in wake flow (TOWNSEND [1.22]) and about 5% of $U_e$ in boundary layer flow (KOVASZNAY et al. [1.32]), showing the rotational zone lagging slightly behind. Velocity and shear stress profiles in the rotational and irrotational zone were measured by KAPLAN and LAUFER [2.52] and BLACKWELDER and KOVASZNAY [2.58], respectively. KLEBANOFF [2.29] showed the free-stream boundary to be at a mean position $\bar{Y}=0.78\delta$ with the standard deviation $\sigma=0.14\delta$ in a zero-pressure gradient boundary layer. These values vary with the state of the boundary layer [2.51]. Though the spreading rates for distinct shear flows are well known (see, e.g., SCHLICHTING [Ref. 1.3, p. 678]) the entrainment process as such is not yet fully understood (see also MOLLO-CHRISTENSEN [2.286] and FAVRE et al. [2.287]).

*The Viscous Superlayer*

The viscous superlayer, as described in Chapter 1, was postulated by CORRSIN and KISTLER [1.30]. It is maintained thin by propagation relative to the fluid with a propagation velocity $V_0$ and by the random stretching of vortex lines in its local velocity gradient. By dimensional reasoning the thickness of the superlayer was estimated to be of the same order as the Kolmogorov length $\eta \equiv (v^3/\varepsilon_0)^{\frac{1}{4}}$, its propagation velocity normal to itself being as proportional to the Kolmogorov velocity $v_\varepsilon \equiv (v\varepsilon_0)^{\frac{1}{4}}$ (see also PHILLIPS [2.59]) where $\varepsilon_0$ is the viscous dissipation within the turbulent fluid. So the propagation of the interface into the irrotational fluid is effected by viscous stresses but at a rate set by the spatial transport of turbulent energy from below by the large eddies. The mean velocity component normal to the mean position of the interface, $y=\bar{Y}(x)$, is called the "entrainment velocity", a notation which is used in practice for the velocity normal to the line $y=\delta(x)$. By calling this velocity $V_E$, application of the principle of conservation of mass to a control volume bounded by $y=0, y=\delta(x), x=x_0$ and $x=x_0+\Delta x$ gives

$$\varrho_e V_E = \frac{d}{dx}\int_0^\delta \varrho U dy = \frac{d}{dx}[\varrho_e U_e(\delta-\delta^*)] \qquad (2.7)$$

where the suffix e denotes conditions at the edge, $y=\delta(x)$, and where $\delta^*$ is the displacement thickness as defined by (2.22).

Note that $V_E$ is orders of magnitude more than the propagation velocity of the interface because the latter is highly convoluted [1.33]. A semi-empirical relationship between the entrainment velocity $V_E$ and the velocity profile has lead HEAD [2.60] to a simple closure for the boundary layer equations (see also HEAD and PATEL [2.61]); details are given in Chapter 5. Theoretical approaches for a description of the evolution of the superlayer were presented by TOWNSEND [2.62] and PHILLIPS [2.59], and further experiments on its structure were published by KOVASZNAY et al. [1.32].

### The Outer-Layer Eddy Structure

So far the entrainment process has been solely ascribed to the normal propagation of the interface as a nonlinear diffusion of vorticity. This is increased as compared with that of a laminar boundary layer by the enlarged free surface of the turbulent boundary layer which is convoluted by the large eddies in the outer layer. The large eddies contribute at least 50% to the turbulent energy associated with the $u$- and $v$-fluctuations and about 80% to the Reynolds stress (BLACKWELDER and KOVASZNAY [2.58]; see also FALCO [2.63]). Whether the large eddies arise in the outer layer because of some form of interfacial instability near the superlayer (KOVASZNAY et al. [1.32]), or result from the ejection of bursts from the inner region (ANTONIA [2.53]), or both, is still an open question.

For a survey on the structure of turbulent shear flows the reader is referred to KOVASZNAY [2.48], or WILLMARTH [2.447].

### The Outer-Layer Velocity Profile

If the mean velocity is averaged over the rotational and irrotational zones of the outer layer, then it is usual to express the velocity distribution in terms of the velocity defect $U_e - U$ (e.g., ROTTA [2.32]):

$$(U_e - U)/u_\tau = f(y/\Delta) \tag{2.8}$$

where the function $f$ depends on the pressure-gradient history and on the bulk Reynolds number, although the direct effect of the latter is small except in small-scale laboratory flows. Here $\Delta$ is a characteristic length scale, introduced by ROTTA [2.64] and defined as

$$\frac{\Delta}{\delta} = \int_0^1 \frac{U_e - U}{u_\tau} \, d\frac{y}{\delta}. \tag{2.9}$$

Equation (2.8) is a good approximation to the velocity distribution (outside the viscous sublayer) in constant-pressure boundary layers, where $f$ is a universal function if $u_\tau \delta / v$ is greater than about 2000 (see Sect. 1.9). The same holds for fully developed flow in parallel-wall ducts. In this case $\delta$ is replaced by the half-height of the duct and $U_e$ by the centerline velocity. A convenient way of rewriting (2.8) for constant-pressure boundary layers (ROTTA [2.65]) is

$$(U_e - U)/u_\tau = -(1/\kappa)\ln(y/\delta) + (B/\kappa)[2 - w(y/\delta)] \tag{2.10}$$

where $w(y/\delta)$, with $w(1) = 2$, is Coles' wake function, and the wake factor $B$ is an absolute constant for $u_\tau\delta/\nu > 2000$.

Equation (2.10) applies only to constant-pressure flows (or, with a change of $B$ or $w$, to the self-preserving boundary layers discussed in the next section). However COLES [2.16] showed that the velocity profile outside the viscous sublayer in almost any turbulent boundary layer could be represented by (2.10) with $w(y/\delta)$ taken as a universal function, independent of pressure gradient or Reynolds number, and $B$ (replaced by $\Pi$ to avoid confusion) taken as a free parameter. It is important to realize that this is merely a convenient data correlation, the effect of pressure-gradient history being submerged in $\Pi$. Equation (2.10), on the other hand, is based on the assumption that pressure-gradient history is unimportant—as indeed it is in a constant-pressure flow. COLES refers to the function $w(y/\delta)$ as the "law of the wake", analogous to the "law of the wall" function $f_1(u_\tau y/\nu)$ in (1.47a), and writes, for the velocity profile outside the viscous sublayer in any turbulent boundary layer at low Mach number,

$$(U/U_e) = (u_\tau/U)[\kappa^{-1}\ln(u_\tau y/\nu) + C + (\Pi(x)/\kappa)w(y/\delta)]. \tag{2.11}$$

The function $\Pi(x)$ is determined by inserting the boundary conditions at the outer edge of the boundary layer into (2.11), and the wake function $w$ is approximated by $w(\eta) = (1 - \cos(\pi\eta))$ according to COLES, or, according to ROTTA [2.65] by

$$w(\eta) = 39\eta^3 - 125\eta^4 + 183\eta^5 - 133\eta^6 + 38\eta^7 \tag{2.12}$$

where in each case $\eta = y/\delta$. For a different formulation of the wake law giving $\partial U/\partial y = 0$ at $y = \delta$ see Section 3.2, and for skin-friction laws derived from the inner and outer laws see Chapter 5.

A family of velocity profiles covering the whole width of the boundary layer has been given by THOMPSON [2.66], taking into account the intermittency in the outer layer and the inner layer by two universal functions. This approach is readily adaptable to account for the effects of wall roughness, or of distributed suction or injection, by adapting the velocity profile in the inner region. It may be regarded as a numerically specified extension of the simple analytic data fit of COLES. The fact that these are merely data fits and not laws of nature is illustrated by the quite large differences between Thompson's "intermittency function" and the true behavior of the intermittency. The main use of these profile families is in checking experimental data and in "integral calculation methods" (Chapt. 5).

### The Turbulent Boundary Layer at Small Reynolds Numbers

Turbulent boundary layers in the range of $320 \leqq Re_\theta \leqq 5000$ show a behavior which deviates from that in higher Reynolds number boundary layers with decreasing values of $Re_\theta$. The following brief discussion is therefore of some importance for incompressible boundary layers with zero or negative pressure gradients and compressible boundary layers at high Mach numbers.

If the ratio $\Delta U/u_\tau$ as given in Fig. 1.6, equal to $2\Pi/\kappa$ in present notation, is used as a measure for the strength of the wake component, it can be shown from measurements (Coles [2.67]) that the wake component disappears[6] at about a value of $Re_\theta \sim 500$, which is not far from that given as a lower limit for turbulent boundary layers ($Re_\theta = 320$) by Preston [2.68]. This leads to a breakdown of the outer law. With decreasing Reynolds number both viscous sublayer and superlayer thicken considerably, so that viscous effects begin to dominate over effects due to turbulence, a process which finally can lead to relaminarization of the boundary layer (see Subsect. 2.3.11). The inner law on the other hand seems to be very little, if at all, affected. Smoke photographs in boundary layers at low Reynolds numbers indicate changes of the shape of the interface and the structure of the large eddies (private communication Head and Falco) but more detailed experimental evidence must be awaited. It is this lack of reliable experiments which makes it impossible at present to decide whether the constants in the law of the wall (2.3) are dependent on the Reynolds number (Simpson [2.69]) or not (Huffman and Bradshaw [2.70]). Further basic experimental evidence about turbulent boundary layers in the Reynolds number range $800 \leqq Re_\theta \leqq 6000$ has just been reported by Murlis [2.71] and appears to support Head and Falco's findings as well as those of Huffman and Bradshaw.

### 2.3.2 Self-Preserving Shear Layers

In the case of a laminar boundary layer the equation of motion can be reduced from a partial to an ordinary differential equation dependent

---

[6] I.e., the velocity never exceeds the logarithmic-law value—not in itself a critical situation.

on $\eta \sim yg(x)/Re_x^{\frac{1}{2}}$ and a shape parameter $m$ if the free-stream velocity $U_e$ is proportional to $x^m$ (FALKNER and SKAN [2.72]). For constant $m$ the mean velocity profile is similar at all distances downstream, say from some leading edge. Since the various values of $m$ cover a range between highly accelerated and separated flows, the variation of the velocity profile is described by these similar solutions which are exact solutions of the boundary layer equations (see LOITSIANSKI [2.73]). In order to obtain dynamic similarity in turbulent shear layers not only the mean velocity profile, but also the profiles of Reynolds stresses and other turbulence quantities, would have to be similar. Such a strict similarity does not exist except in trivial cases. Some turbulent flows, however, have regions where the action of viscosity on the mean flow is negligible and where the motion of the energy containing components of the turbulence is determined by the boundary conditions alone. At sufficiently high turbulence Reynolds numbers the motions of such flows differ, at given $y/\delta$ and outside viscous regions, only in their scales of velocity and length and are then called self-preserving flows (TOWNSEND [1.22]). Clearly, the condition for a flow to be self-preserving is that the equations for the mean velocity and the turbulent energy (see Chapt. 1) and the boundary conditions can be satisfied to an adequate approximation by self-preserving distributions of mean velocity, Reynolds stress, etc. (see, e.g., TOWNSEND [2.74]). For example, both plane and circular jets and wakes are characteristic shear flows of this type, if observed far enough downstream from their origin. It is in this sense that the word "equilibrium" was used above.

The general form of the mean velocity and shear stress profiles in a self-preserving flow is

$$(U_{ref} - U)/u_0 = F(y/l_0) \tag{2.13a}$$

$$-\overline{uv}/u_0^2 = g(y/l_0) \tag{2.13b}$$

where $u_0$ and $l_0$ are the velocity and length scales of the flow, both functions of $x$ in general. $U$ must be referred to $U_{ref}$, usually the free-stream velocity, because the self-preserving region excludes the viscous sublayer and therefore the point where $U=0$. Outside the viscous sublayer, but still in the inner layer, there is a region where the local rates of production and dissipation of turbulence energy are large compared with the rates of advection and diffusion (1.36) so that the latter can be neglected. A shear layer where these local rates of production and dissipation balance each other is called a (local) equilibrium layer (TOWNSEND [1.26]). Local equilibrium can be thought of as a special case of self preservation, with $u_0 = (-uv)^{\frac{1}{2}}$ and $l_0 = y$. However having the scale and the coordinate the same makes this rather confusing and we have therefore considered local equilibrium separately (Sect. 1.8). Local equilibrium is the more important; it is believed that almost any inner layer has a local-equilibrium region, whereas truly self-preserving flows are rare.

Townsend [1.26, 2.74, 76, 77] and Rotta [1.5, 2.75], for example, have investigated self-preserving flows. Rotta determined the conditions under which self-preserving flows can exist by substituting self-similar distributions of mean velocity and Reynolds shear stress.

$$(U_e - U)/u_\tau = F(\eta, \omega, \beta) \tag{2.14a}$$

$$-\overline{uv}/u_\tau^2 = g(\eta, \omega, \beta) \tag{2.14b}$$

into the $x$-momentum equation (1.39) obtaining the following differential equation:

$$\beta(2F - \omega F^2) + [(\omega^{-1}d\Delta/dx) - \beta](\eta - \omega \int_{\eta_1}^\eta F d\eta)F' = g' \tag{2.15}$$

where $\Delta$ is defined by (2.9) and

$$\left.\begin{aligned} \beta &= (\Delta/\tau_w)\omega dp/dx \equiv (\delta^*/\tau_w)dp/dx \\ \eta &= y/\Delta \\ \omega &= u_\tau/U_e \end{aligned}\right\} \tag{2.16}$$

Unfortunately, symbol $\Pi$ is often used for $\beta$, as well as for the wake parameter. Equation (2.15) is only independent of $x$ if $\beta, \omega$ and $\Delta$ or $d\Delta/dx$ are constant. The condition $\beta = $ constant is fulfilled by velocity distributions of the type $U_e \sim x^m$ or $U_e \sim \exp x$ and $\Delta = $ constant. $\omega$ equals constant can be satisfied only if the following relationship holds

$$\omega = (1/\kappa)\ln(U_e\delta^*/\nu) + C_1 + C(k^*) \tag{2.17}$$

where the variation of the displacement thickness $\delta^*$ with $x$ is balanced by the variation of the characteristic roughness $k^* = (k_r u_\tau/\nu)$, $k_r$ being the height of a roughness element.

For a further discussion of the theoretical aspects of self-preserving boundary layers the reader is referred to the above-mentioned investigations of Rotta and Townsend and a paper by Mellor and Gibson [2.78]. The latter authors have supplemented Rotta's work by extending the range of $\beta$ covering values $-0.5 \leqq \beta \leqq \infty$ by using an eddy viscosity relationship for the Reynolds shear stresses and keeping the parameters

$\beta$ and $G$ constant. $G$ is defined as

$$G= \int_0^\delta [(U_e - U)/u_\tau]^2 d(y/\delta)/\int_0^\delta [U_e - U)/u_\tau]d(y/\delta) . \qquad (2.18)$$

The "defect shape parameter" $G$ is connected with $\beta$ and $\omega$ via the momentum integral equation (ROTTA [2.32]).

An approximate relation between $G$ and $\beta$ based on the available, rather scattered experimental data was given by NASH [2.79] (for self-preserving flows only)

$$G \sim 6.10 (\beta + 1.81)^{\frac{1}{2}} - 1.70 . \qquad (2.19)$$

This gives $G = 6.5$ at $\beta = 0$, while the actual value is about 6.8.

It is by no means easy to verify self-preserving boundary layers experimentally. CLAUSER [2.15] kept the velocity profile in the outer layer (2.8) independent of $x$ by adjusting the adverse pressure gradient distribution in the $x$-direction accordingly. In these experiments $G$ was nearly constant while $\beta$ varied, especially in the boundary layer where the pressure gradient was kept high. Both HERRING and NORBURY [2.80] and LAUNDER and STINCHCOMBE [2.81] set up self-preserving flows with favorable pressure gradients. But there are doubts about the applicability of their method to determine skin friction in highly accelerated flows. Therefore it is difficult to decide whether self-preservation was reached at all (cf. COLES and HIRST [2.7]). BRADSHAW and FERRISS [2.82] and BRADSHAW [2.83] have investigated two self-preserving flows in adverse pressure gradients ($U_e \sim x^{-0.255}$ and $x^{-0.15}$) measuring mean velocity profiles and turbulence structure. They found that Reynolds shear stress and fluctuating velocities show self-preservation in the outer layer on the average, without being able to specify the degree of accuracy reached in their measurements of turbulence quantities.

Similarity laws for turbulent boundary layers with suction or injection were given by COLES [2.39] and STEVENSON [2.84].

### 2.3.3 Upstream History, Relaxation Effects and Downstream Stability

A shortcoming of many of the older boundary layer calculation methods consists in their inability to take into account the upstream history of the boundary layer and the relaxation of these upstream effects downstream, since it is assumed that the boundary layer development, and especially the shear stress distribution, is determined by local conditions only, via an "eddy viscosity" relation (Sect. 1.7) or otherwise. Relaxation is not used here in the strict sense of the word as it is defined for example in thermodynamics, but is meant to describe the transition process between two well-defined, possibly equilibrium states of a shear flow. Upstream history and relaxation effects may influence the inner and the outer region of a boundary layer more or less strongly or for a different length of time. In Klebanoff and Diehl's experiment [2.85], where the boundary layer had been disturbed by a spanwise rod, lying on the wall, diameter $d$ about equal to the undisturbed boundary layer thickness $\delta_0$ at the trip position, a downstream distance of $26\,\delta_0$ was not sufficient for the velocity profile to regain similarity. If a much smaller trip rod with $d/\delta_0 = 0.055$ (CLAUSER [2.5]) is placed parallel to the wall and across the boundary layer, but now at distances $y/\delta_0 = 0.15$ and $0.55$ from the wall, the inner layer returns much more quickly to the universal inner law than the outer layer, confirming the strong influence of upstream history and relaxation on the large eddies. One reason for this behavior consists in the rather long lifetime of these eddies. If a typical eddy time scale is defined as (typical wavelength)/(typical velocity fluctuation), one finds a time scale of roughly $\delta/0.003\,U_e$, during which time the large eddies travel a distance of about $30\,\delta_0$. The response of a retarded self-preserving turbulent boundary layer to the sudden removal of the pressure gradient has been investigated by BRADSHAW and FERRISS [2.82] who simulated the pressure distribution near the leading edge of a "peaky" aerofoil. It turns out from measurements of the production and advection terms of (1.43) that the advection is an unusually large fraction of the production and that the decrease of the advection after the release of the pressure gradient is small, indicating a slow approach to the new self-preserving state in zero pressure gradient. Not only is the advection term large compared with that in a self-preserving boundary layer, but also the turbulent energy integrated across the layer is much higher than in the zero-pressure gradient case. This must be attributed to the large-scale turbulence in the outer region of the boundary layer, i.e., to the large eddies. A further flow dominated by its upstream history is a turbulent boundary layer responding to a step change in surface roughness from rough to smooth (JACOBS [2.86] and ANTONIA and LUXTON [2.87]). The more intense initial rough-wall flow dictates the rate of diffusion of the disturbance for a considerable distance, and the turbulent energy budget shows that the advection term is comparable with the production or dissipation terms.

Relaxation phenomena in boundary layers due to the upstream heat transfer history (GRAN et al. [2.88], GATES [2.89]) or to severe pressure gradients in the vicinity of the throat of supersonic nozzles (LEE et al. [2.90]) have been observed to last far downstream into the zero-pressure gradient region. In the same type of flow the laminar-turbulent transition process seems to take much longer in the near wall region than in the outer part of the boundary layer, evidently for similar reasons. As far as downstream stability is concerned CLAUSER [2.15] reported difficulties in establishing self-preserving boundary layers in strong adverse pressure gradients where a small change in pressure gradient upstream produced large changes in the flow downstream. Problems of this sort are often linked to secondary flows or to the unsteadiness of a separation region, but Clauser's results cannot be ruled out completely. A plausible argument for downstream stability was put forward by BRADSHAW [2.83] and a more extensive discussion of the stability of self-preserving flow was given by TOWNSEND [1.26].

### 2.3.4 The Effect of Free-Stream Turbulence

From the aeronautical viewpoint a discussion of the effects of freestream turbulence on a turbulent boundary layer is of little importance. For a large subsonic aircraft at cruise the rms velocity fluctuation over an octave bandwidth centered on a wavelength of 0.3 m (the order of $\delta$ over the rear upper surface of the wing) will be only 0.03% of the free-stream velocity in heavy turbulence (GREEN [2.91]). However in wind tunnel tests of aircraft and other bodies, and in many internal-flow rigs, the influence of free-stream turbulence on the boundary layer must be known in order to interpret or improve the results. The rate of growth of the boundary layer thickness increases by about 50% and that of the skin friction by about 20% (CHARNAY et al. [2.92]) for a change of the turbulence level $T_u = (\overline{u^2})^{\frac{1}{2}}/U_e$ from 0.2 to 5%. As expected from inner-layer similarity arguments the log-law is hardly affected. It is difficult, however, to reconcile the increase in skin friction with Kestin's [2.93] statement that a measurable effect of the free-stream turbulence on heat transfer could not be found. As is to be expected, the effects of free-stream turbulence on the outer part of the boundary layer (here $y/\delta > 0.7$) lead to an increase in the level of $\overline{u^2}, \overline{uv}$ (by about 150%) and of the turbulent kinetic energy, to a sharp decrease of the wake component $\Delta U/u_\tau$ for $T_u \sim 5\%$ and to a slightly fuller mean velocity profile (for the use of the latter effect to increase the "effective" Reynolds number in windtunnels see GREEN [2.91]). HUFFMAN et al. [2.94] state further that at the high turbulence level a highly non-equilibrium boundary layer exists because the advection term is no longer small and therefore production and dissipation are no longer roughly equal. Apart from the measurements by KLINE et al. [2.95], CHARNAY et al. [2.92] and HUFFMAN et al. [2.94] from which the above conclusions were drawn, there are measurements by PICHAL [2.96], KEARNEY et al. [2.97] and ROBERTSON and HOLT [2.278]. Further references together with a discussion of semi-empirical relationships for the outer law, the skin friction, the shape parameters, etc., under the effect of free-stream turbulence were given by BRADSHAW [2.281].

In the above work, the free-stream turbulence was generated by an upstream grid and was not too far from spatial homogeneity and isotropy, having a scale (wavelength) comparable with the thickness of the shear layer. The effects of much larger or much smaller scales are not clear yet. This is especially true for the boundary layer in a turbomachine where the so-called free-stream turbulence is a mixture of roughly homogeneous turbulence from far upstream, highly inhomogeneous turbulence from the wakes of the blades and miscellaneous large-scale unsteadiness, presenting a good example of an ill-posed boundary condition. The influence of free-stream turbulence on the structure of a wake and on the "potential" core region of a jet has been investigated by KOMODA [2.98] and VAGT [2.99], respectively. A thorough survey and experiments on the management of free-stream turbulence have been carried out by LOEHRKE and NAGIB [2.100].

### 2.3.5 The Effect of Streamwise Pressure Gradient

The pressure gradient, adverse (positive) or favorable, is the parameter which has the most far-reaching consequences for a boundary layer, possibly leading to separation in the one case and to supersonic flow and/or reverse transition in the other. More detailed reviews of the effect of the pressure gradient on incompressible turbulent boundary layers were given by CLAUSER [2.5], ROTTA [2.32] and TOWNSEND [1.22], and a data catalogue containing mean velocity profiles, skin friction data and pressure distributions of turbulent boundary layers was published by COLES and HIRST [2.7]. A critical compilation of measurements of turbulence data, updated at the same time, is very much needed as an addition to the Stanford catalogue [2.7]. The interaction between pressure gradient, mean velocity distribution and skin friction can be seen from the integral form of the momentum equation for two-dimensional thin shear layers which is obtained by integrating (1.45) from $y=0$ to $y=\delta$:

$$\frac{d}{dx}\int_0^\delta \varrho U(U_e - U)dy - \frac{dp}{dx}\int_0^\delta \left(1 - \frac{U}{U_e}\right)dy - \tau_w = 0, \qquad (2.20)$$

or, in more common forms, valid also for compressible boundary layers,

$$\frac{d}{dx}(\varrho_e U_e^2 \theta) = \tau_w(1 + \beta) \qquad (2.21a)$$

or

$$\frac{d\theta}{dx} + \frac{\theta}{U_e}\frac{dU_e}{dx}(H+2-M_e^2)=(\tau_w/\varrho_e U_e^2) \equiv \frac{C_f}{2} \qquad (2.21b)$$

where

$$\delta^* \quad \text{or} \quad \delta_1 = \int_0^\delta [1-(\varrho U/\varrho_e U_e)]dy \quad \text{(displacement thickness)} \qquad (2.22)$$

$$\theta \quad \text{or} \quad \delta_2 = \int_0^\delta (\varrho U/\varrho_e U_e)[1-(U/U_e)]dy$$
$$\text{(momentum loss thickness)} \qquad (2.23)$$

$$H \quad \text{or} \quad H_{12}=\delta_1/\delta_2 \quad \text{or} \quad \delta^*/\theta \quad \text{(shape parameter)} \qquad (2.24)$$

and $\beta$ was defined in (2.16). Eq. (2.21a) shows that $\beta$ measures the relative effects of pressure gradient and surface shear stress on the rate of loss of momentum. Since the mean pressure gradient does not appear in the transport equations for the fluctuating quantities (see Chapt. 1), the changes in turbulent intensity, shear stress, etc., which are characteristic of pressure gradient flow are caused by changes of the mean velocity profile. This again is influenced by the pressure gradient and implicitly by the skin friction. It is this rather complex relationship between mean velocity and turbulence quantities which causes the difficulties inherent in the closure problem of the boundary layer equations. The structure of the turbulent boundary layer with longitudinal pressure gradients was investigated by SCHRAUB and KLINE [2.279].

We have seen in Subsection 2.3.1 that the logarithmic law is valid also in boundary layers with adverse pressure gradients (LUDWIEG and TILLMANN [2.13] if the flow is not too close to separation, where its range of validity vanishes. It does not hold, however, for highly accelerated boundary layers (PATEL [2.36]). The outer law (2.8) is strongly influenced by the pressure gradient, exhibiting under certain conditions similar (self-preserving) velocity profiles with $u_\tau/U_e$ and $\beta$ as parameters (see ROTTA or CLAUSER). Typical Reynolds stress $\overline{uv}$, kinetic energy $\overline{q^2}$ and mean velocity profiles are shown in Fig. 2.5. At the wall the gradient of the shear stress profile is determined by the following relation

$$(\partial\tau/\partial y)_w = dp/dx \qquad (2.25)$$

indicating a constant stress layer in zero pressure gradient flow and a maximum of the shear stress in the boundary away from the wall if an adverse pressure gradient is present. Unfortunately there still exist few

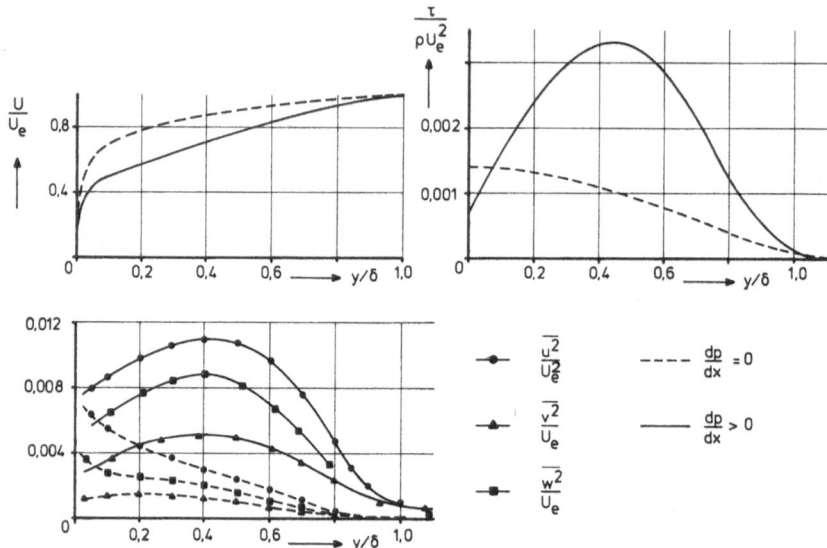

Fig. 2.5. Reynolds stress, kinetic energy and mean velocity profiles in a turbulent boundary layer with zero and medium adverse pressure gradient (from [2.29, 83])

experimental investigations in which even basic turbulence quantities have been measured so thoroughly that all terms in the transport equations can be determined for a wide enough range of pressure gradients, thus allowing assessment of the existing turbulence models. For adverse pressure gradient boundary layers see for example BRADSHAW [2.101] (equilibrium flow), BRADSHAW and FERRISS [2.82] (relaxing to zero pressure gradient), GOLDBERG [2.102], SAMUEL and JOUBERT [2.103], SPANGENBERG et al. [2.104], SANDBORN and LIU [2.105], SCHUBAUER and KLEBANOFF [2.106], NEWMAN [2.107] and for favorable pressure gradients LAUNDER [2.108].

### 2.3.6 Separation, Separation Bubbles and Reattachment Flow

*Separation*

Separation was probably the phenomenon which inspired Prandtl's thinking about those shear flows which are now called boundary layers. There exists a vast body of references on the topic, covering almost every conceivable kind of flow geometry (see, e.g., CHANG [2.109]). Following ROSHKO [2.110] we are concerned here neither with separation at low Reynolds numbers nor with the difficult mathematical questions relating to the singularity at the laminar separation point (see, e.g., KAPLUN [2.111] and BROWN and STEWARTSON [2.112]). Separation may occur at sharp edges, for instance at the shoulder of a wedge, or in boundary layers on smoothly curved surfaces in adverse streamwise pressure gradient. In the first case there is no question as to the point or line of separation, in many other cases the determination of the separation point is part of the problem.

In some flows the wall-streamline leaves the wall at the point where the gradient of the mean velocity at the wall becomes zero, as in Prandtl's [2.113] classical case, followed by a region of reverse flow (Fig. 2.1). The angle of this separating streamline was determined by OSWATITSCH [2.114]. ROSHKO then names secondly the flow with a separation bubble and thirdly the flow with breakaway separation (for the latter see also STEWARTSON [2.115], [2.116]).

All these different types of separation have in common the interaction between the boundary layer and the flow in the free stream. Even in incompressible flow the pressure distribution upstream of separation is influenced by the flow downstream of separation, thus converting the pressure from an independent boundary condition to a variable. This means that the pressure calculated from inviscid flow theory and the measured pressure are no longer consistent. Furthermore the assumptions made to derive the thin shear-layer equations, such as $d\delta/dx \sim O(\varepsilon)$ where $\varepsilon \ll 1$ and $\partial p/\partial y = O(\varepsilon)$, are no longer valid due to the sharp increase of the boundary layer thickness and the curvature of the streamlines close to separation. Streamline curvature will again have a strong influence on the turbulence structure in this region (see BRADSHAW [2.117], and Sect. 3.1).

Most of the criteria for separation are based on determining the position of zero skin friction (see also Chapt. 5), e.g., the "rapid separation criterion" of Stratford as modified by TOWNSEND [2.118], who showed that the pressure distribution near separation depends only on the pressure rise towards separation and on the characteristics of the initial boundary layer and not on the flow geometry. Further investigations were performed by SANDBORN and LIU [2.105] based on the separation model of KLINE [2.119] and SANDBORN and KLINE [2.120].

An important phenomenon related to separation is that the boundary layer close to separation may show a steady, an unsteady or a transient behavior, superimposed on the turbulent motion proper, exhibiting a difficult feedback mechanism. For similar effects of unsteady flow behavior in boundary layers on wings with transonic flow regions where buffeting occurs due to shock-boundary layer interaction see, e.g., MABEY [2.121]. Boundary layers which showed a steady behavior up to separation were investigated by STRATFORD [2.122], FERNHOLZ and GIBSON [2.123], GOLDBERG [2.102] and SPANGENBERG et al. [2.104] among others. In the former two investigations the pressure distributions were set in such a way that the boundary layer remained on the verge of separation, i.e., the flow had nominally zero skin friction. Theoretical discussions about the boundary conditions (pressure distribution) for, and the development of turbulent boundary layers with negligible wall shear stress were given by STRATFORD [2.122], TOWNSEND [2.124], FERNHOLZ [2.125] and ALBER [2.126] for example. Hardly any information is available, however, about the structure of the turbulence in the two extreme cases when a boundary layer reaches separation gradually or is caused to separate rapidly by an obstacle. In the latter case the boundary layer is subject to a strong perturbation (e.g., BRADSHAW and GALEA [2.127]), thus behaving in a distinctly different way from the former case.

Since separation is a phenomenon which normally has detrimental effects on the overall flow, aerodynamicists have developed many ingenious techniques to influence separation, the most important one, of course, being to design the streamwise pressure distribution in such a way that separation is prevented as far downstream as possible. A survey of the techniques to prevent separation by blowing, suction, moving walls or propeller slipstreams was given by GERSTEN [2.128], updating the monograph edited by LACHMANN [2.129]. For discussions of large-scale separated regions and reattachment in internal flows see Sect. 3.4.

### Separation Bubbles (Fig. 2.1c)

After the discovery of the phenomenon of separation bubbles by JONES [2.130], fresh interest in this topic was aroused through the use of thin aerofoil sections for reducing the effects of compressibility. The behavior of separation bubbles is therefore of particular interest both for the stalling characteristics of these aerofoils, showing high suction peaks near the leading edge followed by a sharp pressure rise, and also for the heat transfer on turbine blades or for flow in poorly designed nozzles at low or medium Reynolds numbers. Separation bubbles are also observed in the flow around circular cylinders and in connection with the inter-action of shock waves and laminar boundary layers. There is a whole series of earlier experimental investigations summarized by YOUNG and HORTON [2.131] and followed by those of HORTON [2.132, 133] and DELPAK [2.134]. A survey paper on the subject which is still an excellent introduction was given by TANI [2.135].

At first sight the problem of separation bubbles seems to be one of laminar flow only, for it is the laminar boundary layer which separates: but, under certain conditions not yet clarified, it reattaches in the turbulent state, having undergone transition somewhere between separation and reattachment. The shallow recirculation region shown in Fig. 2.1c is called a separation bubble. The necessary conditions for the formation of a separation bubble are that the pressure gradient shall decrease soon after separation (so that the shear layer does not move rapidly away from the surface) and, except in the rare cases of laminar reattachment, that the Reynolds number and free-stream turbulence level shall be high enough for transition to occur soon after separation but not so high that transition occurs before the expected laminar separation point. At the higher Reynolds numbers the extent of the bubble is small, of the order of 1% chord, and the slight step in the pressure distribution produced by the bubble has a negligible effect on the flow around an aerofoil. It should be noted that the upper limit of the Reynolds number at separation is about 350, where $\theta_s$ is the characteristic length. With an increase of incidence or a reduction of speed (Reynolds number) the short bubble may burst to form either a long bubble or a separated shear layer (GASTER [2.136]) which then leads to a severe reduction of the lift (Fig. 2.1c). No conclusive criterion is known so far, either for the formation or for the bursting of separation bubbles. If the separation region is small, conventional iteration procedures should suffice to calculate the effect of the bubble on the flow around an aerofoil. Promising steps to solve the separation bubble problem via the Navier-Stokes equations for laminar flows have been taken by BRILEY and McDONALD [2.137], LUGT and HAUSSLING [2.138] and MEHTA and LAVAN [2.139]. Some experiments on the turbulence structure of the flow were performed by McGREGOR [2.140] and GASTER [2.136] without conclusive results, and separation bubbles in three-dimensional boundary layers were investigated by EICHELBRENNER [2.141, 142] and HORTON [2.133]. The latter author also presented measurements of three-dimensional turbulent boundary layers re-developing after reattachment behind short separation bubbles.

## Reattachment of Turbulent Shear Flow

The phenomenon of reattachment of a separated shear layer, which we have just met downstream of a separation bubble, has been of special interest in supersonic flows, mainly because of the need to predict base drag and heat transfer for flight vehicles. Let it suffice here to mention the names of the authors who presented the basic theoretical approaches: CHAPMAN [2.143], KORST [2.144] and CROCCO and LEES [2.145]. An eclectic merger of the Crocco-Lees and Chapman-Korst approach to the near wake and reattachment region has been proposed recently by SMITH and LAMB [2.146] including about 70 references. Reattachment is also discussed in [2.9]. Though there are numerous applications in low speed flow such as spoilers, forward- and backward-facing steps, windbreaks, etc., the problem has attracted less scholarly attention than in supersonic flow (see, e.g., the survey paper by BRADSHAW and WONG [2.147]).

Shear layers with reattachment are in most cases highly perturbed flows, the more so if they have two regions of separated flows as in Fig. 2.3 or if in the case of a backward-facing step the thickness of the oncoming boundary layer $\delta_0$ is much smaller than the step height ($h \gg \delta_0$). In the first case the upstream history of the flow is less important due to the severe perturbation by the first separation region, but it is surprising to note that upstream conditions seem to play a minor role also in the latter case (e.g., ABBOTT and KLINE [2.148]). These authors have applied flow visualization techniques to the flow in the separated region downstream from a step, finding a rather complex vortex pattern and a two-dimensional recirculation region. BRADSHAW and WONG [2.147] have noticed large differences between the turbulence structure in the reattachment region and that in a conventional boundary layer; both a marked "dip" below the log-law for the near wall region and persisting characteristics of the mixing layer in the outer part of the boundary layer far downstream.

There are further experimental investigations by ARIE and ROUSE [2.149], MUELLER and ROBERTSON [2.150] and LE BALLEUR and MIRANDE [2.151] (with variable pressure gradients). MUELLER and ROBERTSON also observed a slow redevelopment from the mixing layer to the boundary layer flow. Again there is a need for further measurements in shear layers with reattachment, with more refined measurement techniques. Calculation methods for reattached shear flows were developed by MUELLER et al. [2.152] and LE BALLEUR and MIRANDE [2.151].

### 2.3.7 The Effect of Changing Wall Geometry

In this subsection we shall discuss the effects of two kinds of wall boundary conditions on the turbulent boundary layer, the effects of roughness elements and those of waviness of the wall. Wall curvature will be dealt with in Subsection 2.3.8.

### Wall Roughness

The effect of wall roughness on a turbulent shear layer was first investigated in pipe flow by NIKURADSE [2.153] in 1933. Since then a wide variety of experiments has been performed in shear flows, ranging between the extremes of flows in natural watercourses and over aircraft wings or compressor blades. Due to the multiplicity of geometrical shapes and distribution patterns of roughness elements not even a semi-empirical correlation of the experimental data exists. There are, however, relationships which cover certain types of roughness, such as those of PERRY et al. [2.160; see 2.154] and SIMPSON [2.155]. Naturally it is the boundary layer over the so-called sand-grain roughness covering a surface uniformly which was investigated most thoroughly (PRANDTL [2.156], SCHLICHTING [2.157]). The former showed that results obtained in pipe and duct flows also hold for boundary layers (see, e.g., HAMA [2.158]). If the roughness length scale for this type of roughness $k_r$ is made dimensionless by $v/u_\tau$, one can distinguish between three cases (for nominally zero pressure gradient flow: if $\partial \tau/\partial y \neq 0$, a parameter $(k_r/\tau_w)\partial\tau/\partial y$ must appear but its effects are unknown):

1) Hydraulically smooth ($0 \leqq k_r u_\tau/v \leqq 5$). The roughness elements are submerged in the viscous sublayer and do not affect the boundary layer significantly.

2) Transient roughness ($5 \leqq k_r u_\tau/v \leqq 70$). The velocity profile is affected by the roughness of the wall, and this effect is taken into account by a function $f$ which is substituted into the law of the wall instead of the constant $C$ in (2.3).

$$f(k_r u_\tau/v) = C - \Delta U_k/u_\tau. \tag{2.26}$$

The roughness function $\Delta U_k/u_\tau$ can be expressed according to HAMA [2.158] for natural roughnesses such as wrought or cast iron by

$$\Delta U_k/u_\tau = (1/\kappa) \ln [(k_r u_\tau/v) + 3.30] - 2.92. \tag{2.27}$$

The logarithmic law then reads:

$$(U/u_\tau) = (1/\kappa) \ln (u_\tau y/v) + f(k_r u_\tau/v). \tag{2.28}$$

An analytic approximation for the function $f$ was given by IOSELEVICH and PILIPENKO [2.159] and a skin friction law was given by CLAUSER [2.15]. For flows other than with sand-grain roughness (2.28) can be used also if an "equivalent sand-grain roughness" is determined. PERRY et al. [2.160] have found that for a roughness geometry consisting of rectangular elements of square cross section placed normal to the flow direction $\Delta U_k/u_\tau$ is of the form

$$\Delta U_k/u_\tau = (1/\kappa) \ln (\varepsilon u_\tau/v) + C^+, \tag{2.29}$$

where $\varepsilon$ is called the error in origin and is the distance below the crest of the roughness which locates the origin of $y$ in (2.28). $C^+$ is a constant for the specific roughness geometry.

3) Fully developed roughness ($70 \leqq k_r u_\tau / v$). The velocity distribution in the log-law region is now independent of viscosity because the Reynolds number characteristic of the flow over the rough wall is large. It takes the following form, of which (2.27) yields a special case for large $k_r u_\tau / v$,

$$U/u_\tau = (1/\kappa) \ln (y/z_0). \tag{2.30}$$

This is a relationship often adopted for velocity profiles in atmospheric boundary layers (see Chapt. 4) where $z_0$ is chosen to absorb the constant $C$ in (2.3) altogether.

It has become usual to distinguish between two types of roughness according to $\varepsilon$ in (2.29), or $z_0$ in (2.30): $\varepsilon$ proportional to $k_r$, defining "$k$-type" or sand-grain roughness and $\varepsilon$ proportional to $\delta$, or a function of the distance from the leading edge $x$, defining a "$d$-type" roughness. The latter type is generated for example by the above-mentioned rectangular elements, now spaced at distances such that quasi-stable vortices are formed in the cavity in between (e.g., Liu et al. [2.161], Perry et al. [2.160] and Wood and Antonia [2.154]). For a $d$-type roughness little direct interaction seems to occur between the roughness and the mean flow in the boundary layer (see also Townes and Sabersky [2.162]), so that there are features similar to those of a smooth-wall boundary layer. A calculation method for the flow over a $d$-type roughness was published by Antonia and Wood [2.163]. Further investigations dealing with the structural features of turbulent flow over rough boundaries were performed by Grass [2.164] and by Antonia and Luxton [2.280].

Since the pressure gradient does not influence the flow in the inner layer directly, little effect on the rough wall boundary layer is to be expected (e.g., Perry and Joubert [2.165]). It is important, however, to distinguish secondary effects due to size or spacing of roughness elements, such as displacement effects ($k_r \gg \delta_0$) leading to a change in the overall pressure distribution or to vortex shedding from the crest of the first roughness element. In the experiments of Antonia and Luxton [2.166] the crests of the roughness elements were aligned with the smooth surface upstream to avoid these detrimental effects. Of course upstanding roughness elements, such as belts of trees, may be important in practice. The same authors [2.167] then found an internal layer growing over the rough surface like $\delta. \sim x^{0.80}$, i.e., similar to the growth rate of a smooth boundary layer which is primarily controlled by diffusion. This internal layer denotes the outward limit of the influence of the roughness and has two important consequences: the influence of the "old" upstream boundary layer is confined to the outer part of the composite boundary layer, and, due to the intermittent behavior of the interface between the two layers of the composite boundary layer, the flow switches between regions of the "new" and the "old" turbulence. A self-preserving state is reached about 15 $\delta_0$ downstream from the first roughness element (LUXTON [2.168]). The fast growth together with too crude assumptions about velocity and shear stress distributions may account for the unsatisfactory theoretical results (e.g., TOWNSEND [2.169, 170] and TAYLOR [2.171]). RAO et al. [2.172] obtained better results from a higher-order closure method.

In a fully developed rough-wall boundary layer the turbulent energy production is significantly higher than in a smooth-wall boundary layer and the shear stress has a positive gradient at the wall (ANTONIA and LUXTON [2.173]). The reason for the nonzero shear stress gradient in the absence of acceleration is not wholly clear. The only plausible explanation is that $U$ and $V$ vary cyclically with $x$ as the flow passes over the roughness elements, leading to an apparent "Reynolds stress".

The response of a turbulent boundary layer to a step change in surface roughness from rough to smooth was investigated experimentally by ANTONIA and LUXTON [2.87]. Measurements of the drag of some characteristic aircraft excrescences immersed in turbulent boundary layers were performed by GAUDET and WINTER [2.174]. For the discussion of three-dimensional and of nonuniformly distributed roughness the reader is referred to SCHLICHTING [2.157].

## Wavy Walls

There are three main areas where one is interested in the flow over wavy walls: aeronautical, civil, and chemical engineering. Wave generation by wind and the interaction of wind and sea are outside our scope here. In aerodynamics even very small ratios of amplitude to wavelength of a wavy wall, say 0.005, are of interest, since a surface waviness of such a magnitude can provide a significant contribution to the total drag of a transonic aircraft (ROGERS [2.175]). INGER and WILLIAMS [2.176] showed that large changes in the phase of the pressure and temperature perturbation occur across the highly nonuniform flow of a turbulent boundary layer if the boundary layer thickness is comparable to the surface wavelength. As a result, a wall pressure distribution typical of subsonic flow can exist in the presence of supersonic external inviscid flow (see also KENDALL [2.283] and DAVIS [2.284]). The reader is referred to ASHTON [2.177] and OWEN and THOMSON [2.178] for heat transfer problems along wavy walls. Civil engineers are interested in the formation of bed forms in alluvial channels. For flow over rigid wavy boundaries the objective has been to evaluate the interactions between the boundary layer and the external flow as reflected in pressure distribution and flow resistance (e.g., Ho and Gelhar [2.179]). Turbulent flow in wavy pipes was investigated by Hsu and Kennedy [2.180] who measured both mean flow and fluctuating quantities. Besides boundary layers along wavy walls as such, there exists considerable interest for gas-liquid boundary layer flow where the phase changing interface shows a wavy form (e.g., Kotake [2.181]). In this area little research on turbulent flow has become known.

### 2.3.8 The Effect of Wall Curvature

Dealing with the effect of wall curvature on turbulent shear layers one must distinguish between curvature in the transverse (cross-sectional) and in the longitudinal (camber) direction (the flow along a body of revolution as investigated by Winter et al. [2.182] contains both effects). If for a body of constant radius the boundary layer thickness is small compared with the body radius, then the flow can be treated as if it were two dimensional. This assumption does not hold for long slender bodies of revolution, and the boundary layer equations and turbulence models must contain terms which take into account the transverse curvature (e.g., Cebeci and Smith [1.4]). Effects of transverse curvature on turbulent boundary layers were considered by Richmond [2.183] and Ginevskii and Solodkin [2.184] who derived relationships for the mean velocity and the Reynolds shear stress profile. For the flow in the vicinity of the wall they obtained for the shear stress distribution

$$\tau/\tau_{\mathrm{w}} = [1 + (dp/dx)y/\tau_{\mathrm{w}}]/[1 \pm y/r_{\mathrm{w}}], \qquad (2.31)$$

where $r_{\mathrm{w}}$ is the radius of the transverse curvature and where the positive sign refers to a convex and the negative sign to a concave surface. The reader is referred to discussions of the inner law by Willmarth and Yang [2.185], Bradshaw and Patel [2.186] and of the eddy-viscosity distribution by Cebeci and Smith [1.4]. The boundary layer on air-drawn artificial fibers was investigated by Walz and Mayer [2.187] and that near the tail of a body of revolution by Patel et al. [2.188].

Effects on the turbulent boundary layer due to longitudinal curvature are much more difficult to cope with since the structure of the turbulence is strongly affected, in apparent contrast to the case of transverse curvature. Details are given in Subsection 3.1.1; here we give a few basic references to work on boundary layers. Prandtl [2.189] explained why the flow of a turbulent boundary layer along a convex surface is more stable than that along a concave wall, a result consistent with the instability mode in laminar boundary layers on concave walls represented by Taylor-Görtler vortices (TAYLOR [2.190] and GÖRTLER [2.191]). Such longitudinal vortices in turbulent boundary layers were first found by TANI [2.192]. Finally BRADSHAW [2.193] made quantitative use of the analogy between streamline curvature and buoyancy in turbulent shear flow.

On a convex wall both Reynolds shear stress and turbulent kinetic energy across the boundary layer are strongly reduced as compared with a boundary layer on a plane wall, whereas the inverse effect can be observed in the flow along a concave wall ([So and MELLOR [2.194, 195]). THOMPSON [2.196] provided evidence that even at ratios of $\delta/R = 0.003$ [7] changes in the entrainment rate could be attributed to curvature effects. For $\delta/R = 0.01$ MERONEY [2.197] noticed a steep decrease in Reynolds shear stress outside the near wall region compared with a flat plate boundary layer and a change of the integral properties by about 10%. Ratios of $\delta/R = 0.01$ are often found with thick aerofoils or with turbo-machine blades. In compressible turbulent boundary layers on curved walls with zero longitudinal pressure gradient, changes in heat transfer of $\pm 20\%$ were measured for a concave and a convex wall, respectively (THOMANN [2.198]).

In most of the cases discussed above the wall curvature was so small that changes in static pressure across the boundary layer were still negligibly small. In experiments in curved ducts (So and MELLOR, and PATEL [2.199]) the normal pressure gradient due to the strong curvature must be taken into account. The relevant equations and the importance of these terms with regard to curvature effects were discussed by BRADSHAW [2.117] who gives a large number of references for both boundary layers and wall jets. The reader is referred to NEWMAN [2.200] for a review on curved turbulent wall jets and to FERNHOLZ [2.201, 202] for the influence of upstream history, boundary conditions and effects due to very high curvature.

The effect of longitudinal curvature on eddy viscosity was correlated by CEBECI and SMITH [1.4]. Calculation methods which account for curvature effects on turbulent boundary layers were presented by DVORAK [2.203] and IRWIN and SMITH [2.204].

### 2.3.9 The Effect of Heat Transfer at the Wall

Since Chapter 6 will deal extensively with heat transfer problems in turbulent flows the discussion of "shear flow—heat transfer" interaction will be confined to a few items in preparation for the discussion of

---

[7] $\delta$ is the thickness of the boundary layer and $R$ the radius of the longitudinal curvature.

compressible flow in Section 2.5. We shall refer to turbulent boundary layers at low Mach numbers and at a Prandtl number of about one only, first with small temperature differences and then with temperature differences large enough to affect the velocity field. One of the technically very important relationships is that between skin friction and heat transfer at a fixed wall, the so-called Reynolds analogy which allows heat transfer to be determined by a measurement of the skin friction

$$St \equiv Q_w/[\varrho_e U_e c_p (T_r - T_w)] = F \cdot c_f/2 . \tag{2.32}$$

The analogy in its simplest form applies only to boundary layers with zero pressure gradient along a smooth isothermal wall. $St$ denotes the Stanton number, $c_f$ the skin friction coefficient and $F$ the Reynolds analogy factor (see ROTTA [2.205]). Simple transfer problems could be solved by boundary layer methods but most of the technical flow configurations with heat transfer are extremely complex (e.g., heat exchangers) and many of them in addition ill-posed problems. The Reynolds analogy factor is fairly well documented for constant-pressure boundary layers, or duct flows, with small temperature differences. Data for flows in pressure gradients, and plausible methods of correlating the data, are rarer, and in the case of large temperature differences little information is available about the turbulence structure, which is now influenced also by the fluctuations of density and temperature. In the face of these problems, the usual approach is to transfer first-order closure concepts, such as those of mixing length or eddy viscosity, from the momentum equation to the energy equation and to introduce an eddy diffusivity $\gamma_T$:

$$\gamma_T = -\overline{v\theta}/(\partial T/\partial y) \tag{2.33}$$

where $\theta$ is the fluctuating temperature and $c_p$ the specific heat at constant pressure. In most cases the turbulent Prandtl number $\sigma_t$ defined, analogously to the molecular Prandtl number $\sigma$, as the ratio of eddy viscosity to diffusivity is assumed to be constant across the boundary layer (for $\sigma_t \neq$ constant see MEIER et al. [2.206]). First-order closure schemes were used in the calculation methods of LEONT'EV [2.207], PATANKAR and SPALDING [2.208], GOSMAN et al. [2.209], and CEBECI et al. [2.210] FLAHERTY [2.211] and CRAWFORD and KAYS [2.212] and more modern closure schemes are discussed in Chapter 6.

There exist few experimental investigations of turbulent boundary layers with heat transfer and even fewer with high heat transfer rates. For zero pressure gradient flow see KELNHOFER [2.213] (heated wall), BACK et al. [2.214] (cooled wall) and for accelerated flows BOLDMAN et al. [2.215], BACK and CUFFEL [2.216], and KEARNEY et al. [2.217]. The other experiments known to the author have only moderate to small temperature differences across the boundary layer (e.g., HATTON and EUSTACE [2.218], and THIELBAR et al. [2.219]) and assume constant fluid properties. Little information is available on boundary layers with heat transfer in an adverse pressure gradient. As separation is approached, the heat transfer at the wall falls off, but less rapidly than the skin friction (DVORAK and HEAD [2.220]). It is obvious that the heat transfer does not vanish at separation or reattachment since $\partial T/\partial y$ does not tend to zero (e.g., CHILCOTT [2.221] and FLETCHER et al. [2.222]).

In most of the references quoted above the wall was kept more or less isothermal. Under this condition there is good reason to assume that the influence of the pressure gradient on the temperature profile is weak (COHEN and RESHOTKO [2.223]). However, if there is a streamwise gradient of wall temperature, a strong influence of the heat transfer on the temperature profile must be expected (DIENEMANN [2.224]). Another important problem is the effect of the upstream thermal history of the boundary layer which markedly influences the surface heat transfer characteristics downstream (REYNOLDS et al. [2.225], MORETTI and KAYS [2.226], and KEARNEY et al. [2.217]).

Finally we should mention relationships for the mean velocity and the temperature in turbulent boundary layers. For the inner layer they have the following form (DEISSLER [2.227]):

$$U = u_\tau f(yu_\tau/v_w, \beta) , \tag{2.34}$$

$$T = T_w f^+(yu_\tau/v_w, \beta) . \tag{2.35}$$

Here $\beta = Q_w/(\varrho c_p u_\tau T_w) \equiv T_\tau/T_w$, where $T_\tau$ is Squire's "friction temperature" [1.2, p. 823] which according to [2.228] was introduced independently by LANDAU and LIFSCHITZ in 1944. Relationships for the log-law region were derived by BRADSHAW [2.229] and the author [2.230], for the sublayer by MEEK and BAER [2.231] and for the whole profile by ROTTA [2.205]. Equilibrium profiles were calculated by ALBER and COATS [2.232]. Survey papers on heat transfer problems were published by KESTIN and RICHARDSON [2.233] and SPALDING [2.234].

### 2.3.10 Transition from Laminar to Turbulent Flow

The phenomenon of laminar-turbulent transition was identified as long ago as 1883, by REYNOLDS, and accordingly there exists a large body of theoretical and experimental results, a description of which would go well beyond the scope of this book. What can be done, however, is to warn the reader to be especially critical before applying transition criteria and other research results, usually obtained under quite specific conditions. MORKOVIN [2.235] gives an admirable survey of the problems and of current knowledge, while RESHOTKO [2.236] outlines the most urgent problems and suggests further work. It is generally accepted now that transition is preceded by instability of the laminar shear layer. Since the theory of hydrodynamic stability in shear layers, with boundary layers as a special case, can be considered as a field of its own, it must suffice here to refer to the monograph by LIN [1.10] and the papers by MACK [2.237, 238], CRAIK [2.239] and GASTER [2.240] to name but a few. The more recent stability calculation schemes deal with non-parallel flow models and take into account nonlinear effects.

In the "ideal case" of an incompressible two-dimensional laminar boundary layer with zero pressure gradient which is free of "artifical" disturbances due to boundary conditions, the process of instability and "natural" transition was summarized by STUART [2.241] and MOLLO-CHRISTENSEN [2.242]: "Beyond a certain Reynolds number travelling waves of velocity fluctuations develop and grow with downstream distance. Next, spanwise disturbances cause local free shear layers to form which again cause bursts of high-frequency instability fluctuations with a very high growth rate. The flow then erupts locally into the Emmon's spots (i.e., patches of turbulent flow in a laminar surrounding) which further downstream cause such a confused flow that we call it turbulent". It is obvious from this description, from the basic experimental investigations by SCHUBAUER and KLEBANOFF [2.243] and KLEBANOFF et al. [2.244], and from the smoke photographs by KNAPP and ROACHE [2.245] that transition does not occur instantaneously and that both the location and the extend of the transition region in a specific flow configuration are required. For this ideal case—even with variable pressure gradient—the critical point (denoted by a critical Reynolds number $(Re_\theta)_{crit.}$ as the upstream boundary) of the transition region can be reasonably well predicted (e.g., SMITH and GAMBERONI [2.246], JAFFE et al. [2.247]). A general prediction of transition is impossible at present as we have to cope with an ill-posed problem. It is rarely possible to specify the details of the initial disturbances that exist in a real flow because they arise from a number of boundary conditions such as free-stream turbulence, surface roughness, surface curvature, surface temperature, three-dimensional effects (secondary flow), compressibility, noise and structural vibrations. This means that besides the model and flow conditions, the disturbance environment, i.e., the wind tunnel or flight path conditions, must also be known. Furthermore, though it is generally known whether the above-mentioned parameters have a stabilizing or a destabilizing effect on the boundary layer, little information is available about the mutual interaction of these effects. As for supersonic boundary layers, the reader is referred to Morkovin's survey paper with its comprehensive study of the available literature (345 references) up to 1969. One of Morkovin's conclusions was that the empirical basis for the correlations and the prediction methods of transition in high-speed boundary layers used in industry was severely challenged, with little hope offered for more solid evidence in the near future.

We close with a brief summary of the effects on transition of the parameters and boundary conditions reviewed in previous sections (for more details see MORKOVIN [2.235] and TANI [2.248]).

A favorable pressure gradient increases the stability of a flow in the sense that the critical Reynolds number is higher than for the boundary layer with zero pressure gradient. An unfavorable pressure gradient destabilizes the flow due to the formation of velocity profiles with an inflection point (for a criterion see SMITH and GAMBERONI [2.246]). The transition region can be rather long in a zero pressure gradient with low free-stream turbulence (SCHUBAUER and KLEBANOFF [2.243]) and rather short in the free shear layer of a reattachment region or on a separation bubble, i.e., in case where the pressure gradient is unfavorable.

The transition process is accelerated by the presence of free-stream turbulence which appears to control the rapidity with which the rest of the spectrum feeds on the amplified component. A lower limit for the significant influence of free-stream turbulence seems to be a turbulence level of 0.08% (TANI). The effect of surface roughness on transition has been investigated more than any other parameter. This is due to the multiplicity of two- and three-dimensional roughness elements and configurations and to the impossibility of determining the effect of transition devices by theoretical methods. For incompressible boundary layers the reader is referred to SCHLICHTING [1.3] and TANI [2.248] for further references.

It is known from theoretical studies by GÖRTLER [2.191] and his co-workers that the instability of a boundary layer along a concave surface is enhanced due to the normal pressure gradient, resulting in a configuration of longitudinal vortices. On a convex surface the vortex mode is not excited, and transition proceeds as on a flat surface.

So far as we know today, stabilization of the laminar boundary layer can be achieved only by cooling of the wall, suction or a favorable pressure gradient. But here again a warning is necessary since transition on so-called "heat-sink" noses of ballistic missiles was not delayed by the strong favorable influences of cooling and rapidly falling pressure (MORKOVIN [2.235]).

The problem of the disturbance environment as summarized by MORKOVIN was investigated further by DOUGHERTY [2.249]. For more recent experimental investigations of transition the reader is referred to the papers by Pate [2.250, 251], La Graff [2.252], Owen and Horstman [2.253] and Stainback and Anders [2.254]. Among the attempts to predict the boundary layer behavior through the transition region the investigations by Harris [2.255], McDonald and Fish [2.256] and Cebeci (Sect. 5.8) should be mentioned, where the eddy viscosity is modified according to the statistical distribution of turbulent spots. A paper using a more refined approach is that by Bushnell and Alston [2.257].

Transition in pipe flow was investigated by Smith [2.258], Wygnanski and Champagne [2.259] and Wygnanski et al. [2.260]. A detailed picture of the sequence of instabilities and interactions which lead to transition in free shear layers was given by Miksad [2.261].

## 2.3.11 Relaminarization and Reverse Transition

Reverse transition and relaminarization have been observed in pipe- and duct flows and in boundary layers. Though the terms are used synonymously in most investigations, a distinction will be made in this subsection. An initially turbulent shear flow is in a *state* of relaminarization if the laminar state of the flow is clearly established, such as by the parabolic shape of the mean velocity profile in pipe flow (Sibulkin [2.262], Laufer [2.263], McEligot et al. [2.264] and Badri Narayanan [2.265] ( in duct flow)) or by a typical laminar profile in boundary layers (Launder [2.108] and Badri Narayanan and Ramjee [2.266]). Relaminarization of pipe flow for example was achieved by decreasing the pipe Reynolds number below its critical value of about 2100.

The *process* of reverse transition is much more difficult to define. Some features exist, however, the occurrence of which indicates reverse transition of an initially turbulent boundary layer. The mean velocity profile departs from its characteristic log-law form, being different from a laminar and from a highly accelerated velocity profile (Patel and Head [2.35]); the relevant Reynolds number is comparatively low ($Re_\theta < 2500$) and the sublayer thickness increases. There is some discrepancy between the "reverse transition" mean velocity profiles measured by Blackwelder and Kovasznay [2.268] and Badri Narayanan and Ramjee [2.266], showing a similar behavior to forward-transition profiles, and those of Patel and Head [2.35] which fall below the log-law in the outer layer.

Several parameters controlling the onset of reverse transition have been suggested ($K = (v/U_e^2)dU_e/dx$ by Launder [2.108], $\Delta_p = (v/\varrho u_\tau^3)dp/dx$ and $\Delta_\tau = (v\partial\tau/\partial y)/(\varrho u_\tau^3)$ by Patel and Head [2.35], and combinations thereof (e.g., BRADSHAW [2.269]): $\Delta_\tau$ is the correct inner-layer parameter. More reliable skin friction measurements are needed before distinct values can be allocated to these parameters (NARASIMHA and SREENIVASAN [2.270]).

As far as the turbulence structure in a boundary layer undergoing reverse transition is concerned, the following was observed (BLACKWELDER and KOVASZNAY [2.268]). Measurements of the fluctuating velocity components and the Reynolds shear stress indicated that the absolute levels of the velocities and the shear stress were approximately constant along a mean streamline except near the wall. However, the relative levels were decreasing, as reported already by LAUNDER [2.108], for example. KLINE et al. [2.11] found a decrease in the number of turbulent bursts in the wall region if the pressure gradient parameter $K$ reached a critical value of $3.7 \times 10^{-6}$, i.e., a reduction in the generation of turbulent energy which again leads to a decay of the turbulence in the outer part of the boundary layer. This cessation of turbulent bursts probably accounts for the departure of the velocity profiles in the log-law region. It is not clear, however, why the large eddy structure was not changed significantly after passing through the region where the favorable pressure gradient was strongest (BLACKWELDER and KOVASZNAY [2.268] from space-time autocorrelations). "Sink-flow" turbulent boundary layers between converging planes were investigated theoretically by LAUNDER and JONES [2.271], and the same authors presented a prediction method of reverse transition with a two-equation model of turbulence [2.272]. Reverse transition and relaminarization are phenomena which are of special practical importance for the heat transfer in sub- and supersonic nozzle flows, and several papers deal with this aspect, e.g., O'BRIEN [2.273], MORETTI and KAYS [2.226], BOLDMAN et al. [2.274], BACK et al. [2.275], NASH-WEBBER and OATES [2.276] and PERKINS and McELIGOT [2.277].

## 2.4 Three-Dimensional Boundary Layers

### 2.4.1 Classification of Three-Dimensional Flows

PRANDTL [2.288] opened his lecture on "Boundary Layers in Three-Dimensional flow" by introducing the two phenomena which distinguish three-dimensional from two-dimensional boundary layers:

a) The lateral (spanwise, $z$-wise) convergence or divergence of the velocity components of the potential flow parallel to the wall. In this case only the thickness of the boundary layer is affected and consequently the velocity profile and the wall shear stress. The mean velocity profiles are collateral, i.e., the velocity vector points in one direction throughout the boundary layer. The departure from two dimensionality can be accounted for by a single term in the momentum integral equation, adding $\theta/(x_0 - x)$ to the right-hand side of (2.21b). Here $x_0$ is the virtual origin of the streamlines as seen in plan view. For details, see PIERCE [2.289].

b) The secondary flow due to the lateral curvature of the potential flow. Because of the turning of the external streamlines the velocity vectors in the boundary layer, subjected to the same radial pressure gradient but of smaller size than in the outer flow, follow a turning angle which is a function of $y$ and is usually largest at the surface, that is, the velocity profiles are skewed.

Three-dimensional boundary layers belong to the class of three-dimensional "thin shear layers" where $\partial/\partial y \gg \partial/\partial x \sim \partial/\partial z$ when operating on any velocity component. The boundary layer equations are presented here for reasons of simplicity in a cartesian coordinate system where the $x$-direction can be either surface orientated or be given by the direction of the external streamlines. Then the coordinates on the wall, $x$ and $z$, are the projections of the external streamlines and their orthogonal trajectories. As for the choice of more general, i.e., curvilinear coordinate systems for three-dimensional boundary layers and a detailed discussion thereof, see for example HAYES [2.290], SQUIRE [2.291], RAKICH and MATEER [2.292], CEBECI et al. [2.293], KRAUSE [2.294] and the review papers by EICHELBRENNER [2.295] and BLOTTNER [2.296], the latter especially on calculation methods.

The second kind of three-dimensional shear flow along a wall is typified by the flow in the corner of a duct or of a wing-body junction where $\partial/\partial y, \partial/\partial z \gg \partial/\partial x$ (slender shear flow). The shear stress gradients in both the $y$- and $z$-directions are important in the Reynolds equations. The flow field as mapped by GESSNER and JONES [2.297] or EICHEL-BRENNER [2.298] for example shows a corner vortex in the case of turbulent flow which is called a secondary flow of Prandtl's second kind. With corner-flow configurations we encounter one of the rare cases where laminar and turbulent flow show completely different tendencies in their general behavior. From the considerations and results of ZAMIR [2.299], ZAMIR and YOUNG [2.300], PERKINS [2.301] and BRAGG [2.302] it appears that the secondary flow along the bisector of the corner is directed away from the corner in laminar flow[8] and towards the corner in turbulent flow (e.g., EICHELBRENNER and NGUYEN [2.303], GESSNER and JONES [2.297]). This latter effect has been found to be linked to the turbulence characteristics of the flow and has been investigated by

---

[8] It is just possible that the secondary flow that appears in the laminar case results from a localized instability in the corner, where the velocity profiles have points of inflection.

EICHELBRENNER [2.304, 298], EICHELBRENNER and PRESTON [2.305], EICHELBRENNER et al. [2.306] and MOJOLA and YOUNG [2.307]. ROTTA [1.5] presents the Reynolds equations for the boundary layer in the corner of a square duct and gives the conditions under which no secondary flow would occur. Transitional corner flow was investigated by ZAMIR and YOUNG and corner flow with an adverse pressure gradient by BRAGG, NGUYEN [2.308] and MOJOLA and YOUNG in turbulent flow showing similar characteristics as in two-dimensional retarded boundary layers. For further discussions of secondary flow in ducts see Section 3.1.

The third kind of three-dimensional shear flow, in which $\partial/\partial x \sim \partial/\partial y \sim \partial/\partial z$, needs either higher order boundary layer equations or the full Navier-Stokes equations, as for example in some cases of the flow in turbomachines (Chapt. 3).

Before we discuss some specific problems of turbulent three-dimensional boundary layers it may be advantageous to mention some further review papers; COOKE and HALL [2.309], JOUBERT et al. [2.310], SHERMAN [2.311], WHEELER and JOHNSTON [2.312, 313], NASH and PATEL [2.314] and BLOTTNER (especially on numerical calculations [2.296]), and three progress reports, HORLOCK et al. [2.315], FERNHOLZ [2.316] and EICHELBRENNER [2.295].

### 2.4.2 Three-Dimensional Thin Shear Layers

Under the conditions stated below, the equations for three-dimensional boundary layers as given in Chapter 1 can be simplified further. Since the validity range of the underlying assumptions has often been extended to flow configurations where these additional assumptions are not justified, a brief discussion seems to be appropriate. One can distinguish between three kinds of simplification:

    a) the "independence principle" according to PRANDTL [2.288],

    b) the infinite swept wing or infinite yawed cylinder,

    c) the "small crossflow assumption" or the "principle of prevalence" according to EICHELBRENNER and OUDART [2.317]; the "crossflow" is the velocity component normal to the local free stream.

Since the "independence principle" applies to laminar flow only, the reader is referred to COOKE [2.318]. In this case, however, the underlined terms in (2.36) to (2.39), the equations of mean motion in rectangular cartesian coordinates, vanish.

$$\frac{\partial}{\partial x}(\bar{\varrho}U) + \frac{\partial}{\partial y}(\bar{\varrho}V) + \frac{\partial}{\partial z}(\bar{\varrho}W) = 0, \tag{2.36}$$

$$\bar{\varrho}\left(U\frac{\partial U}{\partial x}+V\frac{\partial U}{\partial y}+W\frac{\partial U}{\partial z}\right)=-\frac{\partial \bar{p}}{\partial x}+\frac{\partial}{\partial y}\left(\mu\frac{\partial U}{\partial y}-\bar{\varrho}\overline{uv}\right),\qquad(2.37)$$

$$\bar{\varrho}\left(U\frac{\partial W}{\partial x}+V\frac{\partial W}{\partial y}+W\frac{\partial W}{\partial z}\right)=-\frac{\partial \bar{p}}{\partial z}+\frac{\partial}{\partial y}\left(\frac{\partial W}{\partial y}-\bar{\varrho}\overline{vw}\right),\qquad(2.38)$$

$$y=0:\ U=V=W=0\ ;\qquad u=v=w=0\ ,\qquad\qquad\qquad(2.39)$$

$$y=\delta:\ U=U_e\ ;\qquad W=W_e\ .\qquad\qquad\qquad\qquad\qquad(2.40)$$

The reason why the independence principle cannot be applied to turbulent boundary layers is that there is interaction between the spanwise and chordwise components of the velocity fluctuations.

Though the turbulent flow on an infinite swept wing is not different from a general three-dimensional boundary layer from a physical point of view, it is easier to treat mathematically (derivatives in $z$-direction must be identically zero and the surface must be developable). There is a further advantage that under infinite swept-wing conditions it is sufficient to perform measurements at one section only, thus reducing the amount of experimental data considerably. These considerations do not hold for finite swept wings or tapered wings where there exists a boundary layer drift towards the wing tip. For calculation methods dealing with the boundary layer on infinite swept wings see for example Cumpsty and Head [2.319], Adams [2.320], Krause [2.294] and Cebeci et al. [2.293].

The "small crossflow assumption" or the "principle of prevalence" due to Eichelbrenner and Oudart [2.317] is applicable to compressible and incompressible laminar and turbulent boundary layers. It amounts to assuming that $W$ is so small that it may be ignored in the momentum equation for $U$. The equation then becomes uncoupled from the equation for $W$. Eichelbrenner and Peube [2.321] have extended the small crossflow assumption to an attenuated "principle of prevalence". Originally the principle of prevalence was deduced for laminar boundary layers and applied to the case of slender bodies with small angles of incidence and consequently moderate crossflow. An extension to turbulent flow is, however, confined to even smaller crossflow due to the strong crosswise momentum and energy transport caused by the transverse turbulent velocity fluctuations. Furthermore Reynolds shear stresses are influenced implicitly by the spanwise flow (see for example the transport equation for $\overline{uv}$ (Chapt. 1)). Lacking a better criterion it may be advisable to follow Cumpsty and Head [2.322] who restrict the small crossflow assumption to angles between the free stream and the limiting streamline smaller than 6 degrees in turbulent boundary layers.

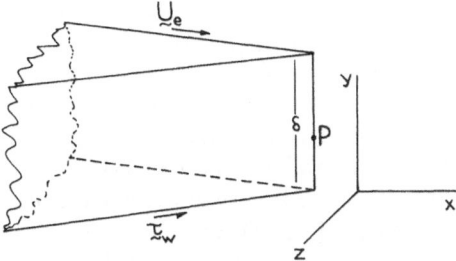

Fig. 2.6. Domain of dependence of a point $P$ in a three-dimensional boundary layer

According to the boundary layer equations, the solution at a given point is affected by upstream conditions only within a wedge-shaped "domain of dependence" whose apex is the point considered [2.323, 324]. This is RAETZ's [2.323] "influence principle". The wedge (Fig. 2.6) is bounded by envelopes of the characteristics (normals to the surface through the limiting streamline and the outer edge streamline) and the subcharacteristics which are the streamlines themselves. Though this model, like the boundary layer equations, neglects diffusion of vorticity in the $x, z$ plane and assumes instantaneous propagation of pressure disturbances normal to the $x, z$ plane, it is of great practical importance since it allows one to determine which upstream conditions must be known to calculate the boundary layer in a specific region (see WANG [2.325] and WESSELING [2.324]).

There are two further flow configurations for which the equations for three-dimensional boundary layers can be simplified (see PIERCE [2.326] and NASH and PATEL [2.314]). In all flows which have a plane of symmetry, the cross-flow components of the mean velocity ($W=0$) and of the shear stress ($\overline{vw}=0$) vanish whereas the gradients in $z$-direction must be retained. The equations are of course the same as derived via the small crossflow approximation and hence may be expected to hold for some short distance on either side of the plane of symmetry. For a more detailed discussion of the turbulent flow at a plane of symmetry the reader is referred to investigations by JOHNSTON [2.327], PIERCE [2.326], MELLOR [2.328] and HEAD and PRAHLAD [2.329]. The second special flow is the flow along an attachment line, for instance along the nose of a swept wing. In this case the equations of the three-dimensional boundary layer are even more simplified by setting all derivatives in the spanwise direction zero (see for example CUMPSTY and HEAD [2.330, 331] and CEBECI [2.332]). However as the chordwise velocity is zero, chordwise differentiation is necessary to give equations which can be solved by a numerical "marching" method.

Now, general three-dimensional thin shear layers may be divided into two groups: firstly, those which are mainly pressure driven such as on swept wings, in curved channels and on a flat plate around an obstacle; and secondly, mainly shear-driven boundary layers, which are more difficult to generate. Flows of both kinds can have very complex shear and velocity distributions (see, e.g., EAST and HOXEY [2.333] and can include separation regions. Experiments of the first type which have been published after Nash and Patel's review [2.314] are those on swept wings by ETHERIDGE [2.334], VAN DEN BERG et al. [2.335], ELSENAAR and BOELSMA [2.336] and EAST [2.337]; in curved channels by KLINKSIEK and PIERCE [2.338], VERMEULEN [2.339] and BANSOD and BRADSHAW [2.340]; and on ship hulls by LARSSON [2.341]. Only three experiments of the second type are known so far, MOORE and RICHARDSON [2.342], BRADSHAW and TERRELL [2.343] and CRABBE [2.344]. In contrast to the case of two-dimensional boundary layers little information about the turbulence structure in three-dimensional thin shear layers and the flow structure in the immediate vicinity of the wall is available. Therefore many investigations have assumed that most of the properties of collateral boundary layers can be carried over to skewed turbulent boundary layers. Prediction methods based on such assumptions may give acceptable agreement with experiments where the three-dimensional boundary layer extends only a few boundary layer thicknesses downstream but may well fail in cases where a three-dimensional boundary layer had time to develop properly. Due to the limited amount of space we cannot discuss here the whole range of semi-empirical relations, but some of the controversial assumptions will be pointed out in the hope to make it clear that more experiments are needed. For a more detailed discussion the reader is referred to NASH and PATEL [2.314].

Semi-empirical relationships for the components of the mean velocity vector as deduced from experiments have been given by many authors, e.g., for the "crossflow" profile by MAGER [2.345] and the polar representation which devides the velocity profile into an inner and outer region by JOHNSTON [2.346]. Both models assume a collateral velocity vector in the wall region, which is very unlikely according to more recent measurements (see for instance PIERCE [2.347]). Neither of these models can take into account velocity profiles with a change of signs of the crossflow component (see also PRAHLAD [2.348]). Downstream of inflection points of the outer flow streamlines, crossflow velocity profiles of multiple sign have been observed by BOUSGARBIES [2.349], EICHELBRENNER and PEUBE [2.350], KLINKSIEK and PIERCE [2.338], ETHERIDGE [2.334] and others. EICHELBRENNER and PEUBE [2.350] developed a relationship which can take account of S-shaped crossflow profiles, but agreement with measurements is not satisfactory.

As with two-dimensional boundary layers the set of equations for the three-dimensional case can only be closed if the Reynolds stresses can be determined either from semi-empirical relations (first-order closure) or from transport equations for turbulence properties (second-order closure). For a first-order closure, mixing-length or eddy-viscosity concepts (Sect. 1.8) are generally carried over from the two-dimensional case, very often even assuming isotropy of the eddy viscosity $v_T$, for instance:

$$-\overline{u_\alpha u_2} = (v_T)_\alpha \frac{\partial U_\alpha}{\partial x_2} \tag{2.41}$$

where the index 2 denotes the coordinate normal to the wall and where $\alpha$ may be either 1 or 3. There are several reasons for criticism here. It is well known that first-order closures have been successful in two-dimensional flow only if the boundary layer was always close to equilibrium. Neither eddy viscosity nor mixing length correlations hold for flows which are strongly influenced by their upstream history. Equation (2.41) implies that the direction of the Reynolds shear stress vector is always the same as the direction of the velocity-gradient vector. This is often expressed by another auxiliary equation

$$\overline{uv}/\overline{vw} = (\partial U/\partial y/(\partial W/\partial y)) \,. \tag{2.42}$$

It has been shown, however (cf. JOHNSTON [2.351] and EAST [2.337]), that the angle through which the shear stress has turned is generally intermediate between the angle of the mean velocity and the mean velocity gradient. Johnston found angular differences between the direction of the turbulent shear stress and the direction of the mean velocity gradient of about 20°. There is a further group of calculation methods where an extended version of Head's entrainment concept is used for three-dimensional turbulent boundary layers. A typical method is described in Subsection 5.7.3. Suggestions for a second-order closure procedure have been made by BRADSHAW [2.352] and SPALDING [2.353]. A version of the law of the wall generalized to three-dimensional boundary layers was derived by VAN DEN BERG [2.354] and EAST [2.355].

There exist few measurements of turbulence quantities which is one of the reason why many of the concepts mentioned above could not be checked properly (see for example ASHKENAS [2.356], BRADSHAW and TERRELL [2.343], JOHNSTON [2.351] and ELSENAAR and BOELSMA [2.336]) One of the unsolved questions remains whether and how much the local flow angle varies down to the wall (PIERCE [2.347], PIERCE and KROMMENHOEK [2.357] and PRAHLAD [2.358]). VERMEULEN [2.339] reported rates of change of the crossflow angle as large as 0.2 deg per unit of $u_\tau y/v$ in the immediate vicinity of the wall. For measurements in compressible turbulent boundary layers in three dimensions the reader is referred to HALL and DICKENS [2.359], RAINBIRD [2.360], FISHER and WEINSTEIN [2.361, 362] and COUSTEIX and MICHEL [2.363].

Three-dimensional turbulent boundary layers on rotating bodies have been dealt with for example by PARR [2.364], STEINHEUER [2.365] and CHAM and HEAD [2.366].

### 2.4.3 Separation in Three-Dimensional Flow

In Subsection 2.3.6 the separation problem in two-dimensional boundary layers has been found to be rather complex, for reasons connected with the structure of the problem (lack of experimental evidence) and with the validity of the boundary layer approximations, and also for mathematical reasons. Beginning with three-dimensional *laminar* boundary layers we find that our understanding of the separation phenomenon is even more incomplete, as can be seen from the investigations by BROWN and STEWARTSON [2.112] and by BUCKMASTER [2.367]. The latter author states a basic difficulty as lying in the very definition of separation in three-dimensional flow: "In general the skin friction only vanishes at singular points—where both components of the wall shear stress vector are zero—although naturally the boundary layer leaves the body along a curve". EICHELBRENNER [2.295] discussed separation in three-dimensional boundary layers in more detail arriving mainly at the definition of HAYES [2.290]: "If a particular streamline is found which separates streamlines coming from an unseparated part of the surface from those coming from a clearly separated part of the surface, this bounding line may be taken as the separation line". This separation criterion is also applicable to turbulent boundary layer, but more careful measurements in the separation region of skewed turbulent boundary layer are necessary (TAYLOR [2.368]) and with special emphasis on aerodynamical problems (SMITH [2.369]).

## 2.5 Two-Dimensional Boundary Layers at High Mach Numbers

Supersonic flight of aircraft and rockets has undoubtedly been the main incentive for the research on compressible turbulent boundary layers, and one of the most important aims has been to determine shear stress and heat transfer at the wall. It is not surprising therefore, to find a large body of investigations in the open literature, and the number of those still classified can only be guessed. The momentum and energy equations for steady two-dimensional compressible turbulent boundary layers are as follows if effects due to dissociation or ionization are excluded. The energy equation is given here in dimensionless form to show the explicit influence of the Mach number $M_\infty$.

$$\bar{\varrho}U \frac{\partial U}{\partial x} + (\bar{\varrho}V + \overline{\varrho'v}) \frac{\partial U}{\partial y} = -\frac{dP}{dx} + \frac{\partial}{\partial y}\left(\bar{\mu}\frac{\partial U}{\partial y} - \overline{\varrho uv}\right) \tag{2.43}$$

$$\hat{\bar{\varrho}}\hat{U} \frac{\partial \hat{h}}{\partial \hat{x}} + (\hat{\bar{\varrho}}\hat{V} + \widehat{\varrho'v}) \frac{\partial \hat{h}}{\partial \hat{y}} = (\gamma - 1)M_\infty^2 \hat{U} \frac{d\hat{P}}{d\hat{x}}$$

$$+ \frac{1}{\sigma_\infty \mathrm{Re}_\infty} \frac{\partial}{\partial \hat{y}}\left(\frac{\bar{\lambda}}{\hat{c}_\mathrm{p}} \frac{\partial \hat{h}}{\partial \hat{y}}\right)$$

$$- \frac{\partial}{\partial \hat{y}}(\hat{\bar{\varrho}}\widehat{v\theta}) + (\gamma - 1)M_\infty^2\left[\frac{\hat{\bar{\mu}}}{\mathrm{Re}_\infty}\left(\frac{\partial \hat{U}}{\partial \hat{y}}\right)^2 - \hat{\bar{\varrho}}\widehat{uv} \frac{\partial \hat{U}}{\partial \hat{y}}\right]. \tag{2.44}$$

Quantities denoted by $\wedge$, unless characteristic parameters, were made dimensionless by flow quantities of the undisturbed flow; overbars on $\mu$, $\varrho$, etc., denote mean values, and some small terms have been neglected. For boundary layers with zero pressure gradient the only direct effect of the Mach number occurs in the term where heat is added to the boundary layer, by direct dissipation and indirectly via production and dissipation of the turbulent fluctuations. The larger the Mach number, the greater is the dissipation. Heat transfer at the wall and heat addition by dissipation influence the velocity field by the distribution of the density $\bar{\varrho}$ and the temperature $T$, the latter via the viscosity $\bar{\mu}$ and the heat conductivity $\bar{k}$. Since dissipation occurs mainly in the vicinity of the wall the maximum of the static temperature is found close to (cooled wall) or at the wall (heated or adiabatic wall). Such temperature distributions again lead to mass-flow profiles where the bulk of the mass flow is in the outer layer of the boundary layer and to regions of low density at the wall ($\varrho \sim T^{-1}$ for $\partial p/\partial y \sim 0$). Measurements performed close to the wall need therefore rarefied gas corrections under certain conditions (see e.g. BECKWITH et al. [2.371]). Typical velocity, temperature and mass-flow distributions are given in Fig. 2.7. Other quantities being equal, the increased temperature at the surface implies a larger viscosity and a smaller density, which cause the ratio $\bar{\mu}/\bar{\varrho}$ to increase and the local Reynolds number to decrease, leading to an increase of the viscous sub-

---

[9] $h = c_\mathrm{p}T$; $\gamma = c_\mathrm{p}/c_\mathrm{v}$; $M_\infty = U_\infty/(\gamma P/\bar{\varrho})_\infty^{\frac{1}{2}}$; $\sigma_\infty = (\mu c_\mathrm{p}/k)_\infty$; $\lambda = k/c_\mathrm{p}$; $\mathrm{Re}_\infty = (U\varrho L/\mu)_\infty$; $\theta =$ temperature fluctuation. For a derivation of this equation for laminar boundary layers see MOORE [2.370].

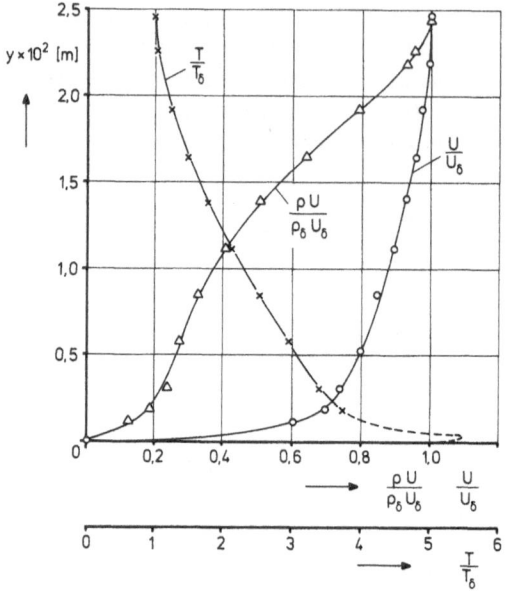

Fig. 2.7. Mean velocity, temperature and mass-flow profile for a $M_e = 7.2$ boundary layer with cooling ($T_w/T_r = 0.51$) at $Re_\theta = 2074$ (taken from [2.393]. Suffix $\delta$ denotes conditions at $y = \delta$

thickness. It may be mentioned in passing that the characteristic Reynolds number

$$Re_\theta = \bar{\varrho}_e U_e \theta / \bar{\mu}_w \tag{2.45}$$

the definition of which arises naturally from the momentum integral relation (WALZ [2.372]), is a sensible compromise taking into account the large variation of $\mu$ and $\varrho$ across the compressible boundary layer. Finally there is, of course, an indirect influence of the Mach number on the boundary layer via the boundary conditions $U_e$, $P_e$, and $h_e$.

A question which cannot be answered from inspection of (2.43) and (2.44) only is whether and how strongly the terms $\overline{uw}$ and $\overline{v\theta}$ are influenced by compressibility effects. Both flow visualization and hot wire measurements have shown that the compressible boundary layer exhibits a large-scale structure and an intermittent flow pattern in the outer layer. Due to the relatively thick sublayer and the location of the maximum of the turbulent fluctuations near the viscous sublayer a similar intermittent behavior between high- and low-intensity fluid has been observed in the inner layer (KOVASZNAY [2.373] and OWEN et al. [2.374]). The latter layer authors showed that the intermittency at the outer edge is more sharply defined in a Mach 7 boundary layer, but does not start until $y/\delta$ much larger than in the low-speed case.

The foundations for fluctuation measurements in compressible boundary layers were laid by KOVASZNAY [2.373, 375] and MORKOVIN [2.376, 377], but there are still only a few data sets available, mainly obtained with hot-wire anemometers (for a recent investigation see DEMETRIADES and LADERMAN [2.378]). In these investigations it could be shown that the three types of disturbance fields or modes (vorticity, entropy and sound wave), all of which obey the Navier-Stokes equations, are only weakly dependent on each other when the fluctuation intensities are small, but interact at larger intensities when linearization is no longer permissible. As long as the temperature fluctuations can be assumed isobaric, the sound wave mode (though important in itself) can be neglected compared with the rms mass-flow fluctuations. It is a lucky coincidence that the hot wire responds mainly to stagnation-temperature fluctuations at low overheat ratios and to mass-flow fluctuations at high overheat ratios, so that these two fluctuation components can be measured almost independently.

It must not be forgotten, however, that—possibly at Mach numbers around 10—the sound mode and its interaction with the other modes mentioned above become important, forming a barrier against the measuring techniques applied so far, so that electron beam or laser-Doppler measuring techniques may have to supplement the hot-wire technique (WALLACE [2.379], HARVEY and BUSHNELL [2.380], PENNER [2.381], and YANTA and LEE [2.382]).

KISTLER [2.383] found that mass-flow fluctuations and total temperature fluctuations $(\theta_t)$ increased with Mach number in adiabatic boundary layers $dp/dx = 0$ and $M_e \approx 4.7$), whereas OWEN et al. [2.374] state that for moderately cooled boundary layers $(dp/dx = 0, M_e \approx 7)$ the mass-flow fluctuations appear to be independent of Mach number while the $\theta_t$-fluctuations decrease with rising Mach number. A relationship between the fluctuating quantities was derived by MORKOVIN [2.384] if second-order terms can be neglected

$$\frac{\theta_t}{T_t} = \frac{\theta}{T}\left(1 + \frac{\gamma-1}{2}M^2\right)^{-1} + \frac{u}{U}(\gamma-1)M^2\left(1 + \frac{\gamma-1}{2}M^2\right)^{-1} \qquad (2.46)$$

or for adiabatic wall conditions $(\theta_t = 0)$

$$\frac{\theta}{T} = -\frac{u}{U}(\gamma-1)M^2. \qquad (2.47)$$

Equation (2.47) was used extensively by BRADSHAW [2.385] for example.

For a further discussion of the mode interaction and the limitations of the mode-concept (e.g., one field-point measurement only) see the excellent survey paper by MORKOVIN [2.384] which forms the base for Morkovin's hypothesis [2.386] where it is stated that "the essential dynamics of adiabatic compressible boundary layers follow the incompressible pattern for flows with $M \leqq 5$", a limit which may now be extended to Mach numbers of about 8 and with cooling of the wall.

For further details of the turbulence structure of compressible turbulent boundary layers the reader is referred to OWEN and HORSTMAN [2.387] and ROSE [2.388]. Calculations of shear stress, eddy viscosity and mixing length distributions were made by MAISE and McDONALD [2.389], MARTELLUCCI et al. [2.388] and BUSHNELL and MORRIS [2.391] from mean velocity and temperature measurements. Prandtl number distributions across the boundary layer have been investigated by MEIER and ROTTA [2.392] and HORSTMAN and OWEN [2.393] for example.

Since the measurements of fluctuating quantities in compressible turbulent boundary layer (e.g., KISTLER [2.383], LADERMAN and DEMETRIADES [2.394], OWEN et al. [2.387, 374] and ROSE [2.388, 395]) are not sufficient yet to develop adequate models of the turbulence structure, information about the distribution of mean velocity and temperature across the boundary layer is still very important. Beginning with the mean temperature distribution, we know of the existence of two energy integrals, i.e., a relation between the static enthalpy $h$ and the mean velocity $U$, which are valid throughout the boundary layer independent of $x$ and $y$. For laminar boundary layers, Prandtl number one and zero pressure gradient, such an energy integral exists if $h_w$ is constant (see MOORE [Ref. 2.370, p. 211]). A second energy integral can be obtained even if the pressure gradient is different from zero, provided that the heat transfer is zero at the wall. In the first case, the existence of an energy integral implies the so-called Reynolds analogy which means that the heat transfer at the wall is proportional to the skin friction. For turbulent boundary layers and Prandtl number close to one—as is the case for the technically important gases—these energy integrals (first derived by CROCCO and BUSEMANN) have been modified by VAN DRIEST [2.396] under the additional condition noted first by MORKOVIN [2.384] that $(\overline{v\theta_i})/(\overline{v\theta}) \ll 1$. The energy integral for a boundary layer along an adiabatic wall, $\sigma \sim 1$ and variable pressure gradient reads:

$$T/T_e = 1 + \tfrac{1}{2}(\gamma - 1)rM_e^2[1 - (U/U_e)^2] \tag{2.48}$$

where $r$ is the recovery factor. $r$ is assumed here to be 0.89 for air as a first approximation or can be expressed as a function of the Prandtl number (MEIER et al. [2.397]).

For an isothermal wall, zero pressure gradient and $\sigma \sim 1$ VAN DRIEST obtained the following relationship:

$$T/T_e = T_w/T_e + (U/U_e)\{1 - (T_w/T_e) + 0.5r(\gamma - 1)M_e^2[1 - (U/U_e)]\}$$

(2.49)

and for the Reynolds analogy:

$$St = F\tau_w/\varrho_e U_e^2$$   (2.50)

where $St$ is the Stanton number and $F$ the Reynolds analogy factor, which unfortunately has been found to vary between 0.8 and 1.2 (CARY [2.398] and COLEMAN et al. [2.399]).

Equations (2.48) and (2.49) are often written for the total temperature or the total enthalpy but it is little known that the form given above permits a more realistic comparison of measured and calculated data. Attempts have been made to extend these relations to flows with pressure gradients and heat transfer but the results are not yet convincing (WALZ [2.372] and KÜSTER [2.400] for example). In order to gain more information about the mean velocity distribution, which must be known to evaluate the energy integrals, it is natural to revert to the multi-layer concept of the turbulent boundary layer, especially since the basic features of the turbulence structure in compressible and incompressible flow can be assumed to be approximately similar, by virtue of Morkovin's hypothesis. Many attempts have been made to scale or transform both abscissa and ordinate of the log law (2.3) or only one of them (COLES [2.401], ROTTA [2.402], BARONTI and LIBBY [2.403] and TENNEKES [2.404] to mention but a few). The best agreement with measurements over a wide range of Mach numbers, heat transfer ratios and even pressure gradients (GRAN et al. [2.405]) has been obtained if $U^*/u_\tau$ is plotted against $yu_\tau/v_w$, where $U^*$ is the mean velocity transformed according to VAN DRIEST [2.396]. It is, of course, important to heed the validity range of (2.48) and (2.49) which are used in the transformation, but the log-law is probably relatively insensitive to adverse pressure gradients, as in incompressible flow. Deviations from the log-law must, however, be expected if the upstream history of the boundary layer—be it a temperature history (e.g., FELLER [2.445]) or the uncompleted transition process (FERNHOLZ [2.406])—still affects the boundary layer. These deviations must not be counted as a failure of the transformation (ROTTA [2.407], FERNHOLZ [2.408] and KEENER and HOPKINS [2.409]).

As for the log-law it has been advantageous to use van Driest's transformation for the outer law, too (FERNHOLZ [2.408]). This relation is valid for zero pressure gradient and constant wall temperature:

$$\frac{U_e^* - U^*}{u_\tau} = -4.70 \ln(y/\varDelta^*) - 6.74 \tag{2.51}$$

where quantities denoted by an asterisk are transformed according to VAN DRIEST. It will be noted that the Rotta-Clauser thickness $\varDelta^*$ (2.9) was used instead of the boundary layer thickness $\delta$. Due to the slower asymptotic approach of the velocity distribution in the boundary layer to the free-stream conditions, the evaluation of $\delta$ is even more liable to the personal touch than in incompressible boundary layers and therefore not a useful length scale. Coles' wake-law was extended to compressible boundary layers by MATHEWS et al. [2.410] and GRAN et al. [2.405].

Since we have just advocated van Driest's transformation, it is appropriate to mention a series of research papers which endeavor to find a transformation allowing a point-to-point mapping between a compressible and an incompressible turbulent boundary layer. This is the transformation concept of COLES [2.67] for adiabatic boundary layers which was extended by CROCCO [2.411] and LEWIS et al. [2.412] to boundary layers with pressure gradient and heat transfer. The development of this transformation theory has not yet been completed.

The majority of compressible two-dimensional boundary layers that have been investigated consists of zero-pressure-gradient flows along adiabatic or cooled walls (hydraulically smooth and rough) generated on surfaces such as contoured plane and axisymmetric nozzles, tunnel side walls, flat plates and axisymmetric bodies. A survey of most of the unclassified experiments is due to be published by FERNHOLZ and FINLEY [2.413]. No experiments have become known so far where the temperature gradient varied along the wall (there exists one experiment with a step change of the wall temperature by GRAN et al. [2.405]). Such a variable wall temperature $T(x)$ is expected to exert a strong influence on the temperature distribution across the boundary layer.

The determination of the skin friction—even for zero pressure gradient flow—still causes problems. The reliability and comparability of floating-element balances and the choice of the correct calibration curve for Preston tubes may serve as examples. These problems are aggravated, of course, in boundary layers with pressure gradients, and it is not surprising to find only a few experiments where mean velocity profiles and values of and skin friction have been measured (e.g., PEAKE et al. [2.414], LEWIS et al. [2.415], VOISINET and LEE [2.416], WALTRUP and SCHETZ [2.417] and ZWARTS [2.418]). Though none of the four adverse pressure gradient flows leads to separation, it appears unusual at first sight that the skin friction can rise in an adverse pressure gradient. One possible reason for this is that the Mach number influence on the skin friction is stronger than that of the pressure gradient. For zero-pressure-gradient boundary layers an increase in Mach number reduces the skin friction considerably, especially at Reynolds numbers $Re\delta_2 < 5000$ (see for example SPALDING and CHI [2.419] or FERNHOLZ [2.420] and for an evaluation of several other theories HOPKINS and INOUYE [2.421] and CARY and BERTRAM [2.422]).

Deviations between calculations of the characteristic flow parameters (especially for $c_f$) and measurements in boundary layers with variable pressure gradient found by BRADSHAW [2.423] were attributed to a direct influence of the mean dilatational rate of strain or compression div $U$ on the turbulence structure. This hypothesis is based on fluctation measurements by BEHRENS [2.424] and ROSE [2.395] in oblique shock/boundary layer interactions and by LEWIS and BEHRENS [2.425] in the shear layer emerging from a Prandtl-Meyer expansion. These measurements have shown unexpectedly large increases in turbulence intensity during compression and large decreases during expansion.

Another flow configuration especially important for air intakes of jet engines is the boundary layer where the adverse pressure gradient is generated by means of a ramp or compression corner (see for example STROUD and MILLER [2.426] or STUREK and DANBERG [2.427]). Boundary layers over curved surfaces are subject to pressure gradients normal to the wall, the influence of which was investigated experimentally by THOMANN [2.198] and theoretically by ROTTA [2.428]. As for compressible turbulent boundary layers with separation the reader is referred to a few of the more recent experimental investigations (THOMKE and ROSHKO [2.429], BATHAM [2.430], ELFSTROM [2.431], COLEMAN and STOLLERY [2.432], ROSE et al. [2.433] and LAW [2.434]).

There exists a sequence of research papers on compressible flow along rough walls which were performed at the University of Texas. Since it is difficult to obtain these reports, details may be taken from FERNHOLZ and FINLEY [2.413], and for another characteristic paper see CHEN [2.435].

For a discussion of transition phenomena in compressible boundary layers the reader is referred to the survey papers by MORKOVIN [2.235] and KISTLER [2.436] and to some more recent experimental investigations (MADDALON [2.437], LA GRAFF [2.438], OWEN [2.439], CARY [2.440], BECKWITH and BERTRAM [2.441], WATSON et al. [2.442], BECKWITH [2.443] and KENDALL [2.444]).

# References

2.1    G. N. Abramovich: *The Theory of Turbulent Jets*. (MIT Press, Cambridge, Mass. 1963)

2.2    S. F. Birch, D. H. Rudy, D. M. Bushnell (Eds.): NASA SP 321 (1972) (2 vols.)

2.3    S. N. B. Murthy (Ed.): *Turbulent Mixing in Non-Reactive and Reactive Flows*. (Plenum Press, New York 1975)

2.4    C. du P. Donaldson, A. J. Bilanin: AGARDograph 204 (1975)

2.5    F. H. Clauser: The turbulent boundary layer, in: *Advances in Applied Mechanics IV*. (Academic Press, New York 1956)

2.6    I. Tani: *Proceedings, Computation of Turbulent Boundary Layers*—1968 AFOSR-IFP-*Stanford Conference*, S. J. Kline et al. (Eds.) (Thermosciences Div., Stanford Univ. 1969), Vol. I, p. 483

2.7    D. E. Coles, E. A. Hirst (Eds.): *Proceedings, Computation of Turbulent Boundary Layers*—1968 AFOSR-IFP-*Stanford Conference*, Vol. II. (Thermosciences Divn., Stanford Univ. 1969)

2.8    L. Ting: J. Math. Phys. **44**, 353 (1965)

2.9    J. E. Green: RAE TR 72050 (1972)

2.10   L. Prandtl: *Ergebnisse der Aerodynamischen Versuchsanstalt zu Göttingen* (Oldenbourg, München-Berlin 1927), vol. III, p. 1
       See also W. Tollmien, H. Schlichting, H. Görtler (Eds.): *Ludwig Prandtl Gesammelte Abhandlungen*. (Springer, Berlin-Göttingen-Heidelberg 1961). For all Prandtl references

2.11   S. J. Kline, W. C. Reynolds, F. A. Schraub, P. W. Runstadler: J. Fluid Mech. **30**, 741 (1967)

2.12   L. Prandtl: *Ergebnisse der Aerodynamischen Versuchsanstalt zu Göttingen* (Oldenbourg, München-Berlin 1932), vol. IV, p. 18

2.13   H. Ludwieg, W. Tillmann: Ingr.-Arch. **17**, 288 (1949), translated as NACA TM 1285

2.14   Th. von Kármán: NACA TM 611 (1931)

2.15   F. H. Clauser: J. Aeronaut. Sci. **21**, 91 (1954)

2.16   D. Coles: J. Fluid Mech. **1**, 191 (1956)

2.17   H. Eckelmann: J. Fluid Mech. **65**, 439 (1974)

2.18   H. Eckelmann: Mitt. No. 48, Max-Planck-Inst. für Strömungsforschung (1970)

2.19   H. P. Bakewell: Thesis, Dept. Aerospace Sci., Penn. State Univ. (1966); see also Phys. Fluids **10**, 1880 (1967)

2.20   S. J. Kline, P. W. Runstadler: Trans. ASME **81E**, 166 (1959)

2.21   E. R. Corino, R. S. Brodkey: J. Fluid Mech. **37**, 1 (1969)

2.22   H. T. Kim, S. J. Kline, W. C. Reynolds: J. Fluid Mech. **50**, 133 (1971)

2.23   A. K. Gupta, J. Laufer, R. E. Kaplan: J. Fluid Mech. **50**, 493 (1971)

2.24   J. M. Wallace, H. Eckelmann, R. S. Brodkey: J. Fluid Mech. **54**, 39 (1972)

2.25   G. R. Offen, S. J. Kline: J. Fluid Mech. **62**, 223 (1974)

2.26   H. Ueda, J. O. Hinze: J. Fluid Mech. **67**, 125 (1975)

2.27   S. Corrsin: *Naval Hydrodynamics*, Publication 515, National Academy of Sciences —National Research Council (1957), Chapt. 15

2.28   J. Laufer: NACA Rept. 1174 (1955)

2.29   P. S. Klebanoff: NACA Rept. 1247 (1955)

2.30   W. W. Willmarth, S. S. Lu: AGARD-CP-93, 3.1 (1972)

2.31   R. F. Blackwelder, R. E. Kaplan: AGARD-CP-93, 5.1 (1972)

2.32   J. Rotta: Prog. Aeronaut. Sci. **2**, 1 (1962)

2.33   J. H. Preston: J. Roy. Aeronaut. Soc. **58**, 109 (1954)

2.34   M. R. Head, I. Rechenberg: J. Fluid Mech. **14**, 1 (1962)

2.35  V.C.PATEL, M.R.HEAD: J. Fluid Mech. **34**, 371 (1968)
2.36  V.C.PATEL: J. Fluid Mech. **23**, 185 (1965)
2.37  G.L.MELLOR: J. Fluid Mech. **24**, 255 (1966)
2.38  H.McDONALD: J. Fluid Mech. **35**, 311 (1969)
2.39  D.E.COLES: AGARD-CP-93, 25.1 (1972)
2.40  R.J.BAKER, B.E.LAUNDER: Intern. J. Heat Mass Transfer **17**, 275 and 293 (1974)
2.41  M.LANDAHL: J. Fluid Mech. **29**, 441 (1967)
2.42  F.H.BARK: J. Fluid Mech. **70**, 229 (1975)
2.43  O.M.PHILLIPS: J. Fluid Mech. **27**, 131 (1967)
2.44  H.K.MOFFATT: Proc. AFOSR-IFP Stanford Conf. **1**, 495 (1969)
2.45  M.J.LIGHTHILL: Proc. AFOSR-IFP Stanford Conf. **1**, 511 (1969)
2.46  A.K.M.F.HUSSAIN, W.C.REYNOLDS: J. Fluid Mech. **41**, 241 (1970)
2.47  A.K.M.F.HUSSAIN, W.C.REYNOLDS: J. Fluid Mech. **54**, 241 (1972)
2.48  L.S.G.KOVASZNAY: AGARD-CP-93, D1 (1972)
2.49  S.CORRSIN: NACA WR W-94 (1943)
2.50  A.A.TOWNSEND: Aust. J. Sci. Res. Ser. A **1**, 161 (1948)
2.51  H.FIEDLER, M.R.HEAD: J. Fluid Mech. **25**, 719 (1966)
2.52  R.E.KAPLAN, J.LAUFER: *Proc. 12th Int. Congr. Appl. Mech.* (Springer, Berlin-Heidelberg-New York 1969), p. 236
2.53  R.A.ANTONIA: J. Fluid Mech. **56**, 1 (1972)
2.54  C.W. VAN ATTA: T.N. F-54, Mech. Engng. Dept., Sydney Univ. (1973)
2.55  J.C.LaRUE: Phys. Fluids **17**, 1513 (1974)
2.56  R.A.ANTONIA, J.D.ATKINSON: J. Fluid Mech. **64**, 679 (1974)
2.57  P.BRADSHAW, J.MURLIS: Aero Rept. 74-04, Imperial College, London (1974)
2.58  R.F.BLACKWELDER, L.S.G.KOVASZNAY: Phys. Fluids **15**, 1545 (1972)
2.59  O.M.PHILLIPS: J. Fluid Mech. **51**, 97 (1972)
2.60  M.R.HEAD: ARC R. and M. 3152 (1960)
2.61  M.R.HEAD, V.C.PATEL: ARC R. and M. 3643 (1969)
2.62  A.A.TOWNSEND: J. Fluid Mech. **26**, 689 (1966)
2.63  R.E.FALCO: AIAA Paper 74–99 (1974)
2.64  J.ROTTA: Ingr.-Arch. **19**, 31 (1951)
2.65  J.C.ROTTA: *Fluid Mechanics of Internal Flow* (G.SOVRAN, Ed.) (Elsevier, Amsterdam 1967)
2.66  B.G.J.THOMPSON: ARC R. and M. 3463 (1967)
2.67  D.E.COLES: Rand Corp. R-403-PR (1962)
2.68  J.H.PRESTON: J. Fluid Mech. **3**, 373 (1958)
2.69  R.L.SIMPSON: J. Fluid Mech. **42**, 769 (1970)
2.70  G.D.HUFFMAN, P.BRADSHAW: J. Fluid Mech. **53**, 45 (1972)
2.71  J.MURLIS: PhD thesis, Imperial College, London (1975)
2.72  V.M.FALKNER, S.W.SKAN: Phil. Mag. **12**, 865 (1931)
2.73  L.G.LOITSIANSKI: *Laminare Grenzschichten* (Akademie-Verlag, Berlin 1976)
2.74  A.A.TOWNSEND: J. Fluid Mech. **22**, 773 (1965)
2.75  J.C.ROTTA: J. Aeronaut. Sci. **22**, 215 (1955)
2.76  A.A.TOWNSEND: J. Fluid Mech. **1**, 561 (1956)
2.77  A.A.TOWNSEND: J. Fluid Mech. **23**, 767 (1965)
2.78  G.L.MELLOR, D.M.GIBSON: J. Fluid Mech. **24**, 225 (1966)
2.79  J.F.NASH: AGARDograph **97**, 245 (1965)
2.80  H.J.HERRING, J.F.NORBURY: J. Fluid Mech. **27**, 541 (1967)
2.81  B.E.LAUNDER, H.S.STINCHCOMBE: Rept. TWF/TN/21, Dept. Mech. Engng., Imperial College, London (1967)
2.82  P.BRADSHAW, D.H.FERRISS: NPL Aero Rept. 1145 (1965)
2.83  P.BRADSHAW: J. Fluid Mech. **29**, 625 (1967)

2.84   T. N. Stevenson: College of Aeronautics Rept. Aero 170 (1963)

2.85   P. S. Klebanoff, Z. W. Diehl: NACA Rept. 1110 (1952)

2.86   W. Jacobs: ZAMM **19**, 87 (1939) and NACA TM 951 (1940)

2.87   R. A. Antonia, R. E. Luxton: J. Fluid Mech. **53**, 737 (1972)

2.88   R. L. Gran, J. E. Lewis, T. Kubota: J. Fluid Mech. **66**, 507 (1974)

2.89   D. F. Gates: TR-73-152, U.S. Naval Ordnance Lab. (1973)

2.90   R. E. Lee, W. I. Yanta, A. C. Leonas: *Proc. 1968 Heat Transfer.* (Fluid Mech. Institute, Stanford: University Press 1968)

2.91   J. E. Green: RAE TR 72201 (1972)

2.92   G. Charnay, G. Comte Bellot, J. Mathieu: AGARD-CP-93, 27.1 (1972) and G. Charnay: PhD thesis, Ecole Centrale de Lyon (1974)

2.93   J. Kestin: Advanc. Heat Transf. **3**, 1 (1966)

2.94   G. D. Huffman, D. R. Zimmerman, W. A. Bennett: AGARDograph 164 (1972)

2.95   S. J. Kline, A. V. Lisin, B. A. Waitman: NASA TN D-368 (1960)

2.96   M. Pichal: ZAMM **52**, T 407 (1972)

2.97   D. W. Kearney, W. M. Kays, R. J. Moffat, R. J. Loyd: NASA CR 11 3590 (1969), Rept. HMT-9 Dept. Mech. Engng., Stanford Univ. (1970)

2.98   H. Komoda: J. Jap. Soc. Aeronaut. Eng. **1957**, 274

2.99   J. D. Vagt: DFG-Bericht zu Schade 154/1 (1968/69)

2.100  R. J. Loehrke, H. M. Nagib: AGARD Rept. R-598 (1973)

2.101  P. Bradshaw: ARC R. and M. 3575 (1969)

2.102  P. Goldberg: Gas Turbine Lab. Rept. 85, MIT (1966)

2.103  A. E. Samuel, P. N. Joubert: J. Fluid Mech. **66**, 481 (1974)

2.104  W. G. Spangenberg, W. R. Rowland, N. E. Mease: In: *Fluid Mechanics of Internal Flow*, G. Sovran (Ed.). (Elsevier, Amsterdam 1967), p. 110

2.105  V. A. Sandborn, C. Y. Liu: J. Fluid Mech. **32**, 293 (1968)

2.106  G. B. Schubauer, P. S. Klebanoff: NACA Rept. 1030 (1951)

2.107  B. G. Newman: Australian Dept. Supply Rept. ACA-53 (1951)

2.108  B. E. Launder: Trans. ASME **31E**, 707 (1964)

2.109  P. K. Chang: *Separation of Flow.* (Pergamon Press, Oxford 1970)

2.110  A. Roshko: Canadian Congr. Appl. Mech. (Univ. Laval Press, Quebec 1967)

2.111  S. Kaplun: *Fluid Mechanics and Singular Perturbations.* (Academic Press, New York 1967)

2.112  S. N. Brown, K. Stewartson: Ann. Rev. Fluid Mech. **1**, 45 (1969)

2.113  L. Prandtl: *Verh. III. Int. Math. Kongr., Heidelberg* (Teubner, Leipzig 1905) p. 484; (see also Gesammelte Abhandlungen [2.10])

2.114  K. Oswatitsch: *Grenzschichtforschung*, (Springer, Berlin-Göttingen-Heidelberg 1958), p. 357

2.115  K. Stewartson: Quart. J. Mech. Appl. Math. **11**, 399 (1958)

2.116  K. Stewartson: Phys. Fluids **12**, II 282 (1969)

2.117  P. Bradshaw: AGARDograph 169 (1973)

2.118  A. A. Townsend: J. Fluid Mech. **12**, 536 (1962)

2.119  S. J. Kline: Trans. ASME **81D**, 305 (1959)

2.120  V. A. Sandborn, S. J. Kline: Trans. ASME **83D**, 317 (1961)

2.121  D. G. Mabey: Aeronaut. J. **77**, 201 (1973)

2.122  B. S. Stratford: J. Fluid Mech. **5**, 17 (1959)

2.123  H. Fernholz, P. Gibson: Gas Turbine Lab. Rept. No. 91, MIT (1967)

2.124  A. A. Townsend: J. Fluid Mech. **8**, 143 (1960)

2.125  H. Fernholz: Ingr.-Arch. **35**, 192 (1966)

2.126  I. E. Alber: AIAA Paper, 71–203 (1971)

2.127  P. Bradshaw, P. V. Galea: J. Fluid Mech. **27**, 111 (1967)

2.128  K. Gersten: AGARD-CP-4, 863 (1966)

2.129  G. LACHMANN (Ed.): *Boundary Layer and Flow Control*. London: Pergamon Press 1961

2.130  B. M. JONES: J. Roy. Aeronaut. Soc. **38**, 753 (1934)

2.131  A. D. YOUNG, H. P. HORTON: AGARD-CP-4, 779 (1966)

2.132  H. P. HORTON: PhD thesis, Queen Mary College, London (1968)

2.133  H. P. HORTON: Aeronaut. Quart. **23**, 211 (1972)

2.134  H. T. DELPAK: PhD thesis, Queen Mary College, London (1973)

2.135  I. TANI: *Progr. Aeronaut. Sci.* **5**, 70. London: Pergamon Press 1964

2.136  M. GASTER: ARC R. and M. 3595 (1967)

2.137  W. R. BRILEY, H. MCDONALD: J. Fluid Mech. **69**, 631 (1975)

2.138  H. J. LUGT, H. J. HAUSSLING: Rept. 3748, Naval Ship Res. Devel. Center (1972)

2.139  U. B. MEHTA, Z. LAVAN: J. Fluid Mech. **67**, 227 (1975)

2.140  I. MCGREGOR: PhD thesis, Queen Mary College, London (1954)

2.141  E. A. EICHELBRENNER: *Jahrbuch der WGL 1959*. (Vieweg, Braunschweig 1960), p. 119

2.142  E. A. EICHELBRENNER: New York Acad. Sci. Ann. **154**, 655 (1968)

2.143  D. R. CHAPMAN: NACA TN-2137 (1950)

2.144  H. H. KORST: Trans. ASME **78**, 593 (1956)

2.145  L. CROCCO, L. LEES: J. Aeronaut. Sci. **19**, 649 (1952)

2.146  J. H. SMITH, J. P. LAMB: Intern. J. Heat Mass Transf. **17**, 1571 (1974)

2.147  P. BRADSHAW, F. Y. F. WONG: J. Fluid Mech. **52**, 113 (1972)

2.148  D. E. ABBOTT, S. J. KLINE: Trans. ASME **84**D, 317 (1962)

2.149  M. ARIE, H. ROUSE: J. Fluid Mech. **1**, 129 (1956)

2.150  T. J. MUELLER, J. M. ROBERTSON: Modern Develop. Theor. Appl. Mech. **1**, 326 (1963)

2.151  J. C. LE BALLEUR, J. MIRANDE: ONERA T.P. No. 1975-16 (1975)

2.152  T. J. MUELLER, H. H. KORST, W. L. CHOW: Trans. ASME **86**D, 221 (1964)

2.153  J. NIKURADSE: VDI Forschungsheft 361 (1933)

2.154  D. H. WOOD, R. A. ANTONIA: Aeronaut. Quart. **26**, 202 (1975)

2.155  R. L. SIMPSON: AIAA J. **11**, 242 (1973)

2.156  L. PRANDTL: *Ergebnisse der AVA Göttingen 4* (Oldenbourg, München 1932) (see also Gesammelte Abhandlungen [2.10])

2.157  H. SCHLICHTING: Ingr.-Arch. **7**, 1 (1936) and NACA TM 823

2.158  F. R. HAMA: Trans. Soc. Nav. Architects Marine Engrs. **62**, 333 (1954)

2.159  V. A. IOSELEVICH, V. I. PILIPENKO: Sov. Phys. Dokl. **213**, 1266 (1973)

2.160  A. E. PERRY, W. H. SCHOFIELD, P. N. JOUBERT: J. Fluid Mech. **37**, 383 (1969)

2.161  C. K. LIU, S. J. KLINE, J. P. JOHNSTON: Rept. MD-15, Mech. Dept. Engng., Stanford Univ. (1966)

2.162  H. W. TOWNES, R. H. SABERSKY: Intern. J. Heat Mass Transf. **9**, 729 (1966)

2.163  R. A. ANTONIA, D. H. WOOD: TN F-71, Mech. Engng. Dept., Sydney Univ. (1974)

2.164  A. J. GRASS: J. Fluid Mech. **50**, 233 (1971)

2.165  A. E. PERRY, P. N. JOUBERT: J. Fluid Mech. **17**, 193 (1963)

2.166  R. A. ANTONIA, R. E. LUXTON: TN F-1, Mech. Engng. Dept., Sydney Univ. (1969)

2.167  R. A. ANTONIA, R. E. LUXTON: Advan. Geophysics **18**A, 263 (1974)

2.168  R. E. LUXTON: TN F-12, Mech. Engng. Dept., Sydney Univ. (1970)

2.169  A. A. TOWNSEND: J. Fluid Mech. **22**, 799 (1965)

2.170  A. A. TOWNSEND: J. Fluid Mech. **26**, 255 (1966)

2.171  R. J. TAYLOR: J. Fluid Mech. **13**, 529 (1962)

2.172  K. S. RAO, J. C. WYNGAARD, O. R. COTÉ: J. Atmosph. Sci. **31**, 738 (1974)

2.173  R. A. ANTONIA, R. E. LUXTON: Phys. Fluids **14**, 1027 (1971)

2.174  L. GAUDET, K. G. WINTER: RAE TM Aero 1538 (1973)

2.175  K. H. ROGERS: J. Aircraft **11**, 382 (1974)

2.176  G. R. INGER, E. P. WILLIAMS: AIAA J. **10**, 636 (1972)

2.177  G. D. ASHTON: Proc. 1972 Inst. Heat Mass Transfer (1972)
2.178  P. R. OWEN, W. R. THOMSON: J. Fluid Mech. **15**, 321 (1963)
2.179  R. T. HO, L. W. GELHAR: J. Fluid Mech. **58**, 403 (1973)
2.180  S. T. HSU, J. F. KENNEDY: J. Fluid Mech. **47**, 481 (1971)
2.181  S. KOTAKE: Intern. J. Heat Mass Transf. **17**, 885 (1974)
2.182  K. G. WINTER, J. C. ROTTA, K. G. SMITH: ARC R. and M. 3633 (1970)
2.183  R. RICHMOND: PhD thesis, Caltech, Pasadena (1957)
2.184  A. S. GINEVSKII, E. E. SOLODKIN: J. Appl. Math. Mech. **22**, 1169 (1958)
2.185  W. W. WILLMARTH, C. S. YANG: J. Fluid Mech. **41**, 47 (1970)
2.186  P. BRADSHAW, V. C. PATEL: AIAA J. **11**, 893 (1973)
2.187  A. WALZ, M. MAYER: Glastechn. Ber. **39**, 359 and 409 (1966)
2.188  V. C. PATEL, A. NAKAYAMA, R. DAMIAN: J. Fluid Mech. **63**, 345 (1974)
2.189  L. PRANDTL: NACA TM-625 (1929)
2.190  G. I. TAYLOR: Phil. Trans. Roy. Soc. A **223**, 289 (1923)
2.191  H. GÖRTLER: Nachr. dr. Wiss. Ges. Göttingen, Math. Phys. **1**, 1 (1940)
2.192  I. TANI: J. Geophys. Sci. **67**, 3075 (1962)
2.193  P. BRADSHAW: J. Fluid Mech. **36**, 177 (1969)
2.194  R. M. C. SO, G. L. MELLOR: J. Fluid Mech. **60**, 43 (1973)
2.195  R. M. C. SO, G. L. MELLOR: Aeronaut. Quart. **26**, 25 (1975)
2.196  B. G. J. THOMPSON: AGARDograph 97, 159 (1965)
2.197  R. N. MERONEY, P. BRADSHAW: AIAA J. **13**, 1448 (1975)
2.198  H. THOMANN: J. Fluid Mech. **33**, 283 (1968)
2.199  V. C. PATEL: ARC R. and M. 3599 (1969)
2.200  B. G. NEWMAN: Canadian Aero. Space J. **15**, 288 (1969)
2.201  H. H. FERNHOLZ: DLR FB 66-21 (1966)
2.202  H. H. FERNHOLZ: Z. Flugwiss. **15**, 136 (1967)
2.203  F. A. DVORAK: AIAA J. **11**, 517 (1973)
2.204  H. P. A. H. IRWIN, P. Arnot SMITH: Phys. Fluids **18**, 624 (1975)
2.205  J. C. ROTTA: Wärme Stoffübertragung 7, 133 (1974)
2.206  H. U. MEIER, D. F. GATES, R. L. P. VOISINET: AIAA Paper 74-596 (1974)
2.207  A. I. LEONT'EV: *Advances in Heat Transfer* (Academic Press, New York 1966), vol. 3, p. 33
2.208  S. V. PATANKAR, D. B. SPALDING: *Heat and Mass Transfer in Boundary Lasers* (Morgan-Grampian, London 1967)
2.209  A. D. GOSMAN, W. M. PUN, A. K. RUNCHAL, D. B. SPALDING, M. WOLFSHSTEIN: *Heat and Mass Transfer in Recirculating Flows* (Academic Press, New York 1968)
2.210  T. CEBECI, A. M. O. SMITH, G. MOSINSKIS: Trans. ASME **92**C, 133 (1970)
2.211  R. J. FLAHERTY: J. Aircraft **11**, 293 (1974)
2.212  M. E. CRAWFORD, W. M. KAYS: Rept. HMT-23, Thermosciences Div., Mech. Engng. Dept., Stanford Univ. (1975)
2.213  W. J. KELNHOFER: DFVLR-FB 70-66 (1970) (see also Trans. ASME **91**A, 281 (1969)
2.214  L. H. BACK, F. CUFFEL, P. F. MASSIER: Intern. J. Heat Mass Transf. **13**, 1029 (1970)
2.215  D. R. BOLDMAN, J. F. SCHMIDT, R. C. EHLERS: Trans. ASME **89**C, 341 (1967)
2.216  L. H. BACK, R. F. CUFFEL: Trans. ASME **93**C, 397 (1971)
2.217  D. W. KEARNEY, W. M. KAYS, R. J. MOFFAT: Int. J. Heat Mass Transf. **16**, 1289 (1973)
2.218  A. P. HATTON, V. A. EUSTACE: Proc. 3rd Int. Heat Transfer Conf. Chicago **2**, 34 (1966)
2.219  W. H. THIELBAHR, W. M. KAYS, R. J. MOFFAT: Rept. HMT-5, Dept. Mech. Engng., Stanford Univ. (1969)
2.220  F. A. DVORAK, M. R. HEAD: Intern. J. Heat Mass Transf. **10**, 61 (1967)
2.221  R. E. CHILCOTT: Intern. J. Heat Mass Transf. **10**, 783 (1967)
2.222  L. S. FLETCHER, D. G. BRIGGS, R. H. PAGE: AIAA Paper 70-767 (1970)

2.223  C. B. COHEN, E. RESHOTKO: NACA R 1293 (1956)

2.224  W. DIENEMANN: ZAMM **33**, 89 (1953)

2.225  W. C. REYNOLDS, W. M. KAYS, S. J. KLINE: NASA Memo 12-3-58 W (1958)

2.226  P. M. MORETTI, W. M. KAYS: Intern. J. Heat Mass Transf. **8**, 1187 (1965)

2.227  R. G. DEISSLER: Trans. ASME **76**, 73 (1954)

2.228  B. A. KADER, A. M. YAGLOM: Intern. J. Heat Mass Transf. **15**, 2329 (1972)

2.229  P. BRADSHAW: AIAA J. **8**, 1375 (1970)

2.230  H. FERNHOLZ: DLR-FB **72-26**, 135 (1972)

2.231  R. L. MEEK, A. D. BAER: Intern. J. Heat Mass Transf. **16**, 1385 (1973)

2.232  I. E. ALBER, D. E. COATS: AIAA J. **8**, 791 (1971)

2.233  J. KESTIN, P. D. RICHARDSON: Forsch. Gebiete Ingenieurw. **29**, 93 (1963)

2.234  D. B. SPALDING: *Convective Mass Transfer*. London: Edward Arnold 1963

2.235  M. V. MORKOVIN: TR-68-149, AFFDL, Wright-Patterson AFB (1969)

2.236  E. RESHOTKO: AIAA Paper 74-130 (1974)

2.237  L. M. MACK: AGARDograph 97 (1965)

2.238  L. M. MACK: AIAA Paper 74-134 (1974)

2.239  A. D. D. CRAIK: J. Fluid Mech. **50**, 393 (1971)

2.240  M. GASTER: J. Fluid Mech. **66**, 465 (1974)

2.241  J. T. STUART: NPL Aero Rept. 1147 (1965)

2.242  E. MOLLO-CHRISTENSEN: AIAA J. **9**, 1217 (1971)

2.243  G. B. SCHUBAUER, P. S. KLEBANOFF: NACA Rept. 1289 (1956)

2.244  P. S. KLEBANOFF, K. D. TIDSTROM, L. M. SARGENT: J. Fluid Mech. **12**, 1 (1962)

2.245  C. F. KNAPP, P. J. ROACHE: AIAA J. **6**, 29 (1968)

2.246  A. M. O. SMITH, N. GAMBERONI: Rept. ES 26388, Douglas Aircraft Co. (1956)

2.247  N. A. JAFFE, T. T. OKAMURA, A. M. O. SMITH: AIAA J. **8**, 301 (1970)

2.248  I. TANI: *Ann. Rev. Fluid Mech.* **1**, 169 (1969)

2.249  N. S. DOUGHERTY: AIAA Paper 74-627 (1974)

2.250  S. R. PATE: AIAA J. **9**, 1082 (1971)

2.251  S. R. PATE: AIAA J. **12**, 1615 (1974)

2.252  J. E. LA GRAFF: AIAA J. **10**, 762 (1972)

2.253  F. K. OWEN, C. C. HORSTMAN: AIAA J. **10**, 769 (1972)

2.254  P. C. STAINBACK, J. B. ANDERS: AIAA Paper 74-136 (1974)

2.255  J. E. HARRIS: NASA TR-R-368 (1971)

2.256  H. MCDONALD, R. W. FISH: Intern. J. Heat Mass Transf. **16**, 1729 (1973)

2.257  D. M. BUSHNELL, D. W. ALSTON: AIAA J. **11**, 554 (1973)

2.258  A. M. O. SMITH: J. Fluid Mech. **7**, 565 (1960)

2.259  I. J. WYGNANSKI, F. H. CHAMPAGNE: J. Fluid Mech. **59**, 281 (1973)

2.260  I. WYGNANSKI, M. SOKOLOV, D. FRIEDMAN: J. Fluid Mech. **69**, 283 (1975)

2.261  R. W. MIKSAD: J. Fluid Mech. **56**, 695 (1972)

2.262  M. SIBULKIN: Phys. Fluids **5**, 282 (1962)

2.263  J. LAUFER: *Miszellaneen der Angewandten Mechanik* (Akademie-Verlag, Berlin 1962), p. 166

2.264  D. M. MCELIGOT, C. W. COON, H. C. PERKINS: Intern. J. Heat Mass Transf. **13**, 431 (1970)

2.265  M. A. BADRI NARAYANAN: J. Fluid Mech. **31**, 609 (1968)

2.266  M. A. BADRI NARAYANAN, V. RAMJEE: J. Fluid Mech. **35**, 225 (1969)

2.267  H. P. HORTON: AGARD-LS-43-71 (1971)

2.268  R. F. BLACKWELDER, L. S. G. KOVASZNAY: J. Fluid Mech. **53**, 61 (1972)

2.269  P. BRADSHAW: J. Fluid Mech. **35**, 387 (1969)

2.270  R. NARASIMHA, K. R. SREENIVASAN: J. Fluid Mech. **61**, 417 (1973)

2.271  B. E. LAUNDER, W. P. JONES: J. Fluid Mech. **38**, 817 (1969)

2.272  W. P. JONES, B. E. LAUNDER: Intern. J. Heat Mass Transf. **15**, 301 (1972)

2.273 R.L. O'Brien: United Aircraft Corp. Rept. (1964)

2.274 D.R. Boldman, J.F. Schmidt, A.K. Gallagher: NASA TN D-4788 (1968)

2.275 L.H. Back, R.F. Cuffel, P.F. Massier: AIAA J. 7, 730 (1969)

2.276 J.L. Nash-Webber, G.C. Oates: Trans. ASME 94D, 897 (1972)

2.277 K.R. Perkins, D.M. McEligot: Trans. ASME 97C, 589 (1975)

2.278 J.M. Robertson, C.F. Holt: ASCE J. Hydraulics Div., 98, HY6, 1095 (1972)

2.279 F.A. Schraub, S.J. Kline: Rept. MD-12, Dept. Mech. Engng., Stanford Univ. (1965)

2.280 R.A. Antonia, R.E. Luxton: TN F-31, Mech. Engng. Dept., Sydney Univ. (1971)

2.281 P. Bradshaw: Aero Rept. 74-10, Imperial College, London (1974)

2.282 H. Fernholz: ARC R. and M. 3368 (1964)

2.283 J.M. Kendall: J. Fluid Mech. 41, 259 (1970)

2.284 R.E. Davis: J. Fluid Mech. 52, 287 (1972)

2.285 A.Y.-S. Kuo, S. Corrsin: J. Fluid Mech. 50, 285 (1971)

2.286 E. Mollo-Christensen: Ann. Rev. Fluid Mech. 5, 101 (1973)

2.287 A. Favre, J. Gaviglio, R. Dumas: Phys. Fluids 10, S 138 (1967)

2.288 L. Prandtl: Festschrift zum 60. Geburtstage von A. Betz (Göttingen 1945) p. 134 (see also Gesammelte Abhandlungen [2.10])

2.289 F.J. Pierce: Trans. ASME 88D, 101 (1966)

2.290 W.D. Hayes: NAVORD Rept. 1313 (1951)

2.291 L.C. Squire: ARC R. and M. 3006 (1957)

2.292 J.V. Rakich, G.G. Mateer: AIAA J. 10, 1538 (1972)

2.293 T. Cebeci, K. Kaups, J. Ramsey, A. Moser: Proc. NASA Langley Conf. Aero Analyses Requiring Advanced Computers (NASA SP-347, 1975) and Douglas Aircraft Co. Rept. MDC J6866 (1975)

2.294 E. Krause: Int. Congr. Aerospace Sci. Paper 74-20 (1974)

2.295 E.A. Eichelbrenner: Ann. Rev. Fluid Mech. 5, 339 (1973)

2.296 F.G. Blottner: AGARD-LS-73, 3-1 (1975)

2.297 F.B. Gessner, J.B. Jones: Trans. ASME 83D, 657 (1961)

2.298 E.A. Eichelbrenner: Rech. Aéro. 104, 3 (1965)

2.299 M. Zamir: Aeronaut. J. 74, 330 (1970)

2.300 M. Zamir, A.D. Young: Aeronaut. Quart. 21, 313 (1970)

2.301 H.J. Perkins: J. Fluid Mech. 44, 721 (1970)

2.302 G.M. Bragg: J. Fluid Mech. 36, 485 (1969)

2.303 E.A. Eichelbrenner, K.T. Nguyen: Compt. Rend. Acad. Sci. A 269, 869 (1969)

2.304 E.A. Eichelbrenner: Rech. Aéro. 83 (1961)

2.305 E.A. Eichelbrenner, J.H. Preston: J. Mécanique 10, 91 (1971)

2.306 E.A. Eichelbrenner, P. Florent, K.T. Nguyen: Compt. Rend. Acad. Sci. Ser. A Paris 274, 1063 (1972)

2.307 O.O. Mojola, A.D. Young: AGARD-CP-93, 12.1 (1972)

2.308 K.T. Nguyen: MSc thesis, Université Laval, Quebec (1968)

2.309 J.C. Cooke, M.G. Hall: Prog. Aero. Sci. 2, 221 (1962)

2.310 P.N. Joubert, A.E. Perry, K.C. Brown: In: Fluid Mechanics of Internal Flow, Sovran, G. (Ed.) (Elsevier, Amsterdam 1967)

2.311 F.S. Sherman: Rept. RM-4843-PR, Rand Corp., Santa Monica (1968)

2.312 A.J. Wheeler, J.P. Johnston: Trans. ASME 95I, 415 (1973)

2.313 A.J. Wheeler, J.P. Johnston: Rept. MD-32, Dept. Mech. Engng., Stanford Univ. (1972)

2.314 J.F. Nash, V.C. Patel: Three-Dimensional Turbulent Boundary Layers (SBC Technical Books, Atlanta 1972)

2.315 J.H. Horlock, J.F. Norbury, J.C. Cooke: J. Fluid Mech. 27, 369 (1966)

2.316 H. Fernholz: J. Fluid Mech. 58, 177 (1973)

2.317  E. A. Eichelbrenner, A. Oudart: ONERA Pub. 76 (1955)
2.318  J. C. Cooke: Aeronaut. Quart. 11, 333 (1960)
2.319  N. A. Cumpsty, M. R. Head: Aeronaut. Quart. 21, 121 (1970)
2.320  J. C. Adams: J. Spacecraft 12, 131 (1975)
2.321  E. A. Eichelbrenner, J. L. Peube: Univ. Poitiers, Contract Number: N 62558-3863 (1966)
2.322  N. A. Cumpsty, M. R. Head: Aeronaut. Quart. 18, 55 (1967)
2.323  G. S. Raetz: Rept. NAI-58-73 (BLC-114), Northrop Aircraft, Inc. (1957) and ARC paper 23634 (1962)
2.324  P. Wesseling: Rept. AT-69-01, NLR, Amsterdam (1969)
2.325  K. C. Wang: J. Fluid Mech. 48, 397 (1971)
2.326  F. J. Pierce: Trans. ASME 86D, 227 (1963)
2.327  J. P. Johnston: Trans. ASME 82D, 622 (1960)
2.328  G. L. Mellor: AIAA J. 5, 1570 (1967)
2.329  M. R. Head, T. S. Prahlad: Aeronaut. Quart. 25, 293 (1974)
2.330  N. A. Cumpsty, M. R. Head: Aeronaut. Quart. 18, 150 (1967)
2.331  N. A. Cumpsty, M. R. Head: Aeronaut. Quart. 20, 99 (1969)
2.332  T. Cebeci: AIAA J. 12, 242 (1974)
2.333  L. F. East, R. P. Hoxey: ARC R. and M. 3653 (1971)
2.334  D. W. Etheridge: PhD thesis, Queen Mary College, London (1971)
2.335  B. van den Berg, A. Elsenaar, J. P. F. Lindhout, P. Wesseling: J. Fluid Mech. 70, 127 (1975)
2.336  A. Elsenaar, S. H. Boelsma: TR-74095U, NLR, Amsterdam (1974)
2.337  L. F. East: ARC R. and M. 3768 (1975)
2.338  W. F. Klinksiek, F. J. Pierce: Trans. ASME 92D, 83 (1970)
2.339  A. J. Vermeulen: PhD thesis, Univ. Cambridge (1971)
2.340  P. Bansod, P. Bradshaw: Aeronaut. Quart. 23, 131 (1972)
2.341  L. Larsson: PhD thesis, Chalmers Univ., Göteborg (1975)
2.342  R. W. Moore, D. L. Richardson: Trans. ASME 79, 1789 (1957)
2.343  P. Bradshaw, M. G. Terrell: NPL Aero Rept. 1305 (1969)
2.344  R. Crabbe: Rept. 71-2, Mech. Engng. Dept., McGill Univ. (1971)
2.345  A. Mager: NACA R 1067 (1952)
2.346  J. P. Johnston: Trans. ASME 82D, 233 (1960)
2.347  F. J. Pierce: AIAA J. 10, 334 (1972)
2.348  T. S. Prahlad: AIAA J. 11, 359 (1973)
2.349  J. L. Bousgarbies: thesis, Univ. Poitiers (1963)
2.350  E. A. Eichelbrenner, J. L. Peube: Univ. Poitiers. Lab. Mech. Fluides (1966)
2.351  J. P. Johnston: J. Fluid Mech. 42, 823 (1970)
2.352  P. Bradshaw: J. Fluid Mech. 46, 417 (1971)
2.353  D. B. Spalding: Paper presented at Euromech 60, Trondheim (1975)
2.354  B. van den Berg: J. Fluid Mech. 70, 149 (1975)
2.355  L. F. East: RAE TR 72178 (1972)
2.356  H. Ashkenas: NACA TN 4140 (1958)
2.357  F. J. Pierce, D. Krommenhoek: Interim Tech. Rept. 2, Dept. Mech. Engng., Virginia Poly. (1968)
2.358  T. Prahlad: AIAA J. 6, 1772 (1968)
2.359  M. G. Hall, H. B. Dickens: RAE TR 66214 (1966)
2.360  W. C. Rainbird: AIAA J. 6, 2410 (1968)
2.361  M. C. Fisher, L. M. Weinstein: AIAA J. 12, 131 (1974)
2.362  M. C. Fisher, L. M. Weinstein: AIAA Paper 73-635 (1973)
2.363  J. Cousteix, R. Michel: ONERA T.P. No. 1975-22 (1975)
2.364  O. Parr: Ingr.-Arch. 32, 393 (1963)

2.365  J. STEINHEUER: AGARDograph 97, 567 (1965)
2.366  T. S. CHAM, M. R. HEAD: J. Fluid Mech. 42, 1 (1970)
2.367  J. BUCKMASTER: Phys. Fluids 15, 2106 (1972)
2.368  E. S. TAYLOR: In: SOVRAN, G. (Ed.): *Fluid Mechanics of Internal Flow* (Elsevier, Amsterdam 1967), p. 320
2.369  J. H. B. SMITH: RAE Memo, 1620 (1975)
2.370  F. K. MOORE: *Theory of Laminar Flows* (Princeton Univ. Press, Princeton 1964)
2.371  I. E. BECKWITH, W. D. HARVEY, F. L. CLARK: NASA TN D-6192 (1971)
2.372  A. WALZ: *Strömungs- und Temperaturgrenzschichten.* (Braun, Karlsruhe 1966)
2.373  L. S. G. KOVASZNAY: J. Aeronaut. Sci. 20, 657 (1953)
2.374  F. K. OWEN, C. C. HORSTMAN, M. I. KUSSOY: J. Fluid Mech. 70, 393 (1975)
2.375  L. S. G. KOVASZNAY: J. Aeronaut. Sci. 17, 565 (1950)
2.376  M. V. MORKOVIN: AGARDograph 24 (1956)
2.377  M. V. MORKOVIN: Proc. Int. Symp. Hot-Wire Anemometry, Univ. Maryland (1967) pp. 38—51
2.378  A. DEMETRIADES, A. J. LADERMAN: SAMSO Rept. 73-129 (1973)
2.379  J. E. WALLACE: AIAA J. 7, 757 (1969)
2.380  W. D. HARVEY, D. M. BUSHNELL: NASA SP 216, 11 A1—11 A10 (1968)
2.381  S. S. PENNER: AIAA Paper 71-283 (1971)
2.382  W. J. YANTA, R. E. LEE: AIAA Paper 74-575 (1974)
2.383  A. L. KISTLER: Phys. Fluids 2, 296 (1959)
2.384  M. V. MORKOVIN: AGARD Wind Tunnel and Model Testing PANEL, London (1960)
2.385  P. BRADSHAW, D. H. FERRISS: J. Fluid Mech. 46, 83 (1971)
2.386  M. V. MORKOVIN: In: *The Mechanics of Turbulence*, FAVRE, A., (ed.) (New York: Gordon and Breach 1964)
2.387  F. K. OWEN, C. C. HORSTMAN: J. Fluid Mech. 53, 611 (1972)
2.388  W. C. ROSE: AIAA J. 12, 1060 (1974)
2.389  G. MAISE, H. MCDONALD: AIAA J. 6, 73 (1968)
2.390  A. MARTELLUCCI, H. RIE, J. F. SONTOWSKI: AIAA Paper 69-688 (1969)
2.391  D. M. BUSHNELL, D. J. MORRIS: AIAA J. 9, 764 (1971)
2.392  H. U. MEIER, J. C. ROTTA: AIAA J. 9, 2149 (1971)
2.393  C. C. HORSTMAN, F. K. OWEN: AIAA J. 10, 1418 (1972)
2.394  A. J. LADERMAN, A. DEMETRIADES: J. Fluid Mech. 63, 121 (1974)
2.395  W. C. ROSE: PhD thesis, Univ. of Washington (1972). See also NASA TN-D 7092 (1973)
2.396  E. R. VAN DRIEST: J. Aeronaut. Sci. 18, 145 (1951)
2.397  H. U. MEIER, R. L. P. VOISINET, D. F. GATES: AIAA Paper 74-596 (1974)
2.398  A. M. CARY: NASA TN D-5560 (1970)
2.399  G. T. COLEMAN, C. OSBORNE, J. L. STOLLERY: J. Fluid Mech. 60, 257 (1973)
2.400  H. J. KÜSTER: Dissertation, Technische Universität Berlin (1972)
2.401  D. COLES: J. Aeronaut. Sci. 21, 433 (1954)
2.402  J. ROTTA: Z. Flugwiss. 7, 264 (1959)
2.403  P. O. BARONTI, P. A. LIBBY: AIAA J. 4, 193 (1966)
2.404  H. TENNEKES: AIAA J. 5, 489 (1967)
2.405  R. L. GRAN, J. E. LEWIS, T. KUBOTA: J. Fluid Mech. 66, 507 (1974)
2.406  H. FERNHOLZ: *Fluid Dynamic Transactions* (Polish Academy of Sciences, Warsaw 1971), vol. 6, pt. 2, p. 161
2.407  J. C. ROTTA: Rept. 64 A 10, AVA-DFVLR, Göttingen (1964)
2.408  H. FERNHOLZ: Ingr.-Arch. 38, 311 (1969)
2.409  E. R. KEENER, E. J. HOPKINS: AIAA J. 11, 1784 (1973)
2.410  D. C. MATHEWS, M. E. CHILDS, G. C. PAYNTER: J. Aircraft 7, 137 (1970)

2.411  L. CROCCO: AIAA J. **1**, 2723 (1963)

2.412  J. E. LEWIS, T. KUBOTA, W. H. WEBB: AIAA J. **8**, 1644 (1970)

2.413  H. H. FERNHOLZ, P. J. FINLEY: To be published by AGARDograph 223 (1977)

2.414  D. J. PEAKE, G. BRAKMAN, J. M. ROMESKIE: AGARD-CP-93, 11-1 (1972)

2.415  J. E. LEWIS, R. L. GRANT, T. KUBOTA: J. Fluid Mech. **51**, 657 (1972)

2.416  R. L. VOISINET, R. E. LEE: TR-73-224, U.S. Naval Ordnance Lab. (1973)

2.417  P. J. WALTRUP, J. A. SCHETZ: AIAA J. **11**, 50 (1973)

2.418  F. ZWARTS: PhD thesis, McGill Univ. (1970)

2.419  D. B. SPALDING, S. W. CHI: J. Fluid Mech. **18**, 117 (1964)

2.420  H. FERNHOLZ: ZAMM **51**, T 146 (1971)

2.421  E. J. HOPKINS, M. INOUYE: AIAA J. **9**, 993 (1971)

2.422  A. M. CARY, M. H. BERTRAM: NASA TN-D-7507 (1974)

2.423  P. BRADSHAW: J. Fluid Mech. **63**, 449 (1974)

2.424  W. BEHRENS: AIAA Paper 71-127 (1971)

2.425  J. E. LEWIS, W. BEHRENS: AIAA J. **7**, 664 (1969)

2.426  J. F. STROUD, L. D. MILLER: Tech. Rept. AFFDL-TR-65-123 and J. Aircraft **3**, 548 (1966)

2.427  W. B. STUREK, J. E. DANBERG: AIAA J. **10**, 475 and 630 (1972)

2.428  J. C. ROTTA: Phys. Fluids **10**, S 174 (1967)

2.429  G. J. THOMKE, A. ROSHKO: NASA CR-73308 (1969)

2.430  J. P. BATHAM: J. Fluid Mech. **52**, 425 (1972)

2.431  G. M. ELFSTROM: J. Fluid Mech. **53**, 113 (1972)

2.432  G. T. COLEMAN, J. L. STOLLERY: J. Fluid Mech. **56**, 741 (1972)

2.433  W. C. ROSE, R. J. PAGE, M. E. CHILDS: AIAA J. **11**, 761 (1973)

2.434  C. H. LAW: AIAA J. **12**, 794 (1974)

2.435  K. K. CHEN: AIAA Paper 71-166 (1971)

2.436  A. L. KISTLER: NASA CR-128540 (1971)

2.437  D. V. MADDALON: AIAA J. **7**, 2355 (1969)

2.438  J. E. LA GRAFF: PhD thesis, Oxford Univ. (1970)

2.439  F. K. OWEN: AIAA Paper 70-745 (1970)

2.440  A. M. CARY: NASA TN-D-5863 (1970)

2.441  I. E. BECKWITH, M. H. BERTRAM: NASA-TM-X-2566 (1972)

2.442  R. D. WATSON, J. E. HARRIS, J. B. ANDERS: AIAA Paper 73-165 (1973)

2.443  I. E. BECKWITH: AIAA Paper 74-135 (1974)

2.444  J. M. KENDALL: AIAA J. **13**, 290 (1975)

2.445  W. V. FELLER: AIAA J. **11**, 556 (1973)

2.446  B. THWAITES (Ed.): *Incompressible Aerodynamics* (Clarendon Press, Oxford 1960)

2.447  W. W. WILLMARTH: Advances in Appl. Mech. **15**, 159 (1975)

# 3. Internal Flows

J. P. JOHNSTON

With 15 Figures

An internal flow is any flow through a (circular) pipe, (non-circular) duct or (open, liquid-flow) channel where confining walls, or a free surface, guide the flow from an arbitrarily defined inlet state to an equally arbitrary outlet state. Laminar flow through a circular pipe (Poiseuille flow) is the simplest example of internal flow, whereas turbulent flow in the impeller (rotor) and diffuser (stator) of a centrifugal compressor stage is one of the most complex of internal flows; see [3.1]. According to this definition even the flow over a model in a wind tunnel is an "internal flow", but the practical distinction between the external flows discussed in Chapter 2 and the flows to be discussed in the present chapter is that in the latter the shear layers on the walls interact strongly with each other or, at least, with the "inviscid" fluid between them. As mentioned in Chapter 2, we include in the present chapter several phenomena which can also occur in external flows, simply because they are more common, or more general, or more easily discussed, in internal flows.

Poiseuille flow and its turbulent counterpart are said to be *fully developed*, an expression used to denote a flow field that is independent, in the ensemble or time-mean, of streamwise position along the duct. Fully developed flow is unique to internal flows. It is by definition a flow type that is independent of inlet or outlet state and one where conditions are statistically identical at each axial position. Because of this simplicity, fully developed duct flows are often favored by turbulence research workers. Section 3.2 reviews some of the published results. In practice many internal flows are far from fully developed, but may contain wall boundary layers, thin interior shear layers (wakes or jets) and large regions of effectively inviscid flow.

## 3.1 Zones of Flow and Basic Phenomena

Figure 3.1 illustrates four basic flow zones for the case of a simple, constant area duct that receives a steady flow from an upstream plenum through a short, smoothly contoured nozzle. In *Zone I* where distance

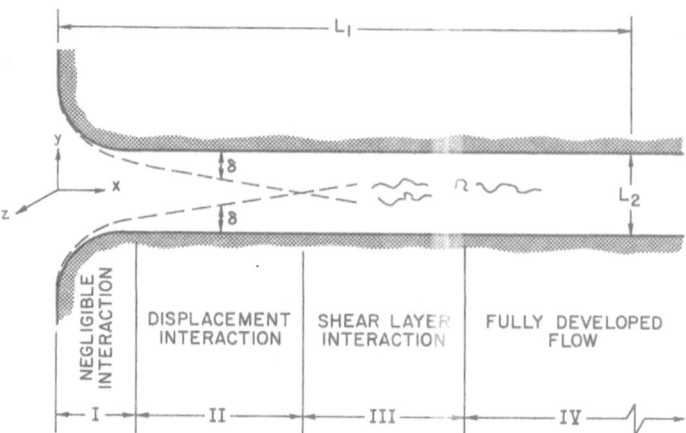

Fig. 3.1. Zones of interaction in duct flow. Note: zone boundaries are in reality poorly defined and diagram is not to scale

from inlet, $x$, is much less than $L_2$, a typical transverse dimension of the duct, the wall boundary layers are very thin and they hardly affect the flow in the interior, inviscid core region. For cases commonly seen in practice where the bulk Reynolds number (based on $L_2$ and mean flow speed, $U_m$), $U_m L_2/\nu$, is of the order $10^4$ to $10^6$, the flow in *Zone II* ($x$ of order 1 to 10 $L_2$) still contains an inviscid core region, but the boundary layers have become thicker so that displacement thicknesses of $\delta^*$ approximately equal to $L_2/10$ are not unexpected. Zone II is called the displacement interaction zone because the blocking effect of the growing wall boundary layers affects the flow speed in the inviscid core. Typically, the displacement interaction effect causes deviations of predicted static pressure change, from inlet to the end of Zone II, that are of the order of 0.1 to 1 times the mean dynamic pressure, $(1/2)\varrho U_m^2$, compared to ideal flow (flow without boundary layers on the duct walls). The flows in Zones I and II are physically identical. The differences are only in the magnitudes of the effects. For further discussion see Section 3.3.

A much more complex interaction starts to occur at the beginning of *Zone III*, the shear layer interaction zone. Here, for example, the boundary layers on opposite walls of a two-dimensional duct start to overlap or interact. At the start of Zone III there is no longer a definable region of inviscid flow or any streamline along which time-mean total pressure, $P$, remains constant. This zone is particularly difficult to analyze, but some progress has recently been made for two-dimensional duct flow (see Subsection 3.1.3).

*Zone IV* is the region of fully developed flow. For Reynolds numbers $U_m L_2/v$ high enough for the boundary layers to go turbulent near the entry, almost fully developed flow may be achieved at $x/L_2 \gtrsim 40$. Fully developed flow is, in fact, the asymptotic result of conditions developed in Zone III, but, as already pointed out, conditions in Zone IV should be independent of upstream conditions. Typical engineering systems where fully developed flow is important are pipe lines and the core tubes in large industrial heat exchangers, surface condensers, superheaters, etc. Very few other devices contain ducts of sufficient axial length, $L_1$, to permit conditions beyond Zone-II or -III type flow. However, large separating zones produced, for example, by sharp-edged contractions at duct inlet may cause wall boundary layers to merge rather near the inlet. For cases of severely disturbed inlet conditions the mean velocity profiles and axial pressure gradients in the regions $x/L_2 \lesssim 10$ to 40 often appear to be fully developed even though turbulent flow structure is still far from fully developed; see Subsections 3.4.1 and 3.4.3 for concrete examples.

The flow in, and between, the blade rows of turbines and compressors is probably the most complex general class of single-phase[1] internal flows. By their very nature, these "ducts" are of complex shape, e.g., both curved and twisted so their main flow field may be fully three dimensional. They are hard to analyze even without the added complexity of turbulence. Furthermore, turbomachine flow is, of necessity, periodically unsteady in either inertial (laboratory or stator) coordinates, or rotating (rotor) coordinates. This periodic unsteadiness is the essential mechanism for high efficiency energy exchange between a rotor and a continuously flowing fluid stream (see DEAN [3.2]). Only in recent years have the instruments and techniques been developed so that turbulence may be properly studied in turbomachine flows where it is necessary to separate the fluctuating signals into three additive parts, (i) $U_i$, the long-time mean over many blade passing[2] cycles, (ii) $\tilde{u}_i$, the component which has the period of blade passing and its harmonics and is a phase average with respect to position between blades, and (iii) the true turbulence $u'_i$, the only part which should display broad-band spectral characteristics. Section 3.5 develops these ideas in greater depth.

Practical internal flows frequently contain more severe examples of effects already introduced in the discussion of external flows in

---

[1] More complex flows are encountered when vapor (or gas) and liquids flow together. Turbulent, multiphase flow is extremely complex; see Chapter 7.

[2] Blade passing frequencies as high as 5000 Hz (periods of $\sim 0.2$ ms/cycle) may be encountered in high speed turbocompressors.

Chapter 2. We list four of the most important, which will be discussed at length in the following subsections.

### a) Curvature and System Rotation Effects (see also Subsection 2.3.8)

Longitudinal curvature of the streamlines or system rotation about an axis perpendicular to the plane of the mean shear has important effects on the structure of turbulent shear flow. The effect of greatest practical importance is the change in Reynolds stresses which may be wrought by curvature and rotation. Subsection 3.1.1 briefly reviews these effects, but a long review paper [2.117] should be studied by serious research workers.

### b) Secondary Flows (see also Section 2.4)

Secondary flows are time mean flows that develop in a plane perpendicular to the main direction of flow, $x$. They are discussed in Subsection 3.1.2 and Section 2.4. Secondary flow of the *first kind* is the result of development of mean streamwise vorticity by exchange of mean vorticity between components as a shear layer is turned about an axis parallel to the plane of the mean shear (i.e., an axis in the $x_\alpha$ or $y$ direction in conventional notation). The exchange is governed by the vorticity/velocity-gradient interaction term in (1.15), $\Omega_l \partial U_i / \partial x_l$. It is commonly seen in boundary layers on the end-walls of bent ducts of rectangular cross section. The *second kind* of secondary flow, the type that develops in corners of straight axis ducts due to cross-stream $(y, z)$ gradients of the Reynolds stresses, is a true turbulence phenomenon that cannot occur in laminar flow of a Newtonian fluid and cannot be simulated by calculations using an isotropic eddy viscosity.

### c) Shear Layer Interaction of Zone II and III Types

Simple displacement interactions, Zone II, are reviewed in Subsection 3.1.3 but most emphasis is placed on the full shear layer interaction zone, Zone III, because of important recent developments in this area, and because engineers have, for some time, recognized the basic importance of full shear layer interaction. The work on displacement interactions is reviewed in Section 3.3 because it is so closely related to extensive basic and applied research on flow in diverging ducts or diffusers.

A diffuser is a duct designed to decrease mean kinetic energy in a flowing stream and to recover this energy as an increase in flow work (static pressure rise). Efficient diffusion with minor loss of available mechanical energy per unit mass of flowing fluid is essential in most internal flow type engineering systems. Because of the need to minimize energy losses, diffusers are usually short (i.e., $L_1 \sim 1$ to $10\, L_2$) to minimize wetted surface. Longer ducts have excessively high total pressure losses, or, in the case of diffusers, loss of available "ideal" pressure rise $(p_2 - p_1)_i$, computed from diffuser geometry, inlet conditions, and the assumption of an inviscid one-dimensional flow from 1 to 2. Well-designed diffusers of high pressure recovery effectiveness $[\eta = (p_2 - p_1)/(p_2 - p_1)_i]$ normally operate with an inviscid core. Zone II in Fig. 3.1.

*d) Flow Separation* (see also Subsection 2.3.6)

To the designer of internal flow systems that are intended on the one hand to minimize thermodynamic losses (total pressure losses) and on the other hand to be simple and compact when executed in hardware, the basic flow phenomenon of separation of a boundary layer has particular importance. Separation also helps to determine the range of stable operation of a flow system (e.g., the advent of surge and rotating stall in compressors). The designer of high-performance flow devices often must push his diffuser designs to the limit of flow separation in order to obtain peak efficiency; as an unwanted consequence he risks sudden, unpredicted stall[3]. Because turbulent boundary layer flow has a central influence on the conditions that lead to flow separation, boundary layers in the various states of stall are discussed in several sections below: basic concepts in Subsection 3.1.4, diffuser stall states in Section 3.3, fully separated flows with large regions of stall in Section 3.4 and some special aspects of turbomachine stall in Section 3.5.

### 3.1.1 Longitudinal Curvature and System Rotation

When two-dimensional, turbulent shear layers flow with mean streamlines that are curved longitudinally (in the $x - y$ plane) in inertial coordinates, the structure of the turbulence is modified by the centrifugal and/or Coriolis accelerations induced by the curvature. The effect, modification of turbulent exchange rates, may be explained in terms of the gain or loss of stability of the flow field with respect to longitudinal-vortex disturbances, a phenomenon known for a long time in curved laminar flows [3.3, 2.190, 191]. The stability of longitudinally curved turbulent flows was first studied by PRANDTL [2.189]. For thin shear layers, he proposed that $F$, a corrective multiplier, be applied to the

---

[3] Stall is a pejorative word often loosely used to denote a state of separated flow coupled with deleterious effects: the disastrous loss of lift of a wing taken to excess angle of attack, the sudden loss of total-pressure rise in a compressor of an aircraft jet engine, etc.

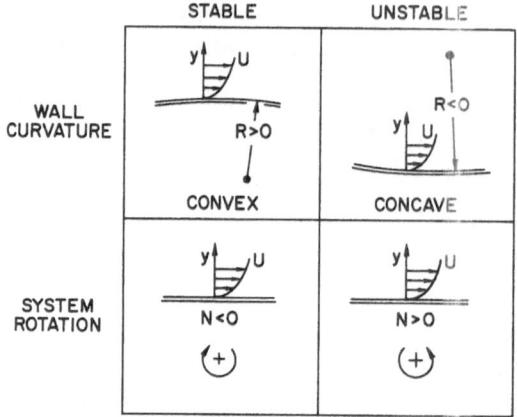

Fig. 3.2. Stability of curved and rotating wall boundary layers

normal mixing length, $l_0$, to account for the local effects of a dimensionless, curvature stability parameter, i.e.,

$$\frac{l}{l_0} \equiv F = 1 - \frac{1}{4}\left(\frac{U/R}{\partial U/\partial y}\right) \tag{3.1}$$

where $R$ is the streamline radius of curvature, positive if $\partial V/\partial x$ is negative. Prandtl's general ideas were supported by some early experiments [3.4, 5], but the data indicated that the order of magnitude of the difference $(F-1)$ should be larger by a factor of ten than that given in (3.1). Many other examples that confirm these results are discussed in [3.4].

Fig. 3.2 illustrates four different simple boundary layer flows which require consideration of the curvature stability effect. Two examples illustrate flow over curved walls where $R$ is the local wall radius of curvature (in thin or fairly thin shear layers the streamlines are nearly parallel to the walls). The other two examples, flow on rotating flat walls, illustrate systems that are in steady rotation at a rate[4] N (radians/second, positive anticlockwise) about an axis perpendicular to the plane of mean shear. Combined cases are possible where fluid flows over curved surfaces in a rotating system. Examples are the boundary layers on the suction and pressure surfaces of curved blades in centrifugal compressor impellers (rotors).

There is at least a qualitative analogy (first invoked by PRANDTL) between the effect of the apparent body force introduced by centripetal or

---

[4] $N$, the system rotation vector, requires inclusion of Coriolis acceleration, $2\varepsilon_{ijk}N_jU_k$, in the equations of motion, e.g., left-hand sides of (1.2), (1.8), etc.

Coriolis acceleration and the effect of a true body force such as buoyancy. Much of the following discussion parallels that of Subsection 4.1.1.

*Stabilizing Curvature* $(R>0$ or $N<0$ if $\partial U/\partial y>0)$ has the effect that Reynolds shear stresses and turbulence energy levels are decreased compared to otherwise equivalent straight shear layer flows. The most striking, and recent, direct measurements are [2.194] and [3.6] for nearly equilibrium boundary layers on convex walls where $\delta/R \lesssim 1/10$, [3.7] for a thicker $(\delta/R \approx 2/10)$, highly curved mixing layer that is not in equilibrium, [3.8] and [3.11] for fully developed flow in long, curved two-dimensional ducts, and [3.9] for fully developed flow in a long, straight, rotating two-dimensional duct. In some of these cases the stabilizing effect becomes large enough to suppress turbulence completely [3.17] or to suppress it in the outer part of the layer [2.194]. Regions of the shear layer develop where turbulence energy, $1/2\overline{u_i^2}$, is produced at a negative rate[5], see Fig. 14 in [3.9]. The latter phenomenon, of course, illustrates the weakness of the gradient-diffusion, or "eddy viscosity", approach $[-\overline{uv}=v_T(\partial U/\partial y)]$ to turbulent flow calculation. A negative energy production rate cannot be predicted by this approach, used in many one-equation or two-equation turbulent stress models (see Chapt. 5).

*Destabilizing Curvature* $(R<0$ or $N>0$ in Fig. 3.2) has effects that are the opposite of those of stabilization. Levels of Reynolds shear stress and turbulence energy are caused to be abnormally high. In addition curved-wall boundary layer flows may become unstable to quasi-steady longitudinal, vortical cells of the order $\delta$ in cross-sectional size. This is the Taylor-Görtler instability originally documented for laminar flows [2.190, 191]. COLES [3.12] has given a very complete experimental account of this effect, including the influence of transitional and turbulent flow, for the case of circular Couette flow. For straight-walled rotating systems in laminar flow POTTER and CHAWLA [3.13] and HART [3.14] examined this instability, and LEZIUS [3.15] has extended the linear stability analyses to turbulent layers. In either case, one sees that these vortical cells are possible where, for given boundary layer Reynolds number, either the curvature parameter $(\delta/R)$ or the rotation parameter $(N\delta/U_\infty)$ exceeds a critical value which is reviewed by So and MELLOR [2.195] for concave wall cases. In the case of fully developed turbulent flow in two-dimensional ducts LEZIUS [3.15] suggests that 0.04 is the minimum critical value of the parameter $2DN/U_m$ ($2D=$ channel

---

[5] The term $(-\overline{uv})\partial U/\partial y$ in (1.43). The existence of negative $(-\overline{uv})$ in regions of positive $(\partial U/\partial y)$ has been discussed by ESKINAZI and ERIAN [3.10] and is seen in other types of data, e.g., fully developed flow in annuli of small radius ratio. The lost energy is supplied by diffusion from other parts of the flow.

width, $U_m$ is area-mean flow speed) where Taylor-Görtler instabilities start to amplify.

The longitudinal vortices formed by this instability affect both transition to turbulence in boundary layers on concave walls [3.16], and also the structure of fully established turbulence itself as can be seen in the data of So and MELLOR [2.195] and PATEL [3.18]. Flow visualization in the unstable side wall layers of a rotating channel [3.9] shows the developed vortex structure clearly. The picture of the flow obtained here, and in the original moving pictures, is not that implied by linear stability theory [3.15]. For one thing, it appears that most of the true turbulence energy and stress may develop and remain in regions where mean motion of two adjacent vortices is away from the wall, an observation partly confirmed for concave wall flow by the data of So and MELLOR [2.195]. Also, the vortex-cell locations may be rather unstable with respect to spanwise location, and they tend to form and decay over time.

The calculation of curved shear layer flow requires quantitative estimates of the effects of curvature, or system rotation, on turbulent shear stress. For example, if a mixing length model is used one can employ (3.1) with the factor 1/4 replaced by a larger, empirical factor. This simple model is only one of several that may be used to close the two-dimensional boundary layer equations. For $\varrho$ and $v$ constant, these equations are, when written in orthogonal, semi-curvilinear coordinates [Ref. 1.9, p. 119], $x$ along the surface and $y$ a series of straight lines normal to the surface:

$$\frac{U}{\alpha}\frac{\partial U}{\partial x} + V\frac{\partial U}{\partial y} + \frac{UV}{\alpha R} - 2NV = -\frac{1}{\alpha\varrho}\frac{\partial p^*}{\partial x} + \frac{1}{\varrho}\left(\frac{\partial \tau}{\partial y} + \frac{2\tau}{\alpha R}\right) \qquad (3.2)$$

$$-\frac{U^2}{\alpha R} + 2NU = -\frac{1}{\varrho}\frac{\partial p^*}{\partial y} \qquad (3.3)$$

$$\frac{\partial U}{\partial x} + \frac{\partial}{\partial y}(\alpha V) = 0. \qquad (3.4)$$

The term $\alpha = (1 + y/R)$ accounts for effects of coordinate system geometry. If the density is constant, centrifugal acceleration can be absorbed in the reduced pressure, $p^* = (p - \varrho N^2 r^2/2)$ where $r$ is radial distance from the axis of rotation. The total shear stress reduces to laminar and turbulent parts, $\tau = \tau_\mu + \tau_t$, where

$$\frac{\tau_\mu}{\varrho} = \alpha v\frac{\partial}{\partial y}\left(\frac{U}{\alpha}\right) \quad \text{and} \quad \frac{\tau_t}{\varrho} = -\overline{uv}. \qquad (3.5)$$

Equations (3.1–5) satisfy the "fairly thin" shear layer approximation, see [2.117], especially when $(\delta/R)$ and $(N\delta/U_e)$ are less than $(1/10)$ in magnitude. When these two system parameters are truly small, say $(1/1000)$, the "simple thin" shear layer approximation may be used and all terms explicitly containing $N$ and $R$ disappear so that the equations reduce to (1.38–41) when the latter are in two-dimensional form.

Examination of equivalent Reynolds stress transport equations (1.35, 36) for the curved case or for the rotating flow case [3.19] shows that the dimensionless quantities[6] $\{2[(U/R)/(\partial U/\partial y)]\}$ and $\{-2[N/(\partial U/\partial y)]\}$ describe the relative importance of the local curvature and rotation effects, respectively. These parameters, which represent the ratio of the buoyant production term in the $v$-component equation to the shear production term in the $u$-component equation, are smallest near the wall in a boundary layer and largest near the free stream. The corresponding effects which are induced on $\tau_t$ vary in magnitude in the same manner across a boundary layer; see [2.194].

For the case of flow in a rotating system where the mean shear layer is assumed purely parallel, and when linearized stability theory is applied, the local quantity (other than $v$) that controls stability is shown by LEZIUS [3.15] to be the square of a Brunt-Väisälä frequency, $-2N[(\partial U/\partial y)-2N]$, analogous to the quantity of the same name in buoyant flows (Subsection 4.1.1). Negative values may lead to growing, oscillatory solutions, but positive values cause damped solutions, i.e., stability. When this parameter is non-dimensionalized by the square of the characteristic turbulence frequency, $(\partial U/\partial y)^2$, we obtain a parameter called, again by analogy with buoyant flows [2.189], a "gradient Richardson number" ($Ri>0$ stable, $Ri<0$ unstable),

$$Ri = -\frac{2N}{(\partial U/\partial y)}\left[1 - \frac{2N}{(\partial U/\partial y)}\right].$$

It is seen that the parameter $[-2N/(\partial U/\partial y)]$ discussed previously is identical to $Ri$ for small values of the parameter, and therefore we see there is a direct link between the turbulent stress effects and the gain or loss of flow stability. Similar conclusions may be drawn for the case of simple curvature; see [2.117].

---

[6] The global system parameters $(\delta/R)$ and $(N\delta/U_e)$ are directly related to these parameters when one takes $U \sim U_e$ and $(\partial U/\partial y) \sim (U_e/\delta)$. Hence, the magnitude and sign of the global parameters, which are directly obtained by dimensional analysis, also are measures of the effects of curvature and rotation.

If current one-equation or two-equation turbulent stress models are used then an empirical constant, $\beta$, is needed in the expression[7]

$$(-\overline{uv})\frac{\partial U}{\partial y}\left[1+\beta\frac{(\partial V/\partial x)}{(\partial U/\partial y)}\right]$$

that must be used to increase, or decrease, the ordinary turbulence energy production rate. As seen in (3.1) $\beta$ is of order 1 if only true production terms are included, but realistic prediction of rotation and curvature effects requires that $\beta \approx 10$ be used. Eide and Johnston [3.22] carried out calculations for a variety of thin, curved and rotating wall boundary layers under various free-stream pressure gradients using a simple mixing length correction factor.

$$l=l_0\left[1+\beta_c\left(\frac{-U/R}{\partial U/\partial y}\right)+\beta_R\left(\frac{2N}{\partial U/\partial y}\right)\right] \qquad (3.6)$$

where $l_0$ is given its normal values; see Chapter 5. Values of $\beta_c = \beta_R = 6 \pm 2$ appear to allow satisfactory engineering prediction of the mean flow measurements of a number of cases, [2.194, 197, 3.23] up to conditions close to flow separation. Related procedures have been used by others, e.g., [3.24], with moderate success. Similar correction factors can be applied to the turbulence length scales used in more advanced turbulence models (Chapt. 5).

Recently Mellor and Yamada [3.20], Launder et al. [3.21], Irwin and Smith [2.204], and So [3.25] suggested that improved modelling of the pressure-strain terms in the full set of Reynolds stress transport equations will give proper values for the curvature effects, at least for equilibrium layers and thin shear layers close to equilibrium, without introduction of a special empirical constant such as $\beta$. The data of Castro and Bradshaw [3.7] do not fall into this category and do not support this conclusion.

### 3.1.2 Secondary Flows

Mean flow fields that are three dimensional may for convenience be divided into three classes. As already noted in Chapter 2, they are:

Class (i) "thin shear layers" and "fairly thin shear layers" where $\partial/\partial y \gg \partial/\partial x \sim \partial/\partial z$. This class includes three-dimensional, unseparated boundary layers on flat, or slightly curved ($\delta/R \leq 0.01$), surfaces, and be-

---

[7] $(\partial V/\partial x)$, the extra rate of strain, is equal to either $2(U/R)$ or $(-2N)$ in cases considered here.

cause it is discussed thoroughly in Section 2.4, class (i) will not be covered here.

*Class (ii)* "slender shear flows" where $\partial/\partial y \sim \partial/\partial z \gg \partial/\partial x$. Flow in the corner of a square duct, intersecting boundary layers at a wing-body junction and flow in the shear layer at the tip of a wing fall in this class.

*Class (iii)* "full three-dimensional flow" where $\partial/\partial x \sim \partial/\partial y \sim \partial/\partial z$ requires solution of the full Navier-Stokes equations. For some special cases inviscid secondary flow theory (see below) may be employed.

"Secondary flow" is a generic term used to denote the mean cross flow field characterized by the velocity components $V$ and $W$ that lie in a plane which is normal to the $x$, "main flow", direction. It is common to define the main flow direction to be the local direction of $U_e$, the free-stream velocity, when one is dealing with boundary layer situations [classes (i) and (ii)] but, for flow along curved ducts or pipes, $x$ is usually defined to follow the duct axis so that the secondary flow plane $(y - z)$ is the normal, cross-sectional plane at each point along the duct. Since the $x$ direction may change along the curved path of the free stream, or duct, analysis is often carried out in curvilinear coordinates that follow the main flow streamlines, or duct axis, e.g., see [Ref. 1.9, p. 101] and [2.314] for equations in general, orthogonal coordinates. Except perhaps close to a solid wall, secondary velocities are usually small compared to $U$, the speed of the main flow. The size of $Q_s = (V^2 + W^2)^{1/2}$, the secondary flow speed, relative to $U$ depends primarily on the physical mechanism that drives the secondary flow. The four principal mechanisms are:

*1) Lateral convergence* (or divergence) thins (or thickens) a shear flow principally by pure geometric effects—consider a boundary layer growing on a cone (or inside a converging nozzle). Here, proper choice of co-ordinates will often reduce the problem to a simpler one in two dimensions. However, the extra rate of strain effect due to transverse stretching (or contracting) of the layer may have important effects on Reynolds stress, similar to those caused by longitudinal curvature [2.117].

*2) Lateral curvature* of the main flow (in the $x - z$ plane for example) subjects the flow field to a pressure gradient in the secondary flow plane $(y - z)$, equal to $\boldsymbol{V}_s p$, where $\boldsymbol{V}_s() \equiv \boldsymbol{i}_y[\partial()/\partial y] + \boldsymbol{i}_z[\partial()/\partial z]$. The vector $\boldsymbol{V}_s p$ points outwards from the center of curvature. If the principal radius of curvature of the main-flow axis is $r_p$, the secondary pressure gradient is balanced by the centripetal acceleration in a path of radius $r_p$ only for fluid elements moving at a certain speed $\tilde{U}$ given to a first approximation by $|\boldsymbol{V}_s p| = \varrho(\tilde{U}^2/r_p)$. Fluid elements moving at a speed $U < \tilde{U}$ drift towards the center of curvature and vice versa. The actual $V - W$ field that results is, of course, constrained by continuity, by the wall boundary shape and by the laminar and Reynolds stress fields.

*3) Lateral wall motion* of a bounding wall or surface relative to the fluid can cause secondary flow by pure viscous drag. Flows typical of this class are seen in pipes that rotate about their central axis, or stationary pipes into which swirling flow is induced. This effect will not be discussed however.

*4) Gradients of the Reynolds stresses*, $\partial(-\varrho\overline{u_i u_j})/\partial x_l$, can produce forces in the secondary flow plane that induce $V$ and $W$. This effect is the only mechanism that can drive the secondary flows in the straight corner region where two flat wall boundary layers interact. It is the type of secondary flow called *secondary flow of the second kind* by Prandtl [3.26]. The mechanism that drives it, unlike mechanisms 2 and 3 above which drive *secondary flow of the first kind*, is not present in laminar flow.

### Curved Ducts and Secondary Flows of the First Kind

Thin developing boundary layers on the walls of curved ducts[8] are usually analyzed by the methods discussed in Chapter 2. The core flow region is considered inviscid and is usually irrotational, so that potential theory can be used for its solution. Where proper correction is made for the displacement interaction effect (see Subsection 3.1.3), it is assumed that the pressure gradients derived from the core solution drive the boundary layer crossflows (secondary flows). Often, in practical devices, the crossflows are strong enough, near the wall in the sublayer at least, to traverse the passage from one side wall to the other in a distance of the order of one passage diameter; see Refs. [3.27, 28]. Very complex flows can result; for example vortices can roll up in the corners where crossflow fluid tends to pile up against a suction side wall. These vortices may have strong effects on the downstream flow; see [3.29]. Complex flows of the latter type are today not amenable to exact analysis. For discussions of the turbulence structure of rotation-stabilized vortex cores, see [2.4, 117].

Because of analytic and computational difficulty, and a desire by engineers to obtain approximate solutions, it is common to analyze pressure driven secondary flows in short, curved ducts dominated by mechanism 2 (above) by neglect of laminar and turbulent shear stresses altogether. Of course shear stresses are necessary to produce a given distribution of inlet mean vorticity, $\Omega_1$, generated in upstream duct elements. Inside the short, curved duct however, vortex lines are "frozen" to fluid particles and we can apply the mean vorticity equa-

---

[8] Zones I and II in the terminology of Section 3.1.

tion for steady "inviscid" flow, a special case of (1.15), with $\Omega$ denoting the mean vorticity vector,

$$U_l \frac{\partial \Omega_i}{\partial x_l} = \Omega_l \frac{\partial U_i}{\partial x_l} \qquad (3.7)$$

subject to satisfaction of the continuity equation and boundary conditions. These are that the normal velocity components are specified (equal to zero on solid walls) and that no flow separation is allowed inside the duct. The velocity field, obtained from the equation defining the vorticity field,

$$\Omega_i = \varepsilon_{ijk} \partial U_k / \partial x_j, \qquad (3.8)$$

is required for the solution, too. Exact analysis by this approach is still difficult because of the nonlinearity of the equations. However, many approximate methods of solution have been devised (see [3.30] and HAWTHORNE in [Ref. 3.31, p. 238]) and are in use today (see HORLOCK in [Ref. 3.32, p. 322]). A very simple approximate method by SQUIRE and WINTER [3.33] is commonly used for a first estimate of the secondary flow in a duct of nearly constant area that turns the main flow (core) through a total angle $\Delta\beta$ in one plane ($x-z$ say), i.e.,

$$\Omega_x - (\Omega_x)_1 = -2(\Omega_z)_1 \Delta\beta \qquad (3.9)$$

$(\Omega_z)_1 = (-\partial U/\partial y)_1$ is the transverse vorticity distribution at inlet, Section 1. The change of streamwise component of vorticity from that given at Section 1 allows one to calculate the change of secondary flow using $\Omega_x = (\partial W/\partial y - \partial V/\partial z)$. In this analysis, streamwise gradients, $\partial/\partial x$, are usually ignored so that the continuity equation at any section along the duct is

$$\frac{\partial V}{\partial y} + \frac{\partial W}{\partial z} = 0. \qquad (3.10)$$

This equation permits definition of a "secondary flow stream function" $\psi_s (V = \partial\psi_s/\partial z, \; W = -\partial\psi_s/\partial y)$. Thus, given $\Omega_x$, the secondary streamline pattern is obtained from solution of Poisson's equation

$$\frac{\partial^2 \psi_s}{\partial y^2} + \frac{\partial^2 \psi_s}{\partial z^2} = -\Omega_x = [2(\Omega_z)_1 \Delta\beta - (\Omega_x)_1]. \qquad (3.11)$$

Implicit in this approximate analysis are the assumptions that the main flow turns the corner, without change, on nearly circular-arc streamlines and that the secondary flows are too weak to distort the Bernoulli surfaces (surfaces of constant total pressure). The secondary flows are therefore only small perturbations, an assumption that is rather poor when $\Delta\beta$ is larger than 10 to 30 degrees. See HAWTHORNE [Ref. 3.31, p. 238] for further discussion, other limiting approximations, and applications to internal flows.

Methods of the type just discussed have often been applied in practice, as have attempts to solve the turbulent three-dimensional boundary layer equations (Section 2.4) near the walls of curved ducts (Zones I and II), but there have, to date, been few attempts at solution of the full turbulent flow problem in a bent duct where the boundary layers interact (Zone III) and where the flow becomes fully developed (Zone IV). PATANKAR et al. [3.34] have provided a solution for the geometrically simple case of flow in a pipe of inside radius a with its central axis bent in one plane[9] on a circle of constant radius R, Fig. 3.3a. They used a two-equation turbulence model (Section 5.3) obtained from [3.35] where the eddy viscosity model

$$\sigma_{ij} = (v_T + v)(\partial U_i/\partial x_j + \partial U_j/\partial x_i) \tag{3.12}$$

is used, and

$$v_T = 0.09(1/2\overline{u_i^2}/\varepsilon). \tag{3.13}$$

Here $1/2\overline{u_i^2}$ or $1/2\overline{q^2}$, the turbulent energy, and $\varepsilon$, the rate of turbulence energy dissipation, are solved for by means of empirical transport equations; see Chapter 5. In addition, the method of GOSMAN and SPALDING (paper 19 in [3.36]) which decouples the main flow pressure gradient from $V_s\bar{p}$, the crossflow gradient, is used to render the equation set parabolic. Several results are obtained with rather good accuracy. In one case, data of [3.37], flow enters a weakly curved pipe, $R/a = 24$, in fully developed turbulent flow from a straight pipe at Reynolds number of $2.36 \times 10^5$. Contours of mean velocity head are, when compared to the data, rather well predicted up to 180° of turn, Fig. 3.3b. Predicted secondary flow speeds are reasonable; $Q_s/\tilde{U}$ is of the order of 1/10. Other comparisons are made to main flow velocity profile data from some fully developed flow measurements in long helically coiled tubes [3.38, 39], and computed friction factors are compared to the extensive measurements of ITO[3.40] in long curved pipes of $R/a$ as low as 16.4. The latter cases showed that the turbulence model

---

[9] The solution is also valid for a long pipe in a shallow helical coil as long as $R/a \gg 1$.

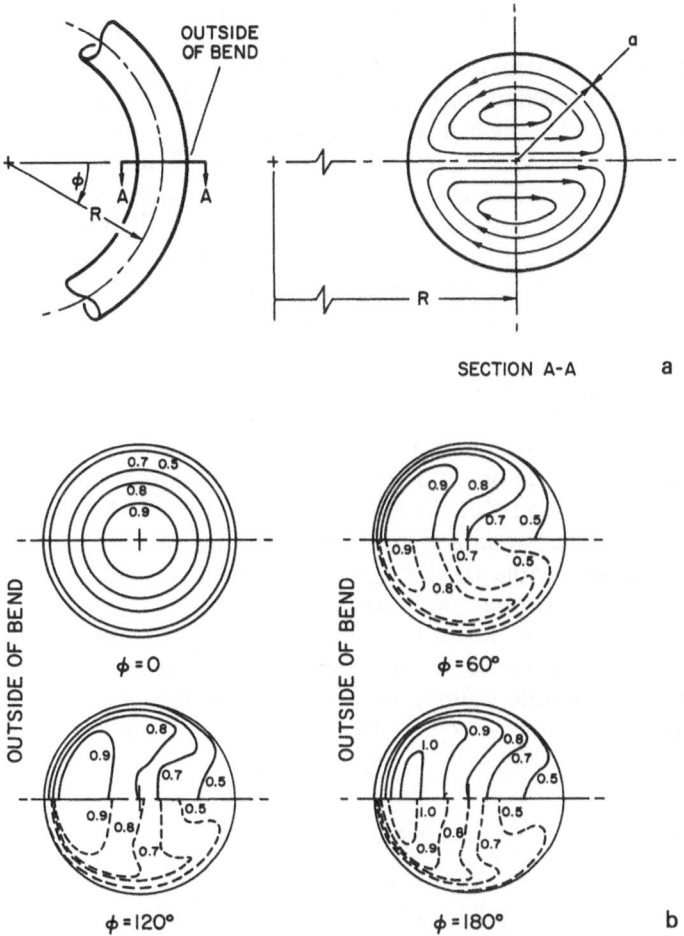

Fig. 3.3 a and b. Flow in a curved pipe. (a) Geometry and secondary flow streamlines, (b) contours of constant mean velocity head: data [3.37] compared to predictions (solid contours) [3.34] at several axial stations $R/a=24$, Reynolds number, $2.36 \times 10^5$

is not altogether satisfactory since computed friction factors were about 8 % lower than measured values for the curved pipes, but very close to the data for straight pipes. The method used here will fail when $R/a$ becomes small, say less than 3 to 6, and flow separation, back flow and reattachment occur. Since calculation methods employing an isotropic eddy viscosity like (3.13) cannot simulate stress-induced secondary flows the latter must have been negligible, compared to secondary flow of the first kind, in the above cases.

*Corner Flows and Secondary Flows of the Second Kind*

Secondary flows of PRANDTL's [3.26] second kind (those driven by gradients of $-\varrho\overline{u_i u_j}$, mechanism 4 above) are important for turbulent flows of class (ii), slender shear flows along straight edges and corners; see Fig. 3.4a. Here, two wall boundary layers on flat planes are interacting in a zone of dimensions about $1\delta$ to $2\delta$. The "streamlines" shown are the projection on the secondary flow plane $(y-z)$ of the actual streamlines. The secondary flow moves inward along the corner bisector and tends to sweep the isovels (lines of constant $U$) toward the corner, as first noted by NIKURADSE [3.41]; see Fig. 3.4b.

There exist many data sets on isovel distortion in turbulent corner layers for boundary layer cases [2.302, 307, 301, 3.42] and for fully developed flow in square and rectangular ducts and channels [3.43, 2.298, 3.44–49]. Fewer workers have tried to measure the very weak, secondary flow velocity vectors. It is seen in all straight corner data that $Q_s/\tilde{U} \lesssim 1/100$. LAUNDER and YING [3.46], using their own data from both rough and smooth walled square section ducts and the data of GESSNER and JONES [3.51], were able to demonstrate that the secondary flow speeds scaled on the average wall friction velocity, i.e., $Q_s/(u_\tau)_{AV} =$ constant, independent of Reynolds number and wall roughness. Thus for smooth wall cases, as Reynolds number increases, one may expect weaker secondary flows and weaker secondary flow effects. In any case, the magnitudes of $Q_s$ by Prandtl's second mechanism are

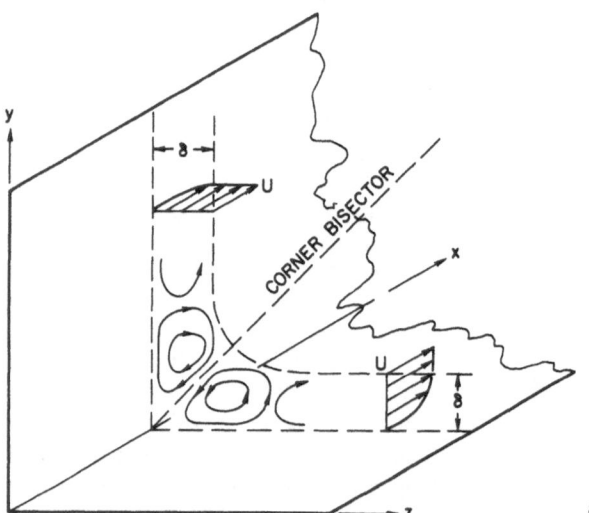

Fig. 3.4a and b. Boundary layer in a 90° straight streamwise corner. (a) Geometry of secondary flow streamlines (b) typical isovel contours

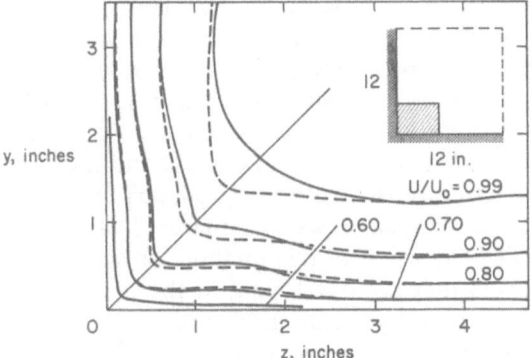

THE CORNER BOUNDARY LAYER (ATMOSPHERIC AIR
$U_0 = 56.70$ ft/sec, $\partial p/\partial x \approx 0$) DATA OF $[2.301]$

―――― TOTAL PRESSURE TUBE ISOVELS
――― HOT-WIRE (X-PROBE) ISOVELS

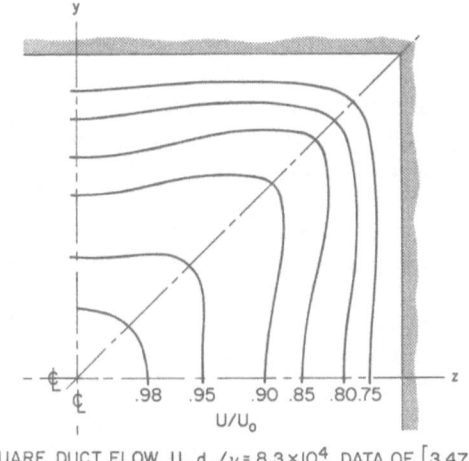

SQUARE DUCT FLOW, $U_m d_e / \nu = 8.3 \times 10^4$, DATA OF $[3.47]$    b

Fig. 3.4b

usually one order less than the $Q_s$ values caused by lateral curvature or lateral wall motion, mechanisms 2 and 3; therefore Reynolds stress driven secondary flows can frequently be neglected when flow proceeds along curved corners.

A few of the references already noted contain partial turbulence data on flow in straight 90° corner regions; the most significant data sets are in [3.44, 48, 50, and 51] for fully developed square and rectangular duct flows. For intersecting turbulent corner boundary layer flows see [2.307,

301, 3.42]. One important feature of these turbulence data is their use in formulation and testing of some modern methods for computing mean main flow, mean secondary flow, and Reynolds stresses; see WILSON et al., paper 11 in [3.36], LAUNDER and SINGHAM, paper 12 in [3.36], and [3.52–55]. All of the theoretical studies except that of Wilson et al. are based, in part, on the model equations of HANJALIC and LAUNDER [3.56]. The most carefully tested of the methods [3.54, 55] appears to give good results for fully developed flows in square [3.44] and in rectangular [3.48] ducts. The extension of this method, or any current corner flow method, to layers that are developing in the main stream direction under imposed free-stream pressure gradients is yet to be developed. It must be remembered too that secondary flow effects of the first kind, lateral curvature effects, are likely to be an order of magnitude larger than the Reynolds stress driven secondary flow. Hence, in many real cases, lateral curvature effects may dominate the corner flow region flow patterns.

### 3.1.3 Interactions in Duct Flows

Here the two principal interaction effects for internal flows i) *displacement* interaction and ii) *shear layer* interaction, will be introduced and illustrated by the steady flow of a constant-density fluid along an impermeable duct with a nearly straight centerline, Fig. 3.5. The cross-sectional shape of the duct is arbitrary, and may change with distance, $x$, along the duct centerline, but the axial rate of change of shape and of cross-sectional area, $dA/dx$, must be small so that cross-stream static pressure gradients are small compared to axial pressure gradients, i.e., $\bar{p}(x, y, z) \approx \bar{p}(x, o, o)$ and $\partial \bar{p}/\partial x = d\bar{p}/dx$. In addition, the mean velocity components are related to each other by the "slender shear layer" approximation, i.e., $|U| \gg |V| \sim |W|$ except at the duct wall. Conditions at the point of maximum flow speed at any section are denoted by subscript c, i.e., $U_c$, $\bar{p} = \bar{p}_c$, $P_c = \bar{p} + (\varrho/2)U_c^2$. The quantity $P_c$, the peak value of total pressure at each section, is constant everywhere inside an inviscid or potential core, such as the core zone between (1) and the merger point upstream of (2) in Fig. 3.5; otherwise $d\bar{P}_c/dx < 0$. From the definition of total pressure, the change of static pressure along the duct is

$$d\bar{p} = d\bar{P}_c - \varrho U_c dU_c . \tag{3.14}$$

We shall use this relationship and some of the ideas developed by SOVRAN and KLOMP on p. 270 of [3.31] to investigate the interaction effects.

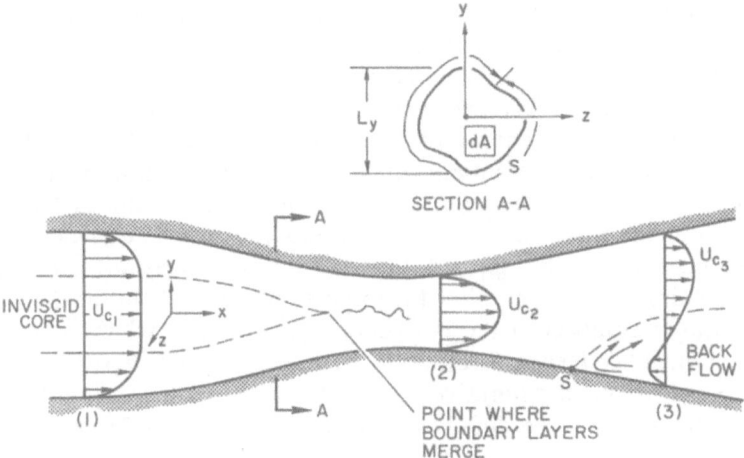

Fig. 3.5. Slender, straight duct flow

The "*blocked area*" is defined in constant density flow by

$$A_{\mathrm{B}} = \int^A \left(1 - \frac{U}{U_{\mathrm{c}}}\right) dA . \tag{3.15}$$

$A_{\mathrm{B}}$ is always less than $A$, the duct cross-sectional area, and is an important generalization of the displacement thickness concept[10]. The volume rate of flow, $Q = U_{\mathrm{c}}(A - A_{\mathrm{B}})$, is constant at all $x$. Therefore, one obtains

$$\frac{dU_{\mathrm{c}}}{U_{\mathrm{c}}} = \frac{dA_{\mathrm{B}} - dA}{(A - A_{\mathrm{B}})} \tag{3.16}$$

and using the above in (3.14) it is also seen that

$$\left[dp - \frac{\varrho U_{\mathrm{c}}^2}{(A - A_{\mathrm{B}})} dA\right] = \left[-\frac{\varrho U_{\mathrm{c}}^2}{(A - A_{\mathrm{B}})} dA_{\mathrm{B}}\right] + (d\bar{P}_{\mathrm{c}}) . \tag{3.17}$$
$$\qquad\quad (0) \qquad\qquad\qquad (i) \qquad\qquad (ii)$$

The purely geometric term caused by duct area change, $dA$, is moved to the left-hand side to make up *term (0)*, the base term which is in-

---

[10] The definition is useful and meaningful even in regions of the duct with backflow, see Zone (3) in Fig. 3.5; if the algebraic sign of $U$ is properly handled and if $S$ is length of the perimeter of the duct at a given section then $\delta^* = A_{\mathrm{B}}/S$ has the well-understood physical meaning of a (mean or average) displacement thickness.

fluenced by the interaction. *Term (i)* is the *displacement* interaction term and *term (ii)* is the *shear layer* interaction term.

In either Zones I or II of generalized duct flow, Fig. 3.1, where wall or interior shear layers have not merged, term (ii) is zero, $dP_c = 0$. The whole effect is caused by change of the blocked area along the duct. The special case of a short constant area duct, $dA = 0$, drawing flow through a nozzle from a large plenum, offers a useful illustration. Initially, (*Zone I* near $x = 0$) in the duct inlet nozzle the effects are small, i.e., $dA_B \ll dA$, and they may usually be ignored, but downstream where $dA = 0$ they become the dominant effect. Now the prediction of $dA_B/dx$ in the $dA = 0$ region requires solution of the boundary layer problem which is itself dependent on $d\bar{p}/dx$. Hence (3.16) should be solved, step-by-step down the duct, in conjunction, and iteratively, with the boundary layer equations. When the boundary layers finally merge, at say $x = x_m$, and in the fully developed region (Zone IV) where $x = x_d$, the interaction analysis is not so easy because term (ii) is no longer zero, i.e., $dP_c < 0$; *shear layer interaction* is also occurring.

To examine *shear layer interaction* in simple terms we shall discuss the special case of two-dimensional flow ($W = 0$, $\partial/\partial z = 0$) that is symmetrical about the duct centerline ($V_c(x, o) = 0$). The special momentum equation applicable to the duct centerline is

$$\varrho U_c \frac{dU_c}{dx} + \frac{d\bar{p}}{dx} = \left(\frac{\partial \tau}{\partial y}\right)_c = \frac{d\bar{P}_c}{dx}. \tag{3.18}$$

In Zones I and II (the inviscid core) the transverse shear stress gradient at the centerline is zero, which is not the case downstream of $x_d$ in Zones III and IV. In Zone IV $dU_c/dx = 0$ however, so conditions are somewhat simplified for this flow region, but in Zone III, all terms are important.

Figure 3.6, where the distribution of centerline speed is shown in relation to the four flow zones, illustrates a basic difference between laminar and turbulent flow. In laminar flow $U_c$ rises monotonically to its final asymptotic value, but in the turbulent case $U_c$ overshoots its fully developed value by a small ($\sim 5\%$), but significant, amount in Zone III. The most recent data on this effect were obtained by DEAN [3.57] at Reynolds number $Qh/Av = 10^5$; he reviews previous work.

According to (3.17), $A_B$ must also attain a peak where $U_c$ attains its peak value. The displacement interaction effect changes sign at this point, a very curious effect. SHARAN [3.58] illustrates similar displacement effects for developing turbulent flows in constant diameter pipes, and also points out the importance of free-stream turbulence and boundary-layer transition.

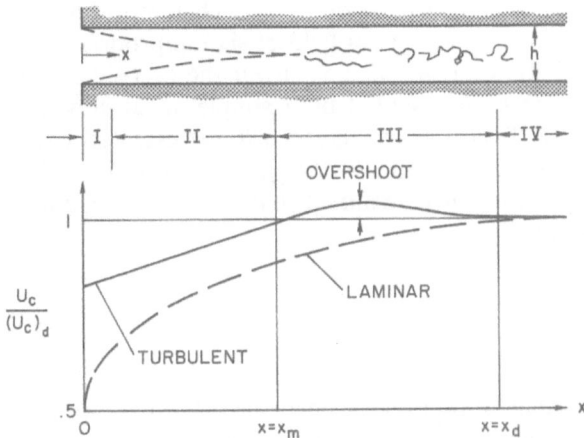

Fig. 3.6. Characteristic core region (centerline) flow speed distributions for inlet regions of constant area ducts: not to scale

Even more curious is the fact that the centerline shear stress gradient and the axial gradient of centerline total pressure, the shear layer interaction effect, also both rise in Zone III to values greater than their final, fully developed values. This can be seen with the aid of (3.18) and the known fact, see [3.57, 58] etc., that the static pressure gradient, $d\bar{p}/dx$, is nearly constant at its fully developed value from the start of Zone III, for all $x \geqq x_m$. Direct measurement of $\tau(y) = -\varrho\overline{uv}$ [3.57] also confirmed these facts.

Simple eddy-viscosity formulae will not predict these effects. To do this, some totally empirical changes would have to be made in the distribution of either eddy viscosity or mixing length at the outer edges of the boundary layer from the start of Zone III and all the way down into Zone IV.

Although the method of [3.21, 56] ought to reproduce the broad features of shear layer interaction, the only theoretical model proposed to date that treats the interaction explicitly is that of BRADSHAW et al. [3.59], and its extension by MOREL and TORDA [3.60] to free turbulent mixing. Problems similar to those outlined above occur in the regions where the free shear layers, on the edges of an inviscid jet core, merge to form the "fully developed" far field jet. The essential feature of this theoretical approach is the hypothesis that overlap of the boundary layers, from the two sides of the duct say, occurs in a "time-sharing" fashion. That is, in Zones III and IV at a fixed measuring station near the duct centerline, one tends to observe large "eddies" or lumps of fluid, from both wall shear layers, pass by in sequence; first a lump from side (+) and then a lump from side (−). Each lump appears to carry, or convect, with it the shear stress field of sign appropriate to its own side, i.e., $\tau_+$ from the (+) side and $\tau_-$ from the (−) side. In the time-mean equations this is represented by a simple linear superposition of the shear stresses from the two layers *without* essential change of the turbulence structure. Thus, even though superposition of the velocity field would be wrong, it is assumed that shear stresses may be added algebraically so that

$$\frac{\partial \tau}{\partial y} = \frac{\partial}{\partial y}(\tau_+ + \tau_-)$$ (3.19)

when used in the equations of motion for each shear layer. DEAN [3.57, 61] provides direct physical evidence for the validity of this model by means of conditionally sampled turbulence measurements. One wall layer was kept slightly warmer than the other so temperature could be used to mark lumps of fluid and allow instantaneous sample identification; eddies from wall (+) were slightly warmer than those from wall (−). Indirect evidence for the appropriateness of this interaction model is also provided in [3.59] by the rather good predictions of the mean flow and wall shear stress for the developing diffuser flows (cases where $dP_c < 0$ everywhere) of [3.62, 63].

### 3.1.4 Separation and Reattachment

The engineering importance of separation in design of systems which must avoid separation while operating close to a separated state, the calculation of effects of separation and separated flow on passage total pressure losses, etc., are well known and are discussed as part of an excellent state-of-the-art (circa 1960) review on separation with emphasis on internal flow, by DEAN, Section 11 in [3.64]. Only a few basic ideas, and a minor update of this field can be developed here. Basic discussions of separation and reattachment phenomena, with particular reference to small separated-flow "bubbles" on airfoils, are given in Subsection 2.3.6. Here we deal with more extensive regions of separation, especially those behind salient edges or sharp corners. (Fig. 3.7).

The separated flow field is sometimes called the "separated zone", "stalled region", "recirculation region" or, misleadingly, "dead-water region".

For three-dimensional flow, there is still considerable controversy about the ways of defining a stalled flow region (see TAYLOR [3.31, p. 320] and [3.65, 66]) and the various means of identifying the kinematic

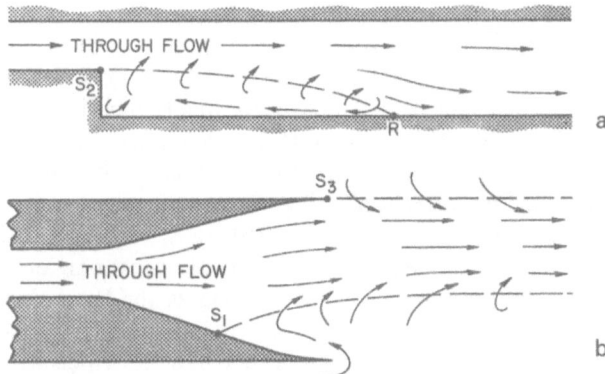

Fig. 3.7a and b. Separated flows. (a) Separation and reattachment in a duct with sudden expansion of area, (b) separated diffuser flow discharging to a large plenum or infinite atmosphere

geometry of the stalled zone (separation and reattachment lines, free-shear layer sheets, etc.) are not today totally settled (see [3.65, 67–69] and LIGHTHILL [Ref. 3.70, p. 60]). We shall not tackle these general questions here, but rather stick to two-dimensional, mean flow, where separated flow fields are easy to define.

Figures 3.7a and b illustrate the several types of steady, separated, two-dimensional internal flows. A separated zone, or stalled region, is that part of the whole flow field adjacent to a free shear layer that separates the zone from the through flow, i.e., the flow that passes through the system from upstream[11]. A free shear layer starts at a separation point (S) and, for practical purposes, may (Fig. 3.7a) or may not (Fig. 3.7b) end at a reattachment point (R). The mean speed of the fluid in a separated zone is usually at least one order of magnitude less than the through flow speed though larger values are found in axisymmetric base flows. The free shear layers which are denoted by dashed lines in the sketches are, in fact, regions of finite width which may often be treated analytically as "thin" or "fairly thin" shear layers. The single most important practical feature of turbulent free shear layers is the rapid rate at which they entrain fluid from the separated zone. The growth rate of a turbulent jet or mixing layer (Fig. 2.2) is of the order $\partial\delta/\partial x \approx 0.1$, many orders of magnitude higher than the rate at which fluid could be captured by a laminar diffusion process. Further discussion of specific separated flows is given in Sections 3.3 and 3.4. Some excellent photo-

---

[11] Flow that enters from downstream (case of no reattachment) is defined to be part of the separated flow even though entrained by the through flow.

graphs that show the complexity of separated and stalled internal flows are shown by DEAN in Section 11 of [3.64] and by ACKERET [3.31, p. 1].

The process of separation is the total of all mechanisms that occur in the wall boundary layer upstream of the separation line (which in two-dimensional flows is called a "point"). In turbulent flow, separation lines, or points, exist either momentarily or as a time-mean or ensemble-mean (average) condition. It is in the latter sense that we use the word point. In Figs. 3.7a and b, separation points are denoted as $S_1$, $S_2$, and $S_3$. Points of type $S_1$ are those which generally cause us the most concern. They occur on flat or slightly curved surfaces. As long as the Reynolds number is high enough to sustain turbulent flow, it is usually easy to predict that separation will always occur at points like $S_2$ which are located on sharp, convex corners or edges. Likewise, flow must separate at a thin cusped trailing edge such as at $S_3$, in Fig. 3.7b. In the latter two cases, $S_2$ and $S_3$, we seldom have to resort to boundary layer theory to locate their position. However, in some cases, the boundary layer properties of the separating flow at the separation point are needed since they play a vital role in the subsequent downstream development of the free shear layers [3.71], wakes and even in the process of reattachment [2.147].

The process of reattachment of turbulent free shear layers has at least as many peculiar effects as the process of separation. The reattachment process entails all mechanisms occurring in the wall-attached shear layer downstream of a reattachment point, R in Fig. 3.7a, until some flow state can be identified where all "memory" of the severe effects of reattachment on turbulence structure is gone. This process has been little understood in the past. In [2.147] most of the previous work is reviewed and it is shown clearly that a distance as long as 25 times the layer thickness at the reattachment point is required before the "memory" of the effects of reattachment on structure is gone.

Most concern and research in internal flow has centered on the process of separation from a flat or slightly curved wall, with a separation point of type $S_1$ in Fig. 3.7a. It is usually accepted by most workers today that many modern methods of calculating turbulent boundary layers, given the adverse gradient type of free-stream pressure field, will allow calculation of boundary layer properties close to, but not quite at, the separation point. However, in the region within one or two layer thicknesses of the point, boundary layer methods fail, and resort to a better approximation, if not full use of the Navier-Stokes equation, is required. In addition, as pointed out by various workers (for example [2.105, 120]) "two-dimensional" turbulent separation on a flat wall is not two dimensional, nor can a definite point of separation be located except in the statistical sense. Local, intermittent backflow occurs near the wall well ahead of the point where the statistical average of zero wall shear stress ($\bar{\tau}_w = 0$) occurs[12]. In any case, methods do exist (see summary in [3.66]) for separation prediction if the free-stream pressure gradients are accurately known.

In internal flows especially, strong interactions due to downstream blockage of a flow passage by a stalled region can influence the inviscid core pressure gradients many $\delta$'s upstream of separation. Hence solution of inviscid core flow (elliptic for subsonic flow) must, in some cases, be carried out simultaneously, and iteratively with the separating boundary layer flow. Some early, but recent attempts at "whole field" solutions of separated, internal flows are discussed in Section 3.4. In many cases the core may not be inviscid and one may have a duct full of shear layer.

Finally, internal flows appear to contain as important problems turbulent separating flows that defy the usual categorization of separation attempted above. These are the flow patterns denoted as *transitory stall*, a highly unsteady and three-dimensional flow that often affects the whole of a diffusing duct. Until twenty years ago flows of this class were not recognized, or were ignored. KLINE [2.119] was the first to describe these flow patterns in a clear and accurate manner and to recognize their technical importance in internal flows. The research from which these ideas come and other work in the area of turbulent diffuser flow (see Section 3.3) testify to the importance of careful, direct study of turbulent flow by flow visualization, i.e., by use of markers such as smoke or dye. In fact, an important breakthrough in basic understanding of the structure of turbulent boundary layers [3.73], the discovery of wall layer streaks and bursting, occurred in the Stanford program as a result of observations and techniques originally developed to study internal flow in diffusers. Even today we see fallout from these past events in the research by FALCO [2.63, 3.74] who uses smoke visualization and hot-wire anemometry for conditional sampling studies of the large eddy structure in boundary layers.

## 3.2 Fully Developed Flow in Pipes and Ducts

Summary discussions and interpretations of known data on fully developed turbulent flow in pipes and ducts may be found in [1.3, 7, 9, 22]. Numerous textbooks on engineering fluid dynamics review this subject,

---

[12] Currently, $\bar{\tau}_w = 0$, appears to be the most common working definition of two-dimensional turbulent separation. Strictly speaking separation is the point where the layer departs from the surface; $\bar{\tau}_w$ is not necessarily equal to zero at separation in three-dimensional flows. SOVRAN [3.72, p. 447] surveyed separation criteria in 1968. His discussion, and the comments of others that follow it are still pertinent today, and the field is still in need of research (see also [3.66]).

Table 3.1. Friction coefficients for fully developed flow in smooth-walled pipes [3.75] and wide, two-dimensional ducts of constant area [3.79] (Aspect ratio > 7)

*Power law form:* $c_f = MRe^{-0.25}$,

*Optimum log-law form:* $c_f^{-1/2} = N_1 \ln(Rec_f^{-1/2}) + N_2$,

  where $c_f \equiv \tau_w / \frac{1}{2}\varrho\bar{U}^2$ and constants $\kappa$ and $C$ from log-law of the wall, (1.49), are matched to values $N_1$ and $N_2$. Also, $Re \equiv \bar{U}L_T/\nu$.

| Quantity | 2-D Duct | Pipe |
|---|---|---|
| $L_T$ | $h$ (depth) | $d = 2a$ (diameter) |
| $\bar{U}$ | $\dfrac{1}{h}\int_0^h U\,dy$ | $\dfrac{1}{a^2}\int_0^a U r\,dr$ |
| *Constants* | | |
| $M$ | 0.073 | 0.079 |
| $N_1$ | 1.72 | 1.74 |
| $N_2$ | 0.47 | $-0.40$ |
| $\kappa$ | 0.41 | $0.40^a$ |
| $C$ | 5.17 | $5.00^a$ |
| *Tested Re range* | | |
| Power law | $6 \times 10^3 - 6 \times 10^5$ | $3 \times 10^3 - 1 \times 10^5$ |
| Optimum log-law | $6 \times 10^3 - 6 \times 10^5$ | $3 \times 10^3 - 3 \times 10^6$ |

[a] Ref. [3.75] recommends these values; the differences from the duct values are probably not significant. According to [3.75], effective values giving the best log-law fit over the whole radius (neglecting the viscous sublayer) are $\kappa_e = 0.406$, $C_e = 5.67$ for the tested Re range.

too. A recent review of the practical aspects of developed turbulent flow is in a book by A. J. REYNOLDS [3.75]. Rather than review the extensive literature on the topic here, we shall note some recent advances and try to guide the reader to material not often discussed in standard texts. For some nomenclature see Table 3.1 and footnote[13].

First, it is useful to be able to known when, and under what circumstances, a state of fully developed turbulent flow is achieved. These are not trivial questions, nor are the answers unique. Workers with different objectives will give different answers. For example, a pipe or duct flow at given Reynolds number $Re = U_m d_e/\nu$ is usually said to be fully developed for $Re > 3000$ and for $x_d/d_e = 20$ to 40 if the sole objective is to know the friction factor, $c_f$ (see Table 3.1 for definition). The variation in reported values of $x_d/d_e$ depends primarily on inlet flow condi-

---

[13] For internal flow in constant area ducts, $Re$ is conventionally based on $d_e$, the equivalent (hydraulic) diameter, $4A/S$, and the mean flow speed $U_m = Q/A$. The speed $\bar{U}$ used in Table 3.1 is identical to $U_m$ for round pipes and for ducts of very large aspect ratio, $AS \rightarrow \infty$. For finite width ducts, $U_m < \bar{U}$ by a small amount due to end-wall layer displacements.

tions (turbulence level and scale) and inlet geometry (smooth or abrupt nozzle). However, as shown by DEAN [3.57] for two-dimensional ducts, a state of fully developed turbulent flow structure may not be achieved until $x/h$ exceeds a value of 80 for the case $\bar{U}h/\nu = 10^5$. In DEAN's experiment, the inlet conditions were ideal: the free-stream turbulence was very low, the flow entered the duct by a smoothly faired nozzle, and wall boundary layers were tripped at the start of Zone II.

It has often been assumed that the fully developed state may be attained for rather short inlet duct lengths if "disturbed" or "tailored" inlet conditions are provided to give high turbulence levels and rapid boundary-layer growth in Zones I and II. This belief is confirmed if either $c_f$ or the mean velocity profile shape is used as the sole criterion for entry into Zone IV, the fully developed region. However, recent data [3.76] on pipe flow and [3.77] on two-dimensional duct flow casts doubt on the inlet "disturbance" method as a means for reducing inlet-length requirements for establishment of fully developed turbulence structure. It appears today that the only safe rule is an extension of one proposed by PATEL [3.76]. Namely, *a fully developed turbulent state is assured when all structure-related statistical quantities ($\overline{u_i^2}$, spectra, skewness, flatness, etc.) at a given x-station do not change with x and, in addition, do not change when the flow system's inlet conditions are altered in a significant way*. In practice, this condition or state may be achieved by strong suppression of large-scale (time and size) upstream fluctuations, by careful attention to inlet-nozzle design, by uniformly tripping inlet wall boundary layers, and by testing at $100 L_T$ downstream of inlet for $Re \lesssim 10^5$.

Flow in long, smooth-walled ducts of high aspect ratio, e.g., two-dimensional ducts, is considered by many workers the sine qua non for study of fully developed turbulence. The 42 references on ducts of aspect ratio greater than 7:1 studied by DEAN [3.78] are believed to be comprehensive as of December 1974. Twenty-six of these references gave DEAN [3.79] useful quantitative results that could be correlated, and from which conclusions were drawn. For example, the two-dimensional friction-factors shown in Table 3.1 were taken from [3.79]. DEAN also correlated shape factor, $H = \delta^*/\theta$ and $U_c$, centerline mean velocity. Both were found to be slowly varying functions of $Re$. In the range $6 \times 10^3 < Re < 6 \times 10^5$, the 26 data sets fit the following formulae to the expected experimental uncertainty. A good fit to the centerline velocity is

$$\frac{U_c}{\bar{U}} = 1.25 \left(\frac{\bar{U}h}{\nu}\right)^{-0.0116}. \tag{3.20}$$

Both velocity defect ratio and Clauser shape factor $G$ (2.18) were found to be nearly constant, i.e.,

$$\frac{U_c - \bar{U}}{u_\tau} = 2.64 , \tag{3.21}$$

$$\frac{U_c}{u_\tau}\left(\frac{H-1}{H}\right) \equiv G = 6.0 . \tag{3.22}$$

As in the case of boundary layers, the complete velocity profile can be fitted by adding a wake function to the universal law of the wall, (1.49), with $\kappa = 0.41$, $C = 5.17$. DEAN deduced a generalized law of the wall-wake,

$$\frac{U}{u_t} = \frac{1}{\kappa}\log_e\frac{u_\tau y}{\nu} + C + g\left(\Pi, \frac{y}{\delta}\right). \tag{3.23}$$

He chose the wake function,

$$\kappa g\left(\Pi, \frac{y}{\delta}\right) = (1+6\Pi)\left(\frac{y}{\delta}\right)^2 - (1+4\Pi)\left(\frac{y}{\delta}\right)^3 \tag{3.24}$$

to satisfy four boundary conditions based on a precise description of the outer layer of a boundary layer and conditions at the duct centerline (where $y = \delta$). This empirical formula appears to be an improvement over the form $(\Pi/\kappa)w(y/\delta)$ discussed in Subsection 2.3.1, because it provides a more accurate fit to the true boundary conditions on the wake function, yielding $\partial U/\partial y = 0$ at $y = \delta$ for all $\Pi$. For two-dimensional duct flows, $\Pi = 0.14$. DEAN also demonstrates that (3.24) provides a good wake function for a variety of boundary-layer cases, for both favorable and adverse free-stream pressure gradients. An identical function has been proposed by FINLEY and by GRANVILLE.

Original data on turbulence-intensity profiles, spectra, and other statistical turbulence quantities for fully developed two-dimensional duct flow are obtainable from a number of sources [3.57, 77, 80–82]. References [3.57] and [3.77], in addition, provide extensive comparisons of data from various sources. Some of these comparisons, particularly profiles of streamwise fluctuation[14] $\overline{u^2}$ and mean velocity, demonstrate possible minor errors in some data sets often taken as standards in the field, e.g., [3.80, 82]. For example, at the channel centerline for $(\bar{U}h/\nu) = 10^5$, values of $(\overline{u^2}/\bar{U}^2)$ from both [3.57] and [3.81] are in good

---

[14] Profiles of $\overline{u^2}$, since they can be taken with a single, normal hot wire, are the commonest turbulence data available, and probably the most accurate.

($\pm 6\%$) agreement, whereas those of [3.80] and [3.82] are 35% lower. Differences in inlet conditions and flow-development lengths, $x_d/h$, does not help account for these discrepancies.

Fully developed turbulent flows in smooth-walled pipes have received nearly as much attention as two-dimensional duct flow. The measurements by LAUFER [2.28], an oft-quoted standard set of turbulence data, have recently been supplemented by those of [3.83–85].

The pipe-flow data by BREMHORST and WALKER [3.85] tend to verify the qualitative "burst" model of the wall-layer flow structure for production of turbulence that was proposed in 1963 by RUNSTADLER et al. [3.87]. This "burst" model was developed from flow visualization experiments since, at the time, conventional $\overline{uv}$ hot-wire data could not be obtained in the inner parts of the wall layer. BREMHORST and WALKER'S $\overline{u^2}$, $\overline{v^2}$ and $\overline{uv}$ profiles and spectra were obtained down to $y^+ = 6.5$ by a new hot-wire anemometer technique described in [3.86].

Pipe and duct flows are considered by many as the easiest and cleanest flows for use as laboratory calibration standards and for use in development work on new measurement equipment. Reference [3.88] illustrates the use of pipe flow as a means to study problems of turbulence measurement in water by laser-Doppler velocimetry. With due precautions and attention to inlet conditions and development length (see discussion above and [3.77]), it is felt that universal standards could be provided by a pipe-flow apparatus.

Most practical engineering use of fully developed flow data has centered on estimations of mean friction factor, $\bar{c}_f$, and mean heat-transfer coefficient, where

$$\bar{c}_f = 2\bar{\tau}_w/\varrho U_m^2 = -(dp/dx)d_e/2\varrho U_m^2 . \tag{3.25}$$

Local values of $\tau_w$, wall shear stress, and $c_f = 2\tau_w/\varrho U_m^2$, local friction coefficient, may vary about the circumference of the duct. Since most real ducts are not smooth-walled, circular pipes or other easily established geometries, engineers rely on extensive sets of carefully established data such as those of KAYS and LONDON [3.89] for design data on specific geometries. The roughness results of Subsection 2.3.7 are applicable to ducts as well. Where high precision is not required, use is often made of circular pipe formulae where $d_e$, the equivalent diameter $(4A/S)$, is used to determine $Re$. Recent methods, such as [3.90], provide better estimates for $\bar{c}_f$ for ducts of complex cross-sectional shape and for flow along rod bundles such as the fuel rods that occur in the cores of nuclear reactors.

An important practical class of flows is fully developed flow through annular passages. REHME [3.91] provides an up-to-date review and some original data on turbulent flow through annuli with concentric core tubes. KACKER [3.92] studied the effects of eccentric core tubes and dual core tubes. He shows the important effects of the weak secondary flows $(Q_s/\tilde{U} \approx 0.01)$ generated by transverse gradients of turbulent shear stresses (mechanism 4, Subsection 3.1.2) in a passage with asymmetric shape. Even though methods such as those of [3.90] give adequate prediction of mean friction factor, $\bar{c}_f$, crossflow momentum transfer causes significant errors in the circumferential distribution of local $c_f$ when the assumptions of these methods, such as the use of $d_e$ as an effective diameter, are used to predict local $c_f$. The data show that even very weak secondary flows cause $c_f$ to be much more uniformly distributed than expected—see distributions of $c_f$ in the corner-flow data in [2.302, 3.45, 47]. Peripheral distribution of surface heat-transfer rates is undoubtedly affected by weak crossflows in the same way, a fact that can have severe consequences in a heat exchanger operating near burn-out heat flux if not properly accounted for.

The effects of rotation of either inner core tube or outer pipe of a long, concentric, annular duct with axial flow offer interesting illustrations of the stabilization effects discussed in Subsection 3.1.1. In a classic experiment on annular flow with a rotating inner cylinder, KAYE and ELGAR [3.93] showed the destabilizing effects of rotation, and they mapped out the four possible flow regimes: laminar, laminar plus Taylor vortices, turbulent, and turbulent plus Taylor vortices. Recent experiments [3.94] in the turbulent regime clearly illustrate destabilization effects by showing that axial component of friction factor increases with inner pipe rotational speed, a condition that would not occur if the destabilization mechanism were not present. The converse case, stabilization and the consequent reduction of turbulence stress, friction factor, and heat-transfer rates, is well illustrated in a number of references on flow through long pipes with rotating outer walls—see [3.95, 96].

As a final illustration of practical, fully developed turbulent flows, consider two-dimensional duct flows where one wall of the duct moves at steady speed relative to the other, either in or against the flow direction. With zero value of $d\bar{p}/dx$, this is the case of turbulent Couette flow. In general, $(d\bar{p}/dx \neq 0)$ the practical application is the analysis of high-speed hydrodynamic and hydrostatic lubrication. Research in this area is briefly summarized in [3.97] where some original $\overline{u^2}$ profiles and spectra data for the case of zero net flow are also presented. A mixing length analysis that requires zero turbulent shear stress at the peak of the velocity profile is also given in [3.97] and is shown to give fair results; but careful examination of computed and measured mean velocity profiles indicates that the basic assumption is not accurate. There are significant differences between the location of the velocity profile peak and the position of zero shear stress, as is the case in asymmetric flow in ducts with one rough and one smooth wall [3.98] and in annuli. The data in [3.97], particularly the profiles of $[\overline{u^2}/(u_\tau)^2_{\text{moving wall}}]$ versus $y/h$ provide an interesting challenge for computation by more advanced methods.

## 3.3 Flow in Diverging Ducts

Flows in diverging ducts or diffusers—ducts with static pressure rise in the flow direction—are not only of great practical importance, but also provide specific examples of some points raised in earlier sections. As noted in Section 3.1, diffusers are usually of modest length ($L_1 < 10L_2$) and thus operate with simple displacement interactions (Zone II) and/or with shear interactions (Zone III). The central problems of diffuser design are prediction and prevention of flow separation.

The literature on diffusers does not abound with results, analytic or experimental, that report the turbulence and its structure in detail. Qualitatively they combine the features of constant-area duct flows and retarded boundary layers. There exist numerous publications of a practical design nature that utilize rough correlations of data and simple analytic models for performance prediction. Many studies have given mean-flow measurements, and some authors have attempted prediction of overall performance of specific, simple geometric classes (conical, annular, or square cross sections) by use of boundary-layer theory. Only a few characteristic references of the latter types will be reviewed. Before this review begins, some general remarks on diffuser flow regimes and performance are in order.

We shall restrict attention to a simple geometric case—symmetric, straight-walled, two-dimensional diffusers, Fig. 3.8. Axisymmetric diffusers are qualitatively similar. Nearly two-dimensional mean flow may be assumed if the aspect ratio at inlet ($AS = b/W_1$) is greater than 4 to 5, and if the inlet flow at station 1 consists of a core of uniform flow at velocity $U_{c_1}$ and moderately thin ($4\delta_1 \lesssim W_1$) wall boundary layers. Under these circumstances, the flow in the diffuser depends on two of three interdependent geometric parameters: $2\theta$, the total opening angle; $AR = W_2/W_1$, the area ratio; and $N/W_1$, the ratio of length to inlet width. Neglecting compressibility effects (subsonic inlet Mach number), the flow is, in addition, defined by the inlet boundary-layer state, shape and thickness, an inlet Reynolds number, $U_{m_1} W_1/\nu$, say, and the turbulence level and structure in the core of the flow. Real diffusers may have a variety of conditions at outlet (station 2). The most common are i) an attached outlet duct, as shown in Fig. 3.8, and ii) dump discharge of the exit flow to a large volume. It is seen that, even for such a simple

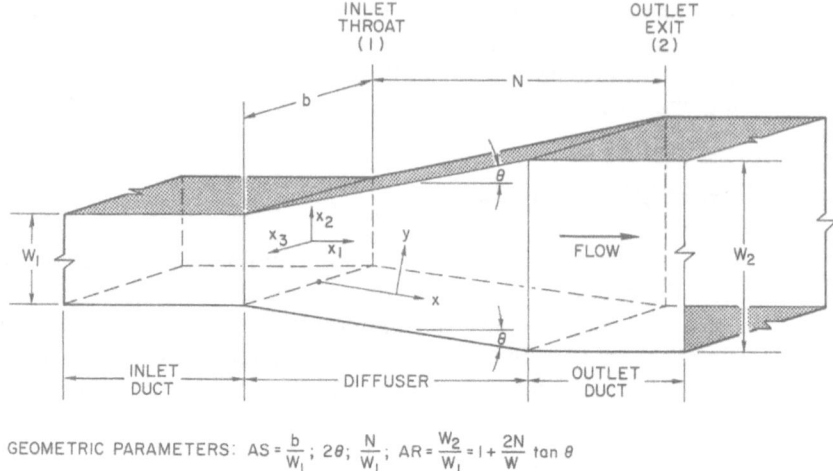

$$\text{GEOMETRIC PARAMETERS: } AS = \frac{b}{W_1}; \ 2\theta; \ \frac{N}{W_1}; \ AR = \frac{W_2}{W_1} = 1 + \frac{2N}{W} \tan \theta$$

Fig. 3.8. Two-dimensional diffuser geometry

case, an amazingly large amount of data is required to specify a diffuser flow field and thus to establish the diffuser's performance.

Figure 3.9, based largely on information from [3.99], illustrates a number of points concerning diffuser flow and performance. First, it was noted a number of years ago[15] that geometry alone is the main controlling factor for establishment of regimes of flow separation and states of stall for situations where inlet flow is subsonic and Reynolds numbers are modest, $Re \gtrsim 10^4$. Even if inlet boundary layers are laminar or transitional, geometry is the main factor that sets flow regimes, because the strong adverse pressure gradients just downstream of station 1 usually cause separation followed a short distance downstream by turbulent reattachment. On Fig. 3.9, the bold lines A, B, C, and D divide the visually observed flow regimes. The zone below line A is called the regime of "no appreciable stall", even though at line A intermittent backflow ($\sim 50\%$ of the time) occurs, by definition, over 20% of wall and corner-region boundary layers at the diffuser exit, station 2. Line B denotes the condition where steady two-dimensional separation occurs very close to the throat, station 1, on one diverging wall. A low-speed, stalled flow region hugs the separated wall, and the through flow leaves the diffuser as a jet along the opposite diverging wall. Line C denotes the upper limit of the "full two-dimensional stall" regime, which

---

[15] Initially by flow visualization work at Stanford University under Professor S. J. KLINE. The numerous references will not be cited. Important contributors to the early work include D. L. COCHRAN, C. A. MOORE, R. W. FOX, and L. R. RENEAU.

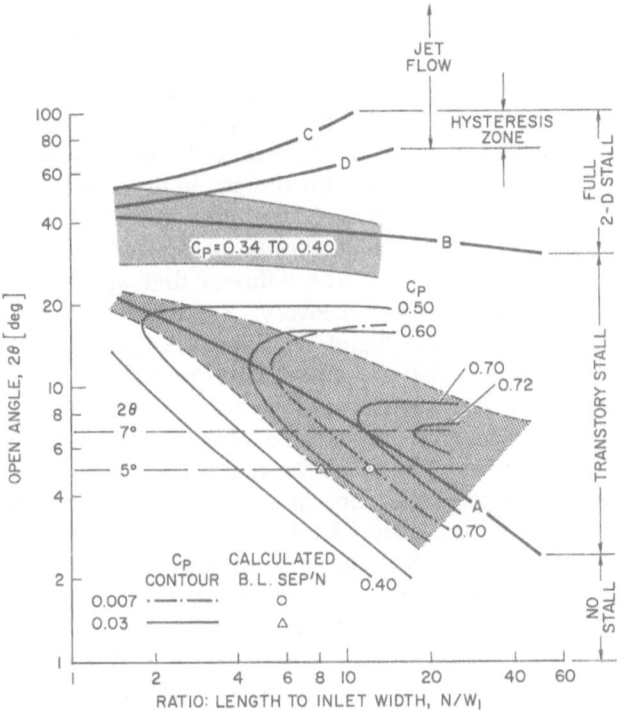

Fig. 3.9. Flow regimes and pressure recovery contours for two-dimensional diffusers of large aspect ratio, $AS > 4$

because of the so-called "Coanda" effect[16] is bi-stable. That is, the through-flow jet may be flipped to the opposite diverging wall by a strong pulse, and once the pulse is removed the through-flow jet sticks in its new location. Line D in Fig. 3.9 is the lower limit of full "jet flow" regime, where flow from the diffuser throat leaves as a jet with flow separation on both diverging walls very near station 1. The region between lines A and B is denoted as the "transitory stall" regime. Here, flow may be very unsteady, particularly in a region well above line A where $2\theta \approx 15$ to 25 degrees. It is sufficient to note that flow in the "transitory stall" regime may range from a condition of slight to moderate unsteady flow (relative to normal turbulence levels) to a state of violently unsteady flow with large-scale wash-out and build-up of stall zones in a quasi-periodic fashion; see [3.100] for details.

———————

[16] This effect is explained as the tendency of a curved flow (at diffuser throat here), with an outer free shear layer, to generate a suction pressure along the wall, to which it adheres. It is basically an inviscid-flow phenomenon. The bi-stable effect is the basis of one type of fluidic control device.

In addition to flow regime lines, Fig. 3.9 contains contour lines of constant static pressure recovery coefficient, $c_p \equiv (p_2 - p_1)/0.5 \varrho U_{m_1}^2$, a common performance parameter in diffuser technology; the effectiveness $\eta$ defined in Section 3.1 is also in common use. Note that maximum pressure recovery occurs in the regimes of "transitory stall", as does the technically important optimum design point—peak $c_p$ at constant $N/W_1$—for cases where $N/W_1 \gtrsim 4$. Technical lore and many engineering textbooks teach that $2\theta = 7°$ affords an optimum design. This indeed is the case here, but only for diffusers that are already stalled, because maximum pressure recovery, $c_p = 0.72$, occurs at $N/W_1 = 20$ in Fig. 3.9. However, the figure also indicates that a variety of design choices is available to reach the same submaximum level of $c_p$ in the unstalled regime. Thinner inlet boundary layers give values higher of $c_p$, as shown by the superposition of one line of $c_p = 0.70$ for which $B_1(= 2\delta_1^*/W_1) = 0.007$ compared to $B_1 = 0.03$ for all other lines. The latter fact illustrates an important technical consequence of the displacement interaction effect discussed in Subsection 3.1.3.

Predictions of $c_p$, up to line A, have been attempted using two-dimensional turbulent boundary-layer theory along the walls, e.g., coordinates $x - y$ in Fig. 3.8. Local blockage corrections $A_B = 2\delta^*(x_1)$ are applied iteratively to the local area $A = W(x_1)$ in order to use a one-dimensional equation such as (3.16) for calculation of local core velocity $U_c$. For low $B_1$ values, $dP_c$ is assumed equal to zero, the simple displacement interaction assumption. Results obtained by this general method, e.g., [3.101–103] are generally able to predict $c_p$ to within the limits of experimental uncertainty. The details of the boundary-layer profiles are predicted to a less satisfactory degree. All methods of the type discussed above tend to predict two-dimensional separation at a location inside the observed "no stall" regime, not at line A. Attempts to carry predictions beyond separation have not been successful. Recent calculations using a program from [3.102] with a modern version of the Hirst-Reynolds integral boundary-layer method [3.72, p. 213] give separation at the points shown on the $2\theta = 5°$ line in Fig. 3.9. Thinner inlet boundary layers separate further downstream, but never anywhere close to the value of $N$ corresponding to line A. If, for a given diffuser, $N$ exceeds the value of $x_1$ where separation is predicted, the simple displacement interaction breaks down because in reality the flow will start to become very unsteady and three dimensional.

Consider now diffusers in the "full, two-dimensional stall" regime where $2\theta = 30°$ to $50°$. Here $c_p$ is nearly constant for all $N/W_1$. The flow appears as sketched in Fig. 3.7b, except that separation point $S_1$ rests very close to the throat. Analysis based on displacement interaction may also be applied in this case. However, because of strong flow curvature in the locality of the throat, the core flow field must be solved by two-dimensional potential theory in order to obtain correct wall-pressure gradients for the boundary-layer calculation. A successful iterative method for this class of flows has been developed [3.102] and has recently been extended to axisymmetric diffusers with axisymmetric flow-separation lines in [3.104]. The same general ideas may be applied to flow in curved diffusers, as demonstrated by the calculations in [3.105], where the flow was not separated but where cross-stream pressure gradients were important. The effects of wall curvature on Reynolds stress (Subsection 3.1.1) are also likely to be important but have to date not been included in curved diffuser calculation methods.

The shaded zone in Fig. 3.9 that straddles line A is an area where simple two-dimensional, steady-flow calculation methods are very unlikely to work. For the case of $B_1 = 0.03$, the lower boundary of this special zone is set where it is estimated that steady, two-dimensional separation will first occur at the diffuser exit, station 2. The upper boundary approximates the line where $c_p$ contours become independent of $2\theta$. Above the upper boundary, it is felt that methods such as that of [3.102] may work for prediction of $c_p$ even if not for prediction of the details of the highly turbulent flow in the region below line B.

Almost all remarks made above apply strictly to flows with thin to moderately thick inlet boundary layers. Experiments show, however, that the lines that demark flow regimes are remarkably insensitive to inlet conditions, at least up to $B_1 = 0.05$, see [3.99]. Nevertheless, for boundary layers this thick, one should be applying models that introduce shear-layer interactions ($dP_c/dx \neq 0$) as well as displacement interactions. Both interaction effects are required to predict and explain performance ($c_p$) trends. With fully developed pipe or duct flow entering a diffuser (see BRADLEY and COCKRELL, [Ref. 3.36, p. A 32]), there is no doubt that shear interactions play a dominant role, as they do in the outlet pipe or duct of a diffuser, where fully developed flow may be re-established if the pipe is of sufficient length. The method discussed in Subsection 3.1.3 is the only direct approach to these types of shear-layer interactions known today. Numerous ad hoc methods for special classes of diffusers, e.g., [3.63] exist, however.

There is little turbulence data for diffusers taken in sufficient quantity and detail to require a lengthy review. The most commonly studied cases are conical diffusers where the inlet flow is fully developed or where the inlet boundary layers are moderately thick [3.106–108] and TRUPP et al. [Ref. 3.36, A 66]. An interesting set of measurements in a two-dimensional diffuser of $2\theta = 2°$ is given in [3.109]. It would be difficult to summarize all these results here. Suffice it to say that the data of several of the sets given above present a significant challenge to developers of full-field prediction methods for turbulent flows.

Finally, some mention must be made of the effects of core (free-stream) turbulence on diffuser flows, extending the more general treatment in Subsection 2.3.4. Early work at Stanford [3.110] and other clues in the diffuser literature indicate that separation may be delayed if core turbulence levels are raised well above the nominal values that are in the range $(\overline{u^2}/U^2)_{c_1} = 1$ to $10 \times 10^{-4}$. However, it is not clear if this was a level effect or if inhomogeneity and eddy-scale (size) effects, caused by the upstream sources of the turbulence, were also important. Recent work [3.111], where careful measurements of separating turbulent boundary layers were carried out under nearly isotropic free-stream flows generated at levels of about $2.5 \times 10^{-3}$, shows that separation may be moved significant distances downstream compared to otherwise equivalent cases with low free-stream turbulence level of $2.5 \times 10^{-5}$. A theoretical model to explain this effect is also presented in [3.111].

## 3.4 Fully Separated Flows

Fully separated internal flows exhibit some exceedingly complex fields that generally have much higher levels of turbulence energy and stress than do boundary layers. The variety of individual cases that could be examined is inordinately large. For a general review, see [2.109]. The main ideas in this area are to be illustrated by use of the cases shown in Figs. 3.10a and b. Another example, the case of full stall in a two-dimensional diffuser, has already been discussed in Section 3.3.

The four cases to be discussed are divided into plane, two-dimensional expansions, Fig. 3.10a, and the axially symmetric flows shown in Fig. 3.10b. All cases have flow separation at a sharp edge and, except for the ejector, have a downstream reattachment point that anchors a mean streamline to the wall. This streamline divides the separated (stalled) region from the through flow (also see Fig. 3.7a and Subsection 3.1.4 for preliminary discussion). In the case of confined jet mixing, the ejector, a separated flow zone with downstream reattachment may also be observed if the ratio of inlet core speeds $(U_{S_2}/U_{S_1})$ drops below a critical value of approximately 0.1; see Subsection 3.4.3.

Free, turbulent shear layers bound the core regions (upstream part of the through flow) in each case, and thus we begin with a brief review of pertinent information on *simple free shear flow*[17]. The upstream and far

---

[17] The term *simple* is used to denote the condition where the environment surrounding the jet or layer is a constant-pressure, infinite atmosphere that may be flowing at constant velocity, parallel to the jet (x) axis. This is consistent with the usage "simple shear layer" in Chapter 2, for nonuniformities in the environment lead to extra strain rates or shear-layer interactions.

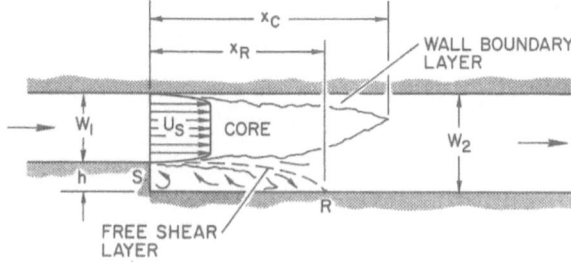

SINGLE-STEP EXPANSION

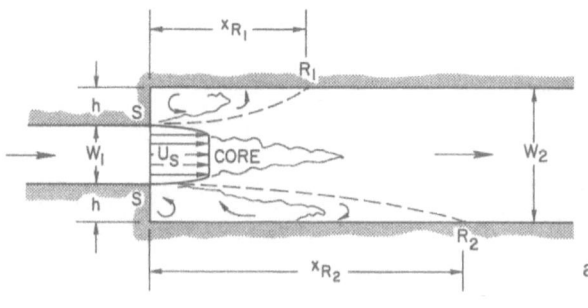

DOUBLE-STEP EXPANSION

NOTE: b  = BREADTH BETWEEN END WALLS
      AS = b/h = ASPECT RATIO
      AR = $W_2/W_1$ = AREA RATIO

Fig. 3.10a and b. Separated flows in sudden expansions. (a) Two-dimensional, fully separated flows

downstream regions of each case shown in Figs. 3.10a and b revert to simple free shear flow in the limit as its area ratio, $AR$, becomes very large. Simple free shear layers grow on the edges of the core, the free shear layers merge and interact until finally, far downstream, a simple, developed turbulent jet emerges. For references to the literature on free shear layers see Section 2.1.

The similarity properties of the mean velocity profiles of shear layers and developed jets have been a cornerstone for analysis in the field of separated internal flows. For example, a fully turbulent, simple free-shear layer with constant mean velocity on either side (Fig. 3.11) has a rapid rate of increase of thickness ($d\delta/dx \approx 1/4$) which is independent of $x$. To a first approximation, this result allows one to estimate the development length of the central core zones in the symmetric cases shown in

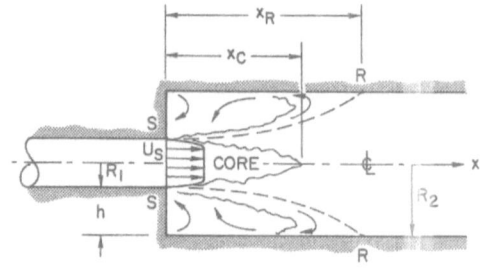

SUDDEN EXPANSION OF A PIPE

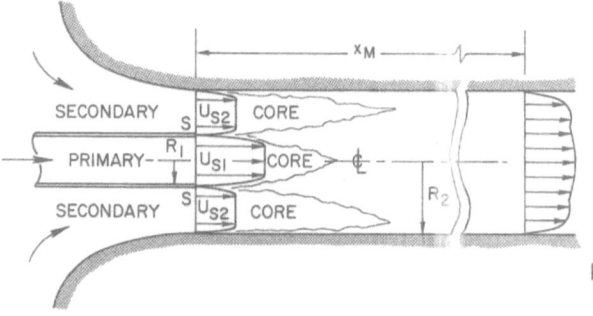

CONFINED JET MIXING (EJECTOR)

Fig. 3.10b. Axially symmetric, fully separated flows

Figs. 3.10a and b, namely, $x_c \approx 4W_1$ or $8R_1$. In many real cases, step height $h$ is of the same order of magnitude as $W_1$ or $R_1$, and therefore reattachment starts to occur at downstream locations not too different from $x_c$. Study of simple free jets shows that the classic, far-field similarity properties of turbulent jets are not attained in distances less than two to ten core-lengths, $x_c$. Thus, because reattachment occurs at $x_r$ close to $x_c$ for the majority of practical cases, one should not expect to encounter, or be able to use, classical simple far-field jet properties. Nevertheless, downstream of reattachment, some evidence of far-field jet behavior is seen in the turbulence characteristics, if not in the mean velocity profiles (Subsection 3.4.1).

Free shear layers are always evident, however, and their general properties are useful to know. First, for simple, developed shear layers and jets, it is a good approximation (attributed to Prandtl) to assume that the eddy viscosity, $v_T = (-\overline{uv})/(\partial U/\partial y)$, is proportional to the shear-layer thickness[18], $\delta$, times the maximum difference in mean velocity

---

[18] The shear layer thickness is based on the intercept points of the maximum velocity gradient line, extrapolated out to $U_h$ and $U_l$; see Fig. 3.11.

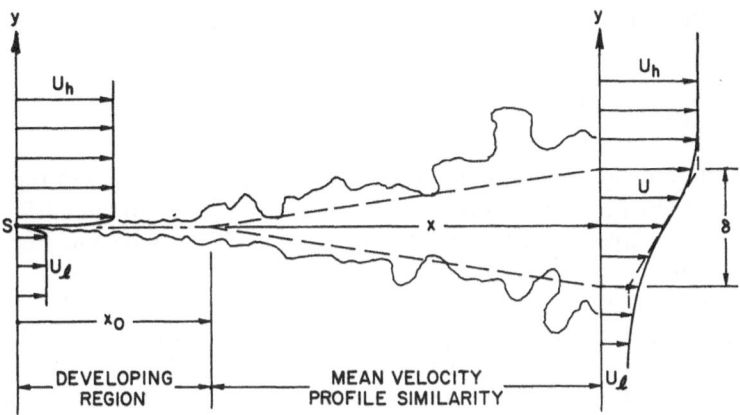

Fig. 3.11. Simple free shear layer (mixing layer)

across the layer. For the simple shear layer (Fig. 3.11), this statement takes the form

$$v_T = C\delta(U_h - U_l). \tag{3.26}$$

From (3.26), the equations of motion for a thin shear layer (1.38–41), and boundary conditions $[U(y \to +\infty) \to U_h,\ U(y \to -\infty) \to U_l]$, one can solve for the mean velocity and shear stress profiles, given a value of the empirical constant, $C$. The solution, like the data in this region, may be represented by a universal, similarity velocity-profile shape, namely

$$\frac{U - U_l}{U_h - U_l} = g\left(\frac{\sigma y}{x - x_0}\right). \tag{3.27}$$

Equations (3.26) and (3.27), with specification of the $g$ function and the "constants" $\sigma$ and $C$, are the starting point for many analyses. $x_0$, the development length, is commonly, but not always properly, set equal to zero. Both the function $g$ and the parameter $\sigma$ are functions of the ratio $r = U_h/U_l$. However $g$ is found to be only weakly dependent on $r$. Also, note that $\sigma$, the diffusion or spreading parameter, is related to layer thickness, $\delta$, by

$$\sigma = 2(x - x_0)/\delta. \tag{3.28}$$

Values of $\sigma$ in the range 9 to 12 are reported for the case $r=0$, and $\sigma$ gets bigger (thus $d\delta/dx$ decreases) as the velocity ratio $r$ increases. Equation

(3.26) and the definition of $\delta$ show that the maximum shear stress, near $y=0$, is

$$\tau_{max} = \varrho C(U_h - U_l)^2 . \tag{3.29}$$

Experiments indicate that $C$ may be as large as 0.010 which shows that the values of the turbulent stress are very large compared to values for wall boundary layers where $C \approx 0.001$ if $U_h$ is taken to be the free-stream velocity.

The parameters $C$, $\sigma$, and $x_0$ are strongly dependent on initial conditions at $x=0$, the separation point. In practical devices $x_0$ may be significant relative to $x_c$ and $x_r$, the distances over which free shear-layer theory is applicable. In fact, there are conditions where initial disturbance levels at the separation point, $S$, are sufficient to cause $x_0$ to be negative. A recent report [3.112] and paper [1.40] clarify some of the effects of initial conditions which are reviewed in [3.71]. BIRCH and EGGERS [2.2, p. 11] and DURÃO and WHITELAW [3.113] provide some recent detailed turbulence data on developing coaxial free jets where initial conditions (inlet flow and lip thickness) are seen to be very important, though in [3.114] it was shown that in simple cases the effect of initial conditions could be adequately predicted by a one-equation turbulence model (see Section 5.2). Finally [3.115] provides an example of a practical ejector flow—an aircraft jet engine thrust augmentor—where the primary jet flow is intentionally disturbed at separation to obtain a so-called "hypermixing" state where $\sigma$ is very low.

Recent research is placing more emphasis on the concept that mixing is heavily influenced by the large, well-ordered eddies that are often observed in turbulent free shear layers. These structures may be of particular importance for the improvement of theories for the developing region of the layer, and in the developed region out to $x < 10 x_0$. Reviews of some of the recent ideas and data in this field are contained in [3.116] and in a number of the papers of a recent Project SQUID workshop [2.4]; see also Section 1.10.

### 3.4.1 Sudden Expansion in a Pipe

In order to illustrate some of the basic phenomena that are particularly important in separated internal flows, a concrete example has been chosen—sudden expansion of a circular pipe of radius ratio $\beta = 0.5$ (see Fig. 3.12). This particular case was chosen because it corresponds to the situation of a rather complete set of turbulence measurements, those of CHATURVEDI [3.117]. CHATURVEDI's flow had high Reynolds number

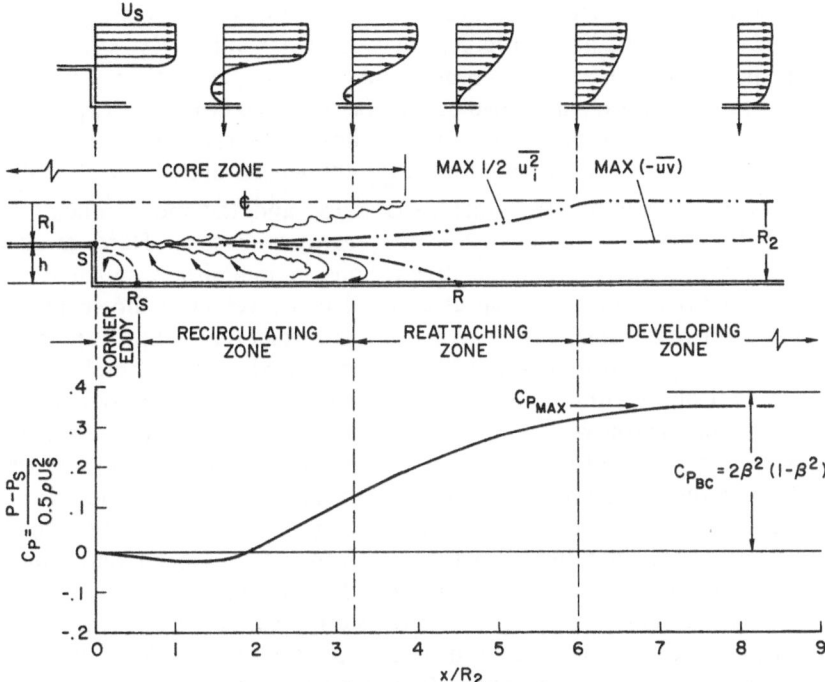

Fig. 3.12. Sudden expansion of a pipe—mean velocity profiles, flow structure, wall static pressure coefficient. Case of $\beta=0.5$, based on data of [3.117, 3.120, 3.121]

$Re_1 = U_s(2R_1)/\nu = 2 \times 10^5$ and very low Mach number at separation, where, in addition, the inlet boundary layer was thin $(\delta_s \ll R_1)$.

In this case, flow reattaches at axial location $x_r \approx 4.5\,R_2$. Downstream of reattachment, the flow redevelops, if the pipe is long enough, to a fully developed state (Section 3.2). The developed flow will be turbulent if the downstream Reynolds number $Re_2 = Re_1(R_1/R_2)$ is greater than $4 \times 10^3$, which is the case in the current example, where $Re_2 = 10^5$. At $x = 8\,R_2$, the last point in Fig. 3. 12, the turbulence structure of the flow is far from fully developed even though the mean velocity profile appears to be.

The zone upstream of reattachment has a number of remarkable features. We have already discussed the phenomena that delimit length of the core in the preceding discussion of the free shear layer that bounds the core. Here, we shall concentrate on a description of the separated region bounded by the wall and the mean reattaching streamline. The largest segment of the separated region is usually called the *recirculating zone*, where mean, back-flow velocities of order $0.1\,U_s$ are seen. Because the back-flow is so slow and because it encounters an adverse pressure gradient (see the $C_p$ line in Fig. 3.12 between $x=0$ and $1.5 R_2$), a reseparation point, $RS$, is always present. A corner eddy of exceedingly slow speed ($\leqq 0.01 \, U_s$) may also be seen between the face of the expansion and $RS$, but other types of complex flows of very low speed were observed in this narrow region for the two-dimensional step cases [2.148].

The most complicated zone to describe and the most difficult in which to make measurements is the *reattaching zone*. Here, at any instant, the velocity has large fluctuations in magnitude and may be positive or negative with respect to local mean velocity (shown in the profiles above the zone in Fig. 3.12). Also, the exact reattachment point $R$ is elusive to measure. The most commonly used method employs the surface oil-film technique (see [3.118] for details), but recently another method that employs a basic definition of reattachment[19] has been employed by the authors of [3.119]. Very little is today known of the details of the structure in this zone, although BRADSHAW and WONG [2.147] have presented interesting conjectures concerning the length scales of eddies that flow downstream from reattachment and which have profound effects on flow in the *developing zone*.

The lower portion of Fig. 3.12 shows a typical curve of $C_p$, the wall static pressure coefficient versus distance from the step face. Wall pressure data are the most commonly obtained results for sudden expansions (see ACKERET [3.31], and [3.120, 121]). LIPSTEIN [3.120] obtained $C_p$ profiles for a wide range of $\beta$ ratios (0.133 to 0.900) and inlet Reynolds numbers $Re_1$ ($4.3 \times 10^4$ to $2.5 \times 10^5$), all for thin inlet boundary layers. Most of the profiles of $C_p$ were nearly universal when $C_p$ was normalized on $C_{p_{max}}$ (the maximum or plateau value of $C_p$) and when $x$ was normalized upon the expansion height $h = R_2 - R_1$. The minimum value of $C_p$ ($-0.01$ to $-0.03$) occurred at $x/h = 2$ to 3 and $C_p$ passed through zero close to $x/h = 3$ to 4. $C_p$ approaches 95% of $C_{p_{max}}$ over a wider range of $x/h$ values, but for the cases $0.40 \geqq \beta \geqq 0.75$ and for $Re_1 \approx 10^5$, this point was seen to lie between $x/h = 11$ and 16, a point set well downstream of the reattachment point, which occurred at $x/h = 8$ to 9. Effects of initial conditions caused either by extreme values of $\beta$ or by inlet Reynolds number effects, which in this case were reflected in separation point boundary layer thickness changes, showed up as minor deviations from the universal $C_p$ curve.

It is also interesting to note that, because the exceedingly high turbulence levels cause the mean velocity profiles to be close to uniform at

---

[19] Two-dimensional reattachment (or separation) points in turbulent flow are the geometric locations where the instantaneous velocity, very near the surface, is in back-flow (forward-flow) 50% of the time.

the beginning of the *developing zone*, the values of $C_{p_{max}}$ were only about 5% lower than those predicted by the Borda-Carnot pressure rise coefficient[20]

$$C_{p_{bc}} = 2\beta^2(1 - \beta^2). \tag{3.30}$$

$C_{p_{max}}$ fell well below $C_{p_{bc}}$ at values of $\beta$ greater than 0.625 because the increasing influence of wall friction was felt in the zone downstream of reattachment, where the core was still decaying and where $C_p$ was still rising to its maximum value. The maximum or plateau value of $C_p$ continues as far downstream as $x = 25h$, but eventually $C_p$ must begin to decrease in the developing zone, where the slow conversion to a state of fully developed pipe flow would occur.

In the case we are examining in Fig. 3.12, transverse (radial) static pressure gradients are negligible compared to axial gradients. Streamline curvature in downstream parts of the recirculating zone suggests, however, that transverse pressure gradients may be important in some flow situations of this sort—the limiting case where $\beta \to 1$ and the core length is much greater than reattachment length. This limiting case also corresponds to constant pressure free-stream flow over a two-dimensional backward-facing step, or to the case of flow near the flat base of a long, slender body aligned with its axis parallel to an infinite stream. In the latter case, base pressures, $C_{p_B}$ ($C_p$ evaluated on the base or step-face) is known to assume values as low as $-0.15$ to $-0.20$ because of streamline curvature in the through flow outside of the reattaching streamlines. However, pressure changes across the separation region itself ($h\partial C_p/\partial y$) are small; see [3.132]. Exact values of $C_{p_B}$ are sensitive to many factors, including i) the state (laminar versus turbulent) and relative thickness of the layer at separation ($\delta_s/h$); ii) the end-wall effects in a two-dimensional geometry for aspect ratio, $b/h$, less than 4; iii) the leakage into, or suction of fluid from, the recirculating zone, etc. (see experimental results in [3.118], TANI [Ref. 3.122, p. 377], ROSHKO and LAU [Ref. 3.123, .157], [3.124] and the theory of GREEN [Ref. 3.125, .393]).

There appears to be only one set [3.117] of detailed measurements of turbulence quantities for fully separated, sudden expansion in a pipe flow. We shall discuss a few of the details from this data set. The normalizing velocity for the r.m.s. fluctuation levels is taken as $U_s$, the core speed at flow separation. Local levels are generally higher, since local mean velocities (see profiles in Fig. 3.12) are lower than $U_s$.

---

[20] The sudden-mixing pressure rise without wall friction is based on the momentum theorem and mass-flow conservation for a control surface $R_2$ in radius. Velocity profiles are assumed uniform and parallel. For noncircular ducts, replace $\beta^2$ with (area ratio)$^{-1}$.

In the shear layer, the peak streamwise and transverse turbulence levels were $\sqrt{\overline{u^2}}/U_s = 0.21$ and $\sqrt{\overline{v^2}}/U_s = 0.14$, respectively (at $x/R_2 = 2$ and $r/R_2 = 0.5$), values that are of the expected magnitude for simple free turbulent shear layers. At the end of the reattaching zone, $x/R_2 = 6$, peak levels occurred close to the centerline, $r = 0$, and the values were still very high, $\sqrt{\overline{u^2}}/U_s = 0.17$ and $\sqrt{\overline{v^2}}/U_s = 0.13$. The line where maximum values of total turbulence energy $0.5\,\overline{u_i^2}$ were measured corresponds to the location of the peak r.m.s. levels. It is seen (Fig. 3.12) to move to the centerline at the beginning of the developing zone. However, the region of maximum turbulence stress ($-\overline{uv}$) remained close to the radial location of the free shear layer, $r/R_2 = 0.5$.

At the beginning of the developing zone, $x/R_2 = 8$, maximum values of turbulence energy and stress are very high relative to developed pipe flow values, i.e., $(0.5\,\overline{u_i^2})_{max} = 0.03\,U_s^2$ and $(\overline{uv})_{max} = 0.0025\,U_s^2$. The individual component r.m.s. turbulence levels based on local mean velocity, rather than $U_s$, are about 0.40 at the same station, $x/R_2 = 8$.

The divergence of the lines of maximum turbulence energy and maximum stress in the reattaching zone resembles the transition seen as a simple turbulent jet develops its far-field characteristics. However, here the jet is fully bounded by the pipe wall downstream of reattachment, and the mean velocity profile at $x/R_2 = 8$ appears to be that of a fully developed pipe flow, not that of a simple jet. Only many pipe radii downstream will the turbulence structure eventually approximate that of fully developed pipe flow with maximum values of turbulence energy and stress located near the pipe walls and a centerline r.m.s, intensity, $\sqrt{\overline{u^2}}/U_{CL}$, of the order of 0.01 compared to values 40 times as large seen (in the data of [3.117]) at $x/R_2 = 8$. It is hoped that these few remarks highlight the complexity of turbulent, fully separated internal flows.

### 3.4.2 Two-Dimensional Step Expansions

Much less is known about the details of the flow for two-dimensional step expansions (Fig. 3.10a) than for the case of the axially symmetric expansion. We have already reviewed some step results in connection with our comments (Subsection 3.4.1) on base pressure for cases where step height is small, $h \ll W_1$. ABBOTT and KLINE [2.148] present the only known comprehensive study on two-dimensional geometries where $h$ is of the order of $W_1$. In their study, the boundary layer at separation was turbulent and about one step-height thick. Most of their investigation (accomplished by flow visualization) concerned the structure of the separated zone and distance to reattachment, although a few mean velocity and turbulence-level profiles were also obtained. The latter results are qualitatively the same as those discussed in Subsection 3.4.1, but in the single-step case, their results show that interaction of the core with the unseparated wall boundary layer may also be of some importance.

In two-dimensional, single-step expansions, the distance to reattachment is remarkably constant—$x_R = 7h$ for the observed range of cases where $h/W_1$ was varied from 2 to 0.2. However, for double-step geometries of area ratio $(W_2/W_1)$ greater than 1.5, the mean flow became asymmetric. Reattachment distances and stalled region volumes were different on the two sides of the expansion; $x_{R_1} < 7h$ and $x_{R_2} > 7h$. Small changes of upstream conditions could cause the stalled region sizes to change and could cause a shift in location of reattachment points from one wall to the other, which indicates, again, the complexity of displacement and shear interactions in purportedly very simple internal flow situations.

Aspect ratio, $b/h$, was not high for all tests of [2.148]. More recent results [3.119] on the single-step configuration where $h/W_1 = 1$ confirmed the basic reattachment distance data, $x_R/h = 6$ to 7.5, independent of aspect ratio for the test range $b/h = 2$ to 15. Other results [3.118] indicate that end-wall effects (reflected as aspect ratio effects) have little influence on mean reattachment location if flow leaves the step as a turbulent boundary layer. However, if the separation-point layer is laminar, $x_R/h$ may be profoundly different from the turbulent case because the upstream laminar flow at separation will affect the free shear-layer development [3.71] and flow in the developing zone [2.147] as well.

The reattachment studies of [3.119] were conducted in a duct that could be rotated about an axis set parallel to the edge of the step. The results of this study with system rotation provide some thought-provoking, if not completely understood, data on effects stabilizing and destabilizing curvature (see Subsection 3.1.1) on free shear-layer turbulence structure. It appears that total flow entrainment in a free shear layer may be partly caused by the "gulping" action of spanwise vortices that grow rapidly from the basic, two-dimensional instability seen already at the start of the free-shear-layer development zone $(x < x_0)$. The remainder of the total entrainment results from the full three-dimensional turbulence field. The stabilizing effects of rotation were observed to decrease, or increase, the magnitude of the three-dimensional turbulence field, without apparent effect on the spanwise vortices. Concomitant changes in reattachment location with changes of rotation rate indicated the effects of stabilization or destabilization on entrainment, and seemed to show that "gulping" by the two-dimensional vortices is very important even when the rotation was zero.

### 3.4.3 Confined Jet Mixing (Ejectors)

Removal of the step-face and its replacement with a smooth nozzle converts the sudden expansion configuration to that of a simple ejector (Fig. 3.10b). If secondary flow enters at speed $U_{S_2}$ equal to or greater than 0.1 to 0.2 times the primary jet speed, $U_{S_1}$, the mean axial velocity will be positive everywhere and the reattachment is replaced by a gradual mixing of the primary and the secondary flow.

Because the coaxial jets are not expanding into a constant-pressure atmosphere, but are confined to a duct or mixing tube, the mass-flow-rate-averaged, mean axial momentum is not conserved and static pressure may vary with $x$. For a constant radius, $R_2$, mixing tube, so long as $U_{S_1} \geqq 2U_{S_2}$, there will be an increase of pressure in the $x$-direction in the zone where the cores are being consumed by rapid shear-layer mixing. Pressure may also continue to rise in the developing flow zone just downstream of the disappearance of the cores. This pressure rise is the source of the pumping effect of the ejector (aspirator, eductor, etc.).

The so-called mixed-out condition occurs at $x_M \approx 10$ to $20 R_2$; here the mean velocity profile is generally quite flat, but turbulence levels are rather high and the structure of turbulence is far from that seen in fully developed pipe flow. Because of the uniformity of mean velocity at $x_M$, it is often assumed that an upper limit on the pressure rise in the mixing zone of a constant-radius duct may be obtained by a Borda-Carnot analysis (see footnote [20], page 151, Subsection 3.4.1) as

$$\frac{p_M - p_S}{\frac{1}{2}\varrho U_{S_1}^2} = \frac{2}{AR}\left(1 - \frac{1}{AR}\right)\left(\frac{U_{S_2}}{U_{S_1}} - 1\right)^2. \tag{3.31}$$

However, because of neglected wall-friction effects, (3.31) provides results that are too large by a factor of two or more except when $U_{S_2}/U_{S_1} \ll 1$, a condition close to that of the sudden expansion [see (3.30)].

Because pressure starts to decrease downstream of $x_M$, a primary concern of the ejector designer is prediction of mixing tube length, $x_M$. It is usually desired to terminate the mixing tube at this point and, if space permits, to add a diffuser in order to recover a fraction of mean flow kinetic energy at $x_M$ as an additional pressure rise. Note that because of the high levels of turbulence and its non-equilibrium structure at $x_M$, normal diffuser design correlations such as those shown in Fig. 3.9 may be of questionable value here (see comments in Section 3.3).

The literature on ejector analysis and experiment is large, but little of it concerns direct investigation of the turbulence field. Some of the most recent and better known analytic methods for low-speed flows are contained in [2.1, 3.126–128]. All of the more rigorous methods

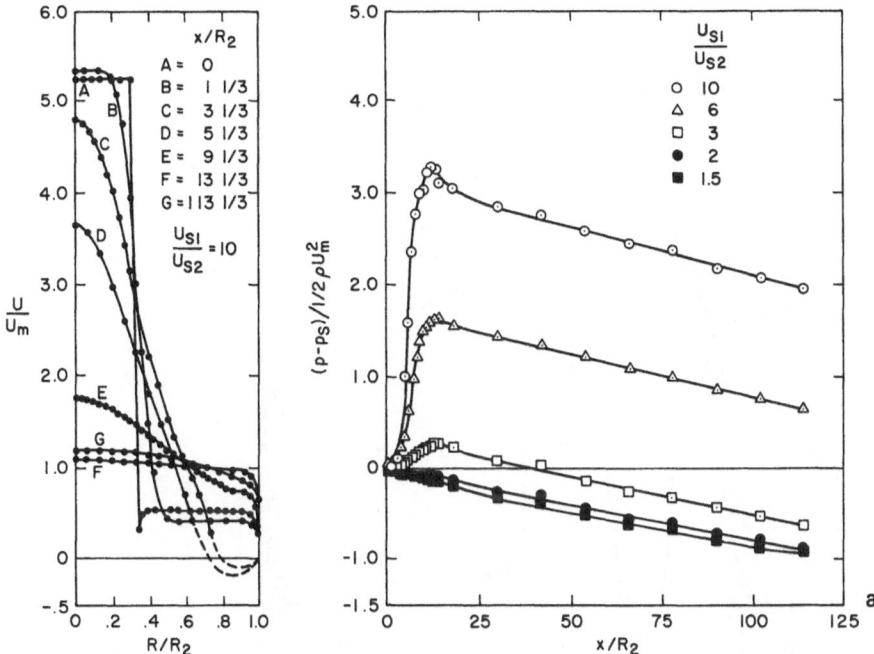

Fig. 3.13a—c. Confined jet mixing in a constant-diameter pipe; data from [3.130] for case $U_{S_1}/U_{S_2}=10$, $\beta=R_1/R_2=1/3$. (a) Mean axial velocity profiles and axial distribution of static pressure

depend, in one way or another, on the properties of turbulent shear layers and jets which expand freely into a constant-pressure atmosphere (see earlier discussions of simple, free shear layers). An interesting recent example of this class of analysis is [3.129] for the difficult case of $U_{S_2}=0$, the sudden expansion of a pipe. The results shown in [3.129] give remarkably accurate profiles of $C_p(x)$ when compared to the data of [3.120], and when compared to the data [3.117] on return-flow velocity in the recirculating zone and the position of the mean reattaching streamline the predictions are also good. None of the analytic methods, however, provide predictions of the details of the turbulence structure.

The work of RAZINSKY and BRIGHTON [3.130] provides a rather complete set of time-mean and turbulence measurements over a range of system parameters for confined jet mixing in constant-area pipes. A few of their results for the case $\beta=R_1/R_2=1/3$ and $U_{S_1}/U_{S_2}=10$ are shown in Figs. 3.13a, b and c. The static pressure profiles for other, lower values of $U_{S_1}/U_{S_2}$ are given in Fig. 3.13a.

This flow had a small separated zone (see mean velocity profiles C and D) because the wall boundary layer of the secondary flow could not

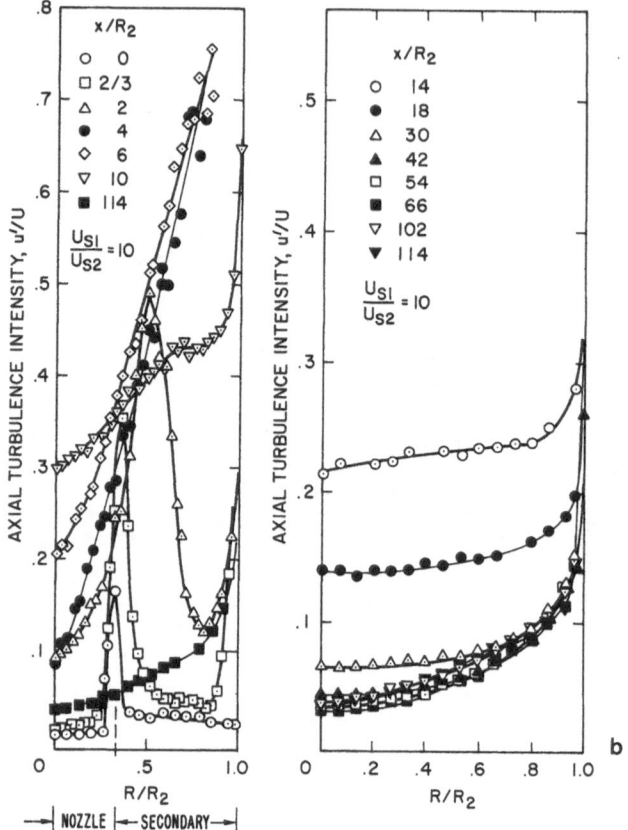

Fig. 3.13b. Axial turbulence intensity profiles

negotiate the adverse pressure gradient. The core of the central jet has disappeared, because of the spreading free shear layer, by $x/R_2 \leqq 3$, as is seen from profile C, Fig. 3.13a, and from the shear-stress profiles at $x/R_2 = 2$ and 4, Fig. 3.13c.

R.m.s. axial fluctuation levels become very large in the free shear layer (see $x/R_2 = 2/3$ and 2 in Fig. 3.13b) and exceedingly large in the separated zone. Near the wall, downstream of reattachment, $x/R_2 = 10$, these levels remain high; in the central region of the pipe, the decay of axial fluctuations is slow, and fully developed pipe-flow values are not attained until $x/R_2 > 42$.

The turbulent shear stress (Fig. 3.13c) achieves its greatest values in the free shear layer, just downstream of the end of the central core zone, $x/R_2 = 4$. The position of the maximum value on a given profile remains near $r/R_2 \approx 0.4$ even in the region where the peak axial intensity moves

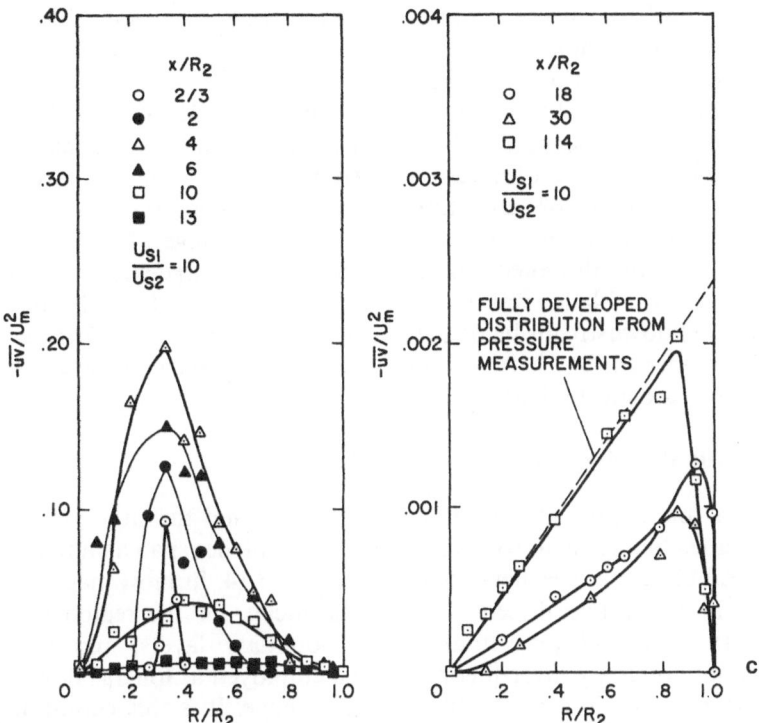

Fig. 3.13c. Turbulent shear stress profiles

close to the tube wall, $x/R_2 = 10$ to 14. This is also the region where the mean velocity profile is the flattest and where static pressure is a maximum. From a practical point of view, $x_M = 10$ to 14, and one would terminate the device there if one were using it as an ejector. It is worth noting that, except for the flow in the free shear layer upstream of $x/R_2 = 4$ to 6, no simple flow model used in calculation, to date, can represent the true complexity of the situation in this practically important zone.

## 3.5 Turbomachinery

Knowledge of the structure of turbulence, and theories of the turbulent boundary layer, have applications in the analysis of turbomachinery flow and performance, even though flow through turbines and compressors is often too complex for most calculation methods to be applied directly in the design process. In this section a few examples of the current state of turbulence knowledge in the field of turbomachinery flow will be illustrated. Explicit recognition of the importance of turbulence in applications may be seen in the papers of several recent conferences on internal flow and turbomachinery [3.31, 32, 133, 134]. A few of the key problem areas that one may identify are: i) the behavior of the boundary layers on the central portions (profile boundary layers) of the blades of the rotors and stators of axial flow compressors, ii) the end-wall boundary layers over the tip and hub casings of axial flow machines, and iii) the effects of turbulence in the free-stream (core region) and the blade wakes on the performance of (and noise produced by) downstream blade rows. There are many other items that could be discussed, but it is in these areas where explicit research on turbulence is the most advanced.

### 3.5.1 Profile Boundary Layers in Axial Flow

The rotors and stators of axial flow compressors and turbines contain airfoil-shaped blades set in rows or arrays[21]. The blades span an annular duct from hub to tip. For new notation see Fig. 3.14. In most machines the flow is nearly two dimensional, in the mean, for the region outside of the end-wall (or annulus-wall) boundary layers, and two-dimensional boundary layer theory may be employed as long as flow is not separated too close to the blade leading edge. It is this condition

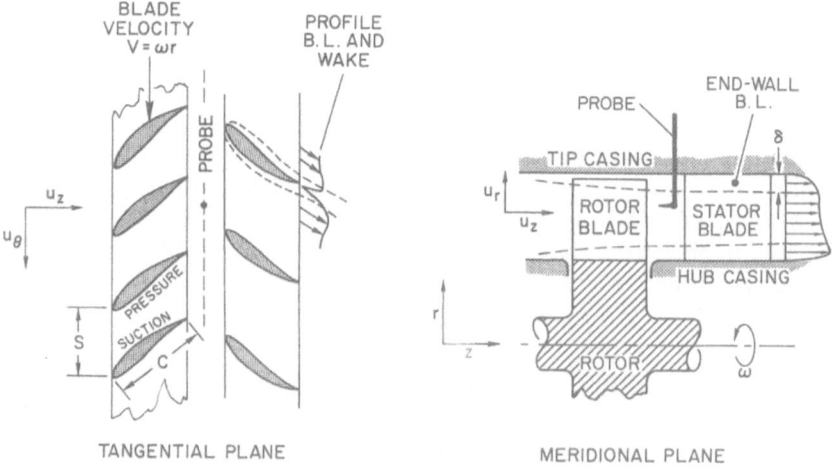

Fig. 3.14. An axial flow stage

---

[21] Called "cascades" when installed in static research facilities where air is simply blown, or sucked, through an annular or linear array.

which is called *profile* boundary layer flow. It represents the simplest real, turbulent flow situation one can find in a turbomachine.

Profile boundary layers are affected by many agents. The most important is the streamwise, adverse, surface-pressure gradient imposed on the suction surface by the airfoil shape (camber, thickness distribution), by the proximity of the other blades in the circumferential array (spacing to chord ratio, S/C, stagger angle) and by inlet conditions (Mach number, incidence angle, etc.). Excessive pressure rise along the suction surface causes thick profile boundary layers and high performance losses due to mixing of wakes downstream. In the worst case, flow separation occurs near the blade leading edges.

For some years it has been known that the pressure gradient is the main determinant of performance losses, and its influence is mainly felt as a diffusion ratio $U_{MS}/U_2$, the ratio of maximum flow speed[22] on the blade suction surface to speed at the trailing edge. LIEBLEIN [3.135] was able to correlate this diffusion ratio with measured values of $\theta_2/C$, the ratio of blade wake momentum thickness to blade chord length, for a very large number of cascade tests with turbulent profile boundary layers over blades of many different shapes and configurations. At optimum incidence, between positive and negative stall angles, all data fell close to a universal curve. When $U_{MS}/U_2$ rose to values greater than 1.8 to 2.0 (LIEBLEIN's stall criterion) $\theta_2/C$ also became large with small additional increases of $U_{MS}/U_2$ indicating that thick wakes, which came from regions of separation, were being shed downstream.

The use by designers of the LIEBLEIN stall criterion is elegant in its simplicity because it requires so little detailed knowledge—nothing more than the assurance that the profile boundary layers are turbulent. Optimum design of blade profile shape requires more detailed analysis, however. Many workers have attempted to apply two-dimensional boundary layer methods (Chapter 2 and [3.72]) to predict profile boundary layers. SCHLICHTING reviewed some early work in this area [3.136] and summarized recent advances in [3.137]. HORLOCK [Ref. 3.32, p. 331] commented on the state-of-the-art as did PAPAILIOU et al. [Ref. 3.134, p. 5] where they improved upon the modern integral equation method of LE FOLL [3.138]. In almost all cases discussed in these references the pressure gradients must be known a priori. To date, few attempts have been made to develop a displacement interaction method (see Subsection 3.1.3) for simultaneous solution of the inviscid external flow and the boundary layers, e.g., see paper I-9 in [3.134]. However, a number of attempts (papers I-1, I-8, in [3.134] and HUO [3.139]) have been made to find "optimum" blade surface

---

[22] Velocities are taken relative to the moving blades in the case of a rotor blade.

pressure distributions on the basis of turbulent boundary layer properties. These "optimum" surface pressure fields are then used in inverse potential flow solutions for blade shapes to be applied in high performance compressors.

It is widely recognized that low blade chord Reynolds numbers[23], $U_1 C/v \leqq 10^5$, may cause the suction surface boundary layer to be laminar over a substantial fraction of the chordwise distance from the leading edge. This is a severe problem in the first stages of axial compressors of aircraft engines operating at high altitude; see [3.137], SCHLICHTING and DAS [Ref. 3.32, p. 243] and HORLOCK [Ref. 3.32, p. 325]. Turbulent flow theory cannot be applied upstream of the final transition point which may, for cases of negative incidence angles, be close to the blade's trailing edge [3.140]. Even if the layer is turbulent over most of the suction surface, BRADSHAW [Ref. 3.133, p. 251] notes that low Reynolds numbers will have important effects on boundary layer turbulence (Subsection 2.3.1).

In his review BRADSHAW examines a number of effects—such as longitudinal curvature (Subsection 3.1.1)—which influence profile boundary layer growth. His comments on the problems of unsteadiness are relevant to turbomachinery applications. As discussed below (Subsection 3.5.3) unsteadiness in turbomachines is basically a superposition of true turbulence and the periodic unsteadiness of moving wakes from upstream blade rows. The periodically unsteady part of the flow over a stator (see Fig. 3.14) has a frequency of $V/S$ which is much lower than the frequencies of the large, stress-carrying eddies in the profile boundary layers, $U_e/\delta$, because to a first approximation $V \approx U_e$ and $\delta \ll S$. BRADSHAW shows that the effects of the periodic component may be important when its r.m.s. intensity level relative to $U_e$ reaches the order of 3 %, and he points out that boundary layer calculation methods such as [3.141] which include models for the effects of streamwise and temporal history of the large eddies may be the most appropriate to account for the effects of upstream blade wake disturbances.

This conclusion does not necessarily hold for the case of the true turbulence moving downstream over a profile boundary layer. However, if the frequency of the turbulence is higher than the typical eddy frequency, $U_\infty/\delta$, the steady-flow turbulent boundary layer equations may still be used with some modifications (see, for example, effects of free-stream turbulence on surface shear stress in Subsection 2.3.4). Recent references pertinent to the problem of calculation of profile boundary layers with high frequency turbulence in the external flow are [3.111, 142, 143].

---

[23] $U_1$ is usually defined to be the inlet relative velocity of the mean flow approaching a blade row.

Profile boundary layers are not always two dimensional. The main cause of three-dimensional effects is the displacement interaction between the end-wall layers and the core flow between blades. When end-wall layers thicken there is a spanwise convergence of the profile boundary layers which, as a result, grow thicker than predicted by two-dimensional theory. DUNHAM [3.144] investigated this effect and found that simple convergence corrections to two-dimensional methods were adequate if the end-wall regions were not stalled. However, in the presence of end-wall stall, direct flow visualization showed much stronger spanwise contraction of the suction surface "streamlines" than could be accounted for by simple blockage—an effect probably associated with the strong corner vortices (secondary flows) seen in the region of the suction surface and casing wall when flow enters the blade row at high positive incidence. Further details of three-dimensional flow on stalled airfoils are given in [3.131].

### 3.5.2 End-Wall Boundary Layers

Turbulent, three-dimensional boundary layer concepts (Section 2.4) have been intensively applied by aerodynamicists to predict end wall (annulus) boundary layers in axial flow compressors and in turbine-nozzle blade rows. These layers are a primary source of loss of stage efficiency. The main source of this loss is the mean kinetic energy transferred to the secondary flow from the main flow in its passage through a rotor or stator. The secondary flow is caused by the mechanisms 1, 2, and 3 discussed in Subsection 3.1.2, and, in addition, by the leakage of flow over the rotor blade tips (or roots in some stators) where they clear the stationary casing.

Until the early 1960's most analysts ignored the viscous nature of the boundary layers and proceeded with inviscid secondary flow methods like those in [3.7–11]. These inviscid theories were often sufficient for estimating mean air flow angles as a function of radial position at blade row exit [3.145], but they were not able to account for losses in stage efficiency.

Two basically different approaches have been advanced. The first, most commonly taken for turbine-nozzle end-walls, attempts solution of the full, three-dimensional boundary layer on the end-wall in the region between blade pressure and suction surface—the region is treated as a curved duct. Modern methods of this type for turbine-nozzles [3.146, 147] usually use momentum integral type methods with assumed mean velocity profiles for the main-flow and the crossflow directions, but otherwise they are quite general and are not restricted to small crossflows or the restrictions placed on methods developed for infinite swept wings (see Subsection 2.5.2).

The second approach to the end-wall problem is that used for flow over the hub and tip casings of axial flow compressors; see Fig. 3.14. Here, the dependent variables ($u_i$ and $p$) are tangentially (in the $\theta$-direction) averaged across one full blade pitch, $S$. Therefore, both the inviscid external flow and the boundary layer problems are reduced to equivalent "axially symmetric" flows where variation of the new, tangentially averaged variable occurs in the axial, $z$, and radial, $r$, directions only. As a result, the method has distinct computational advantages (one independent variable eliminated). MELLOR and WOOD [3.148] have developed this idea to a high level. Their approach permits prediction of boundary layer integral parameters ($\theta$, $\delta^*$, etc.) and also gives estimates of the casing-wall blockage and losses in stage efficiency as is shown in the comprehensive application of the method to prediction of actual multi-stage compressor performance; see [3.149]. Recently, HIRSCH [3.150] has contributed to the development of this method too. Nearly all past references to earlier work on the method are summarized by SMITH (p. 275) and HORLOCK (p. 321) in [3.32] and in the papers by MELLOR and BALSA (p. 363) and DANESHYAR et al. (p. 375) in [3.134].

There are two distinct problems that arise because of the tangential averaging employed by the methods described above. The first is the problem of the compatibility of the boundary layer assumption that static pressure is constant from layer edge to wall, and the observed fact that lift forces per unit span resulting from pressure differences across the blade are different in the boundary layer and at the free-stream edge of the layer. The discussion on pages 381 and 471 of [3.134] clarifies this issue. In essence, the "effective" blade force (a fictitious body force) introduced into the momentum integral equations [3.148] is required to satisfy this compatibility condition.

The second problem caused by tangential averaging arises because of the nonlinear inertia terms, $u_j \partial u_i / \partial x_j$, in the left-hand side of the basic equations of motion; see (1.8). The spanwise, periodically repeating variations of the mean velocity components, defined in coordinates fixed with respect to the blades, are not uncorrelated. Hence product terms $[(u_i u_j)_{av\ over\ S}]$ caused by tangential averaging are, like true turbulent stresses, not necessarily zero in the equations of motion, and some assumptions need to be made concerning these terms. In many analyses, the correlations have been neglected as second order—a correct approximation when end-wall boundary layers are very thick relative to the blade spacing, i.e., when $\delta \gg S$. This is not the case in most real compressor flows where $\delta \leq S$, and where the neglect of the tangential correlations is only justified in cases of very light blade loading. When the boundary layer equations are solved and integrated from inlet to outlet of a single blade row (see approach of [3.148]) then arguments are often given that purport to show the net effects of the correlation terms may be neglected. However, RAILLY in a discussion of [3.149] indicates that this is not always a reasonable assumption and it appears that the issue is still open.

### 3.5.3 Periodic Flow, Turbulence and Blade Wakes

As noted above, there is often a substantial blade-to-blade variation of $u_i$ and $p$, and in coordinates stationary with respect to a given blade row this variation is nearly[24] periodic with a spatial period of $S$. A stationary observer taking readings of velocity or pressure (say at the probe shown in Fig. 3.14) downstream of a rotor row would perceive this effect as a temporally periodic signal with frequency $V/S$. Superposed on this periodic signal would be a random fluctuation $f'(t)$ which is due to the true turbulence; see Fig. 3.15. The unsteady flows induced by both the periodic fluctuation and the turbulence may affect flow in the boundary layers of a downstream blade row (the stator in Fig. 3.14); see discussion in Subsection 3.5.1. Measurement and analysis of the separate effects of both unsteady components is becoming increasingly important in turbomachinery research.

The periodically fluctuating signal $\langle f(t) \rangle$ must be measured by *ensemble averaging* (1.17) of the total signal $f(t)$ at each point in the

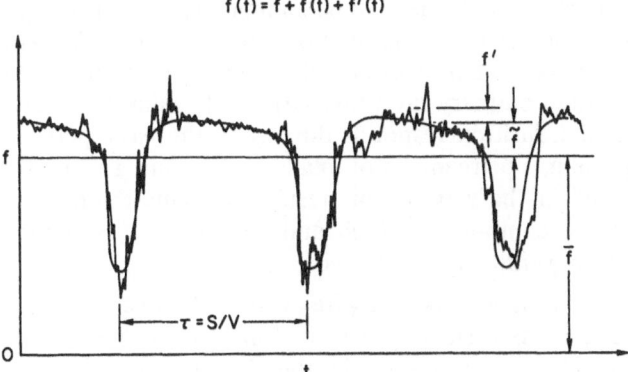

Fig. 3.15. Components of a fluctuating signal from a probe. $f$ may represent velocity $u_i$, static pressure, $p$, total pressure, $P$, etc.

---

[24] In real machines there will be some jitter in period (frequency) and wave shape due to variations (accidental or intentional) in $S$ or blade shape about the circumference.

phase angle for at least one blade space ($0 < t < \tau$). Phase angle (time) is usually determined relative to a fixed point on the rotor. It is useful to refer to this type of average as a *phase average*.

The *long time average* (1.16) of the total signal is $\bar{f}(t)$. It is easy to see (Fig. 3.15) that a periodically fluctuating component $\tilde{f}(t)$ and the turbulence $f'(t)$ may be determined once $\bar{f}(t)$ and $\langle f(t) \rangle$ are known, i.e.,

$$\tilde{f}(t) = \langle f(t) \rangle - \bar{f}$$

$$f'(t) = f(t) - \langle f(t) \rangle .$$

Evans [3.151] used this signal separation method with a stationary hot-wire probe to obtain data on the turbulence about 1/2 a chord length downstream of an axial-flow rotor. He showed that r.m.s. turbulence levels due to $u_i'$ are of the same magnitude as levels due to the periodic part $\tilde{u}_i$ for design point conditions. However, when the rotor profile boundary layers were heavily stalled the true turbulence levels, $\sqrt{\overline{u_i'^2}}/U_i$, rose to double or triple their original values whereas the periodic levels, $\sqrt{\overline{\tilde{u}_i^2}}/U_i$, stayed nearly the same. A few other measurements of this type [3.154, 155] are starting to appear in the technical literature, but much remains to be accomplished.

For flows where turbulence levels were measured without regard to separation of components, the net effects of turbulence and periodic unsteady flow have been shown to be important determinants of performance, particularly in cases where low Reynolds numbers may cause profile boundary layers to have long, laminar separation bubbles on their suction surfaces; see Schlichting, [Ref. 3.32, p. 243], and [3.137, 152]. The work of Walker [3.153], whose detailed measurements were conducted in the boundary layers on large-scale axial-flow stator blades, suggests that transition responds directly to the basic periodic fluctuations of passing, upstream rotor blade wakes and not to the turbulence convected in the wakes. This result highlights the need to separately measure the components $U_i$, $\tilde{u}_i$, and $u_i'$ so that one may intelligently sort out the separate effects of each component.

Reference [3.154] using a triple hot-wire probe technique developed in [3.155] studied the wakes shed by a row of axial rotor blades. The results show that the rate of decay, in the downstream direction, of the wake velocity deficit in the periodic component is faster than of the same wake shed by an equivalent stationary blade row. The reasons for this effect are not yet understood, but practical consequences are important to improved prediction of turbulence effects in turbomachinery. The results of this study also support those of Evans [3.151].

The flow field in the wakes leaving a typical, high speed centrifugal compressor impeller is usually much more complex than that of an axial flow machine. Hot-wire velocity data for real machines have yet to be developed, in part because of the large total temperature fluctuations in the blade wakes. However ECKARDT [3.156], using special pressure and temperature probe techniques [3.157], was able to deduce the details of the periodic components of the wake velocity and pressure leaving a typical centrifugal impeller that had a tip speed $V = 300$ m/s and a blade passing frequency of 4.78 kHz. The distribution and intensity of the random fluctuations of the total pressure, $\overline{P'(t)^2} = \overline{(P(t) - \langle P(t) \rangle)^2}$ were also deduced. Finally, SENOO and ISHIDA [3.158] used hot-wires to measure the complex wake flow patterns leaving a low speed centrifugal impeller. They included no true turbulence data, but their results represent an important step toward settling a controversy surrounding the rate of decay of turbulent wakes from centrifugal impellers; see the discussion that follows [3.158].

*Acknowledgements*

The Affiliates Program of the Thermosciences Division of the Department of Mechanical Engineering, Stanford University, and the National Science Foundation, Fluid Mechanics Program of the Engineering Division, provided partial support for the preparation of the chapter.

# References

3.1    R. C. DEAN, JR: NASA SP-304, p. 301 (1974)
3.2    R. C. DEAN, JR: Trans. ASME 81 D, (1959)
3.3    J. W. S. RAYLEIGH: Proc. Roy. Soc. A **93**, 148 (1916)
3.4    H. WILCKEN: NASA TTF-11421 (translation from Ing.-Arch. **1**, 357 (1930)) (1967)
3.5    H. SCHMIDBAUER: NASA T.M. 791 (translation from Luftfahrtforschung **13**, 160 (1936) (1936)
3.6    S. HONAMI, I. ARIGA, T. ABE, I. WATANABE: ASME Paper No. 75-FE-32 (1975)
3.7    I. P. CASTRO, P. BRADSHAW: J. Fluid. Mech. **73**, 265 (1976)
3.8    S. ESKINAZI, H. YEH: J. Aeronaut. Sci. **23**, 23 (1956)
3.9    J. P. JOHNSTON, R. M. HALLEEN, D. K. LEZIUS: J. Fluid Mech. **56**, 533 (1972)
3.10   S. ESKINAZI, F. F. ERIAN: Phys. Fluids **12**, 1988 (1969)
3.11   L. B. ELLIS, P. N. JOUBERT: J. Fluid Mech. **62**, 65 (1974)
3.12   D. COLES: J. Fluid Mech. **21**, 385 (1965)
3.13   M. POTTER, M. D. CHAWLA: Phys. Fluids **14**, 2278 (1971)
3.14   J. E. HART: J. Fluid Mech. **45**, 341 (1971)
3.15   D. K. LEZIUS: PhD dissertation, Mech. Engg. Dept., Stanford University (1971)
3.16   H. W. LIEPMANN: NACA Wartime Rep. W-107 (1943); also NACA ACR-4J28 (1945)
3.17   J. P. JOHNSTON: Trans. ASME **95**I, 229 (1973)
3.18   V. C. PATEL: ARC CP 1043 (1968)

3.19   J. P. Johnston: NASA SP-304, p. 207 (1974)
3.20   G. L. Mellor, T. Yamada: J. Atmosph. Sci. **31**, 1791 (1974)
3.21   B. E. Launder, G. J. Reece, W. Rodi: J. Fluid Mech. **68**, 537 (1975)
3.22   S. A. Eide, J. P. Johnston: Rept. PD-19, Thermosciences Division, Mech. Engg. Dept., Stanford University (1974)
3.23   J. Moore: Trans. ASME **95**A, 205 (1973)
3.24   K. A. Rastogi, J. H. Whitelaw: ASME Paper 71-WA/FE-37 (1971)
3.25   R. M. C. So: J. Fluid Mech. **70**, 37 (1975)
3.26   L. Prandtl: *Verhandlung des 2. Internationalen Kongresses für Technische Mechanik*, Zürich (1926): see also *Gesammelte Abhandlungen* [2.10]
3.27   M. J. Tunstall, J. K. Harvey: J. Fluid Mech. **34**, 595 (1968)
3.28   A. G. Hansen, H. Z. Herzog, G. R. Costello: NACA T.N. 2947 (1953)
3.29   P. Bansod, P. Bradshaw: Aeronaut. Quart. **3**, 131 (1972)
3.30   W. R. Hawthorne: Proc. Roy. Soc. A **206**, 374 (1951)
3.31   G. Sovran (ed.): *Fluid Mechanics of Internal Flow*. (Elsevier Publ. Co., Amsterdam 1976)
3.32   L. S. Dzung (ed.): *Flow Research in Blading*. (Elsevier Publ. Co., Amsterdam 1970)
3.33   H. B. Squire, K. G. Winter: J. Aeronaut. Sci. **18**, 271 (1951)
3.34   S. V. Patankar, V. S. Pratap, D. B. Spalding: J. Fluid Mech. **67**, 583 (1975)
3.35   B. E. Launder, D. B. Spalding: *Mathematical Models of Turbulence*. (Academic Press, New York 1972)
3.36   J. L. Livesey (Symp. Chairman): *Salford Symposium on Internal Flows* (Inst. Mech. Engrs., London 1971)
3.37   M. Rowe: J. Fluid Mech. **43**, 771 (1970)
3.38   G. W. Hogg: PhD thesis, University of Idaho (1968)
3.39   Y. Mori, W. Nakayama: Int. J. Heat. Mass Transfer **10**, 37 (1967)
3.40   H. Ito: Trans. ASME **82**D, 123 (1959)
3.41   J. Nikuradse: Ingr.-Arch. **1**, 306 (1930)
3.42   S. D. Veenhuizen, R. M. Meroney: ASME Paper 72-WA/FE-34 (1972)
3.43   J. W. Delleur, D. S. McManus: Proc. 6th Midwestern Conf. on Mechanics, Austin, Texas (1959)
3.44   F. B. Gessner: PhD dissertation, Purdue University (1964)
3.45   L. C. Hoagland: ScD thesis (Dept. Mech. Engg.), Mass. Inst. Technology (1960)
3.46   B. E. Launder, W. M. Ying: J. Fluid Mech. **54**, 289 (1972)
3.47   H. J. Leutheusser: ASCE J. Hydraulics Div. **89**, HY3 (1963)
3.48   H. J. Tracy: ASCE Hydraulics Div. **91**, HY6 (1965)
3.49   R. H. Pletcher, H. N. McManus: Proc. 8th Midwestern Conf. on Mechanics (1963)
3.50   E. Brundrett, W. D. Baines: J. Fluid Mech. **19**, 375 (1964)
3.51   F. B. Gessner, J. B. Jones: J. Fluid Mech. **23**, 689 (1965)
3.52   B. E. Launder, W. M. Ying: Rept. TM/TN/A/11, Dept. Mech. Engg., Imperial College of Sciences and Technology, London (1971)
3.53   D. Naot, A. Shavit, M. Wolfshtein: Wärme u. Stoffübertragung **7**, 151 (1974)
3.54   F. B. Gessner, A. F. Emery: Trans. ASME **98**I, 261 (1976)
3.55   F. B. Gessner, J. K. Po: Trans. ASME **98**I, 269 (1976)
3.56   K. Hanjalic, B. E. Launder: J. Fluid Mech. **52**, 609 (1972)
3.57   R. B. Dean, P. Bradshaw: J. Fluid Mech. **78**, 641 (1977)
3.58   V. K. Sharan: ASME Paper No. 72-WA/FE-38 (1972)
3.59   P. Bradshaw, R. B. Dean, D. M. McEligot: Trans. ASME **95**I, 214 (1973)
3.60   T. Morel, T. P. Torda: AIAA J. **12**, 533 (1974)
3.61   R. B. Dean: 5th Australasian Conf. on Hydraulics and Fluid Mech. (University of Canterbury, Christchurch, New Zealand, 1974)

3.62  J. P. Milliat: Pubs. Sci. et Tech. du Min de l'Air 335 (1957)
3.63  O. J. McMillan, J. P. Johnston: Trans ASME **95**I, 393 (1973)
3.64  V. L. Streeter (ed.): *Handbook of Fluid Dynamics*. (McGraw-Hill, New York 1961)
3.65  K. C. Wang: AIAA J. **10**, 1044 (1972)
3.66  M. Hahn, P. E. Rubbert, A. S. Mahal: Rept. AFFDL-TR-72-145, Wright-Patterson AF Base, Ohio (1973)
3.67  E. C. Maskell: Rept. Aero 2565, RAE Farnborough (1955)
3.68  E. A. Eichelbrenner, A. Oudart: Rech. Aero **40** (1954)
3.69  E. A. Eichelbrenner, A. Oudart: Rech. Aero **47** (1955)
3.70  L. Rosenhead (ed.): *Laminar Boundary Layers*. (University Press, Oxford 1963)
3.71  P. Bradshaw: J. Fluid Mech. **26**, 225 (1966)
3.72  S. J. Kline, M. V. Morkovin, G. Sovran, D. J. Cockrell (ed.): *Computation of Turbulent Flow* – 1968 AFOSR-IFP-*Stanford Conference Proceedings*, Vol. 1, Thermosciences Division, Mech. Engg. Dept., Stanford University (1969)
3.73  P. W. Runstadler, S. J. Kline: Trans. ASME **26**E, 2 (1959)
3.74  R. E. Falco, G. R. Newman: Coherent, Repetitive Reynolds Stress Producing Motions in a Turbulent Boundary Layer, APS Division of Fluid Dynamics Meeting (Nov. 25–27, 1974)
3.75  A. J. Reynolds: *Turbulent Flows in Engineering*. John Wiley & Sons, London 1974)
3.76  R. P. Patel: Aeronaut. J. **78**, 93 (1974)
3.77  A. K. M. F. Hussain, W. C. Reynolds: Trans. ASME **97**I, 568 (1975)
3.78  R. B. Dean: Aero Rept. 74-12, Imperial College, London (1974)
3.79  R. B. Dean: J. Fluids Engg. **100** I (June 1978)
3.80  J. Laufer: NACA Report 1053 (1951)
3.81  G. Comte-Bellot: PhD thesis, University of Grenoble (1963); translation as ARC 31609, FM 4102 (1969)
3.82  J. A. Clark: Trans. ASME **90**D, 455 (1968)
3.83  C. J. Lawn: J. Fluid Mech. **48**, 477 (1971)
3.84  K. Bremhorst: PhD thesis, University of Queensland, St. Lucia (1969)
3.85  K. Bremhorst, T. B. Walker: J. Fluid Mech. **61**, 173 (1973)
3.86  T. B. Walker, K. J. Bullock: J. Phys. E. **5**, 1173 (1972)
3.87  P. W. Runstadler, S. J. Kline, W. C. Reynolds: Rep. MD-8, Thermosciences Division, Mech. Engg. Dept., Stanford University (1963), or P. W. Runstadler: PhD thesis
3.88  N. S. Berman, J. W. Dunning: J. Fluid Mech. **61**, 289 (1973)
3.89  W. M. Kays, A. L. London: *Compact Heat Exchangers*, 2nd ed. (McGraw-Hill, New York 1964)
3.90  K. Rehme: Int. J. Heat Mass Transfer **16**, 933 (1973)
3.91  K. Rehme: J. Fluid Mech. **64**, 263 (1974)
3.92  S. C. Kacker: J. Fluid Mech. **57**, 583 (1973)
3.93  J. Kaye, E. C. Elgar: Trans. ASME **80**, 753 (1958)
3.94  M. H. Abdul Khader, H. Suresh Rao: ASCE J. Hydraulics Div. **100**, 25 (1974)
3.95  J. M. Cannon, W. M. Kays: Trans. ASME **91**C, 135 (1969)
3.96  A. White: J. Mech. Engg. Sci. **6**, 47 (1964)
3.97  L. J. Huey, J. W. Williamson: Trans. ASME **41**E, 885 (1974)
3.98  K. Hanjalić, B. E. Launder: J. Fluid Mech. **51**, 301 (1972)
3.99  L. R. Reneau, J. P. Johnston, S. J. Kline: Trans. ASME **89**D, 141 (1967)
3.100 C. Smith, S. J. Kline: Trans. ASME **96**I, 11 (1974)
3.101 L. R. Reneau, J. P. Johnston: Trans. ASME **89**D, 643 (1967)
3.102 R. L. Woolley, S. J. Kline: Rep. MD-33, Thermosciences Division, Mech. Engg. Dept., Stanford University (1973)
3.103 W. W. Bower: AIAA Paper No. 74-1173 (1974)

3.104 J. W. White, S. J. Kline: Rep. MD-35, Thermosciences Division, Mech. Engg. Dept., Stanford University (1975)

3.105 D. J. Parsons, P. G. Hill: Trans. ASME 95 I, 349 (1973)

3.106 D. R. Boldman, H. E. Neumann: NASA TN D-7486 (1973)

3.107 P. A. C. Okwuobi, R. S. Azad: J. Fluid Mech. 57, 603 (1973)

3.108 J. M. Robertson, G. L. Calehuff: Proc. ASCE 83, HY5, Paper 1393 (1957)

3.109 J. R. Ruetenik, S. Corrsin: 50 Jahre Grenzschichtforschung (Vieweg, Braunschweig 1955)

3.110 B. A. Waitman, L. R. Reneau, S. J. Kline: Trans. ASME 83 D, 349 (1961)

3.111 D. Arnal, J. Cousteix, R. Michel: Rech. Aerospatiale no. 1976–1, p. 13 (1976)

3.112 O. K. Oseberg, S. J. Kline: Rep. MD-28, Thermosciences Division, Mech. Engg. Dept., Stanford University (1971)

3.113 D. Durão, J. H. Whitelaw: Trans. ASME 95 I, 467 (1973)

3.114 E. E. Chriss, P. T. Harsha: AEDC-TR-75-54, AEDC, Tullahoma (1975), and AD-A009149

3.115 P. M. Bevilaqua, H. L. Toms, Jr.: ARL TE-74-0006, Aerospace Research Laboratories, Wright-Patterson AF Base, Ohio (1974)

3.116 M. Van Dyke, W. G. Vincenti, J. V. Wehausen (Eds.): Annual Review of Fluid Mechanics (Annual Reviews Inc., Palo Alto 1975)

3.117 M. C. Chaturvedi: ASCE J. Hydraulics Division 89, 61 (1963)

3.118 V. de Brederode, P. Bradshaw: Aero Rep. 72-19, Imperial College London (1972)

3.119 P. H. Rothe, J. P. Johnston: Rep. PD-17, Thermosciences Division, Mech. Engg. Dept., Stanford University (1975)

3.120 N. J. Lipstein: ASHRAE J. 4, 3 (1962)

3.121 G. Heskestad: Trans. ASME 92 D, 437 (1970)

3.122 H. Görtler (ed.): Grenzschichtforschung Symposium, Freiburg i. Br., Aug. 1957. Berlin-Göttingen-Heidelberg: Springer 1958

3.123 A. F. Charwat (ed.): Proc. 1965 Heat Transfer and Fluid Mechanics Institute. Stanford: University Press 1965

3.124 M. A. Badri Narayanan, Y. N. Khadgi, P. R. Viswanath: Aero. Quart, 25, 305 (1974)

3.125 Separated Flow, Part I, AGARD CP No. 4 (1966)

3.126 R. Curtet, F. P. Ricou: Trans. ASME 86 D, 765 (1964)

3.127 P. G. Hill: J. Fluid Mech. 22, 161 (1965)

3.128 P. G. Hill: Trans. ASME 89 D, 210 (1967)

3.129 R. G. Teyssandier, M. P. Wilson: J. Fluid Mech. 64, 85 (1974)

3.130 E. Razinsky, J. A. Brighton: Trans. ASME 93 D, 333 (1971)

3.131 V. de Brederode: PhD thesis, Imperial College, London (1975)

3.132 C. Chandrsuda: PhD thesis, Imperial College, London (1975)

3.133 B. Lakshminarayana, W. R. Britsch, W. S. Gearhart (Eds.): Fluid Mechanics, Acoustics, and Design of Turbomachinery, parts I and II, NASA SP-304 (1974)

3.134 J. Surugue (Ed.): Boundary Layer Effects in Turbomachines, AGARDograph No. 164 (1972)

3.135 S. Lieblein: Trans. ASME 81 D, 387 (1959)

3.136 H. Schlichting: Trans. ASME 81 D, 543 (1959)

3.137 H. Schlichting: AIAA J. 12, 427 (1974)

3.138 J. LeFoll: Proc. of Seminar on Advanced Problems in Turbomachinery, VKI, March 23–30 (1965)

3.139 S. Huo: Trans. ASME 97 A, 195 (1975)

3.140 N. J. Seyb: ARC Rept. 27214 (1965)

3.141 P. BRADSHAW: NPL Aero. Rept. 1288, ARC 30912 (1969)
3.142 R. L. EVANS, J. H. HORLOCK: Trans. ASME 96 I, 348 (1974)
3.143 J. H. HORLOCK, R. L. EVANS: Trans. ASME 97 I, 126 (1975)
3.144 J. DUNHAM: Aeronaut. J. **78**, 90 (1974)
3.145 J. HORLOCK: ASME Paper No. 62-Hyd-11 (1962)
3.146 T. C. BOOTH: ASME Paper No. 75-UT-23 (1975)
3.147 R. P. DRING: Trans. ASME 93 A, 386 (1971)
3.148 G. L. MELLOR, G. M. WOOD: Trans. ASME 93 D, 300 (1971)
3.149 T. F. BALSA, G. L. MELLOR: Trans. ASME 97 A., 305 (1975)
3.150 CH. HIRSCH: Trans. ASME 96 A, 413 (1974)
3.151 R. L. EVANS: Trans. ASME 97 A, 131 (1975)
3.152 R. KIOCK: ASME Paper No. 73-GT-80 (1973)
3.153. G. J. WALKER: ASME Paper No. 74-GT-135 (1974)
3.154 R. RAJ, B. LAKSHMINARAYANA: Trans. ASME 98 A, 218 (1976)
3.155 B. LAKSHMINARAYANA. A. PONCET: Trans. ASME 96 I, 87 (1974)
3.156 D. ECKARDT: Trans. ASME 97 A, 337 (1975)
3.157 H. WEYER, R. SCHODLE: Trans. ASME 93 D, 603 (1971)
3.158 Y. SENOO, M. ISHIDA: Trans. ASME 97 A, 375 (1975)

# 4. Geophysical Turbulence and Buoyant Flows

P. Bradshaw and J. D. Woods

The behavior of turbulence can be greatly affected by fluctuating body forces if the latter are correlated with the velocity fluctuations. The most common example is the large effect of gravity on flows with density fluctuations. If the density fluctuations arise because there is a mean density gradient in the same direction as the mean velocity gradient (as in a boundary layer on a heated or cooled horizontal surface) or if the flow is actually driven by the mean density differences (as in buoyant plumes in still air) then the density and velocity fluctuations are highly correlated and buoyancy can have a large effect. If the density increases upward (heavy fluid on top of light fluid) the flow is "unstable" and the density-velocity correlation can convert potential energy into turbulent kinetic energy. Conversely if the density decreases upward at a faster rate than is expected for fluid in hydrostatic equilibrium, existing turbulent energy can be converted into potential energy (because turbulent mixing tends to reduce the density gradient and thus raise the center of gravity of the fluid). We shall see below that a convenient parameter for correlating the effects of a density difference $\Delta \varrho$ across a fluid layer of thickness $h$ with a typical velocity $U$ is

$$\frac{\Delta \varrho g h}{\varrho U^2}$$

which is the ratio of the hydrostatic pressure difference across the layer to (twice) a typical dynamic pressure. It is also related to the Froude number, $U^2/gh$ or its square root. Complete suppression of turbulence (or a very large increase in intensity, according to the sign of $\Delta \varrho$) is expected if this parameter is roughly of the order of unity. Suppose for simplicity that $\Delta \varrho/\varrho$ is of order unity: then large effects require $gh \approx U^2$. If $U = 1\,\mathrm{m\,s^{-1}}$ then $h = 0.1\,\mathrm{m}$, so that to obtain really large buoyancy effects in a laboratory boundary layer 0.1 m thick in an air stream of $1\,\mathrm{m\,s^{-1}}$ we must heat or cool the air by 200° C or so. On the other hand if $h = 1\,\mathrm{km}$, $U = 10\,\mathrm{m\,s^{-1}}$, large buoyancy effects can appear if the temperature difference is only a few degrees C, as in the boundary layer of the earth's atmosphere. In pure water $\Delta \varrho/\varrho$ cannot exceed about 0.05 (temperature difference $100°$ C) and larger values, whether caused by temperature differences or concentration differences, will be rare in liquid flows, so if $U \approx 1 \text{ m s}^{-1}$ large buoyancy effects will be found only if $h > 2$ m, outside the normal laboratory range. Alternatively if $h = 0.1$ m large buoyancy effects can be obtained in liquids if $U$ is smaller than about $20 \text{ cm s}^{-1}$, which brings us down to coffee-cup proportions.

The rather large range of physical sizes over which significant buoyancy effects can occur is paralleled by a large range of configurations, from heated boundary layers (on laboratory or planetary scale) to plumes from isolated sources of buoyancy, again on laboratory or planetary scale. However, for an introduction to turbulent flows with buoyancy, it is sufficient to consider only a few simple types of boundary condition, leading to external or internal flows not unlike those discussed in the two preceding chapters. The purpose of this short chapter is to review the subject for workers concerned with buoyant flows in the laboratory or in engineering. It is *not* primarily intended for meteorologists, astrophysicists or oceanographers although they may find the discussion of small-scale problems helpful; this book's main contributions to geophysics are the theoretical discussions of buoyancy effects on heat and pollutant transfer in Chapter 6 and of particle-laden flows in Chapter 7.

Numerous reviews of the specialized buoyancy problems of geophysics, notably the earth's boundary layer, are already available and we do not seek to compete with them in this chapter. However it is convenient to start with a description of geophysical boundary-layer turbulence because most of the research work on buoyant turbulence has been done for geophysical purposes and because the boundary conditions of meteorological problems are simpler—or at least better understood—than those of some engineering buoyancy problems. Also, the effects of buoyancy are most easily discussed for shear layers whose mean motion is normal to the gravitational vector and is therefore not affected by gravitational body forces as such. Later, we consider shear layers whose mean motion is parallel to the gravitational vector, such as the free-convection layer on a heated vertical plate in still air, and buoyant turbulence fields with more complicated boundary conditions. We anticipate some of the results of Chapter 6 by using Reynolds' analogy between heat and momentum transfer in shear layers (which is a consequence of the high correlation, mentioned above, between velocity fluctuations and fluctuations of density or other scalars). Effects of condensation of water vapor are almost entirely ignored in the meteorological discussion and the adjective "dry" applied to a formula or definition is a warning that the quantity concerned is different in the real atmosphere. Note that dry formulae apply to liquids if the density rather than the temperature is used. Detailed discussion of the prediction of buoyant flows is postponed until Chapter 6 because it is closely related to the prediction of transfer of heat or concentration which are the sources of buoyancy.

The Reynolds-stress transport equation for buoyant flow is (1.35) with the buoyancy (body force) term as given under (5) on p. 26.

Specialized papers are quoted in context below. In addition LINDEN and TURNER [4.1] report on a Euromech meeting devoted to buoyant turbulence and covering a very wide range of geophysical and laboratory problems; their review is a good entry to the recent research literature. Again for the sake of providing a good entry to the literature, the papers quoted in the present chapter are mainly the latest available, rather than more balanced but slightly outdated reviews; the former lead to the latter. No attempt is made to cover the astrophysical literature. A useful introduction to buoyant convection with application to stars is given by SPIEGEL [4.2].

Notation may cause a little confusion: shear-layer aerodynamicists use $x_2$ or $y$ for the direction normal to the shear layer, notionally the vertical direction, but geophysicists denote the upward vertical co-ordinate by $x_3$ or $z$ (meteorologists) and $-x_3$ or $-z$ (oceanographers) while students of free convection on vertical surfaces use $x_1$ or $x$ for the upward vertical. We follow these conventions here, rather than shielding the reader from reality.

Another confusion in the literature is caused by the use of the word "convection". To some non-geophysicists it is what we call "transport", while to geophysicists it is what we call "buoyant convection". "Free (buoyant) convection" is a flow driven principally by buoyancy forces, while "forced convection" relies on a pre-existing velocity field.

## 4.1 Geophysical Boundary Layers

### 4.1.1 The Atmospheric Boundary Layer

The first kilometer or so of the atmosphere approximates to a boundary layer, whose external stream is the high-altitude wind and whose study is called micrometeorology. Mainly because of the earth's rotation, the high-altitude wind undergoes large changes in direction over horizontal distances of a few hundred km, forming the swirling patterns seen in weather maps and satellite pictures, and the boundary layer thickness does not grow indefinitely. An idealized flow which demonstrates this is the Ekman layer ([1.8], Section 4.4). Wyngaard et al. [4.3] briefly review work on the turbulent Ekman layer and then discuss the effects of buoyancy.

For the purposes of this review we ignore the three dimensionality of the atmospheric surface layer; its consequences are generally the same as those of three dimensionality in laboratory flows (Chapter 2). The effects of superimposed rotation on turbulent flow were discussed in Chapter 3. The reader may find it interesting to compare the following analysis with that of Subsection 3.1.1. Frequently the growth of the unstable boundary layer is inhibited by an "inversion layer", a shallow region with a highly stable density gradient which suppresses the turbulence at the upper edge of the boundary layer. The atmosphere above the boundary layer is most often weakly stable, with relatively little turbulent mixing, although there are frequent exceptions.

The inner ("surface") layer of the atmospheric boundary layer—the first 50 m or so above the surface—is a nearly two-dimensional constant-stress layer, which in the absence of buoyancy effects or sudden changes of surface roughness would obey the "local-equilibrium" inner-layer scaling outlined in Chapters 1 and 2. Meteorologists tend to assume, on rather slender evidence, that the inner-layer scaling is valid throughout the "boundary layer", but this point need not be discussed here. The existing inner-layer formulae such as (1.50) can be formally extended to buoyant flows simply by inserting a dimensionless buoyancy parameter, based on local scales and assumed to be unique. Suppose that at height $z$ above the surface the temperature gradient is $\partial T/\partial z$ and the absolute temperature is $T$. Now in a still atmosphere (implying no net buoyancy effects and a pressure gradient $\partial p/\partial z = -\varrho g$), $\partial T/\partial z$ would be equal to the so-called "dry adiabatic lapse rate", $-g/c_p$, which is the rate at which the temperature of a fluid element would decrease if it were moved in the positive $z$ direction without heat transfer or condensation. If the actual lapse rate $\partial T/\partial z$ equals the adiabatic lapse rate, then a displaced fluid element automatically adjusts to the temperature of its new sur-roundings. Buoyancy effects, therefore, will depend on the "effective" temperature gradient $\partial T/\partial z + g/c_p$, and $T + gz/c_p$ is sometimes called the "potential temperature", $\theta$. This slight difficulty arises because of the compressibility of air; it can usually be neglected in the ocean or the laboratory. If the lapse rate differs from the adiabatic value, a displaced fluid element will have a different temperature and density from its surroundings. A displacement $\Delta z$ implies a density difference of $\bar{\varrho}(\partial T/\partial z + g/c_p)\,\Delta z/\bar{T}$, where $\bar{T}$ is the absolute temperature, and a body force per unit mass of $-g(\partial T/\partial z + g/c_p)\,\Delta z/\bar{T}$, relative to the gravitational force on the surrounding fluid. If the potential temperature gradient is positive, the body force will tend to restore the fluid element to its former position. In the absence of aerodynamic forces a displaced fluid element in a dry atmosphere would oscillate vertically at a frequency

$$N = [g(\partial T/\partial z + g/c_p)/\bar{T}]^{1/2} \equiv [g(\partial\theta/\partial z)/\bar{T}]^{1/2} \qquad (4.1a)$$

or

$$N = [-g(\partial\bar{\varrho}/\partial z)/\bar{\varrho}]^{1/2} \qquad (4.1b)$$

in an incompressible fluid. $N$ is called the Brunt-Väisälä or "buoyancy" frequency. An interesting account of the rather complicated motion of a displaced fluid element in real life is given by McLAREN et al. [4.4]: oscillations at a frequency close to $N$ indeed occur (among others) but the initial injection of potential energy radiates away as internal waves. One way of writing down the dimensionless buoyancy parameter we seek is to divide $N^2$ by the square of a typical turbulence frequency. For the inner, constant-stress layer of a turbulent wall layer which has primary velocity and length scales $u_\tau$ and $z$, a suitable choice is $(u_\tau/z)^2$ or, equivalently in a neutral atmosphere, $(\partial U/\partial z)^2$. We define the gradient Richardson number for dry conditions as

$$\mathrm{Ri} = \frac{-N^2}{(\partial U/\partial z)^2} = \frac{-(g\,\partial\theta/\partial z)/\bar{T}}{(\partial U/\partial z)^2}. \qquad (4.2)$$

L. F. RICHARDSON, in 1920, suggested this first buoyancy parameter. Ignoring the distinction between $\theta$ and $T$ and replacing $\partial/\partial z$ by $1/h$, it is the parameter mentioned in the Introduction. Its attraction is that it is fairly simple to measure the variation of $T$ and $U$ with height. An alternative definition, the flux Richardson number, is based on the rates of heat and momentum through the surface or, more generally, through the plane $z = $ constant. It is proportional to the gradient Richardson number in cases of validity of the inner-layer formula (1.50), and of the analogous formula for the upward heat transfer rate, $Q \equiv \varrho c_p \overline{\theta' w}$ (where $\theta'$ is the fluctuating part of the temperature) namely

$$\frac{\partial\theta}{\partial z} = \frac{-\overline{\theta' w}}{\kappa_\theta(-\overline{uw})^{1/2} z} \qquad (4.3)$$

where $-\varrho\overline{uw}$ is the Reynolds shear stress and $\kappa_\theta$, the analogue of $\kappa$, is about 0.45 compared to about 0.41 for $\kappa$ (both values from laboratory data). Substitution of (1.50) and (4.3) in (4.2) gives

$$\mathrm{Ri} = \frac{\kappa}{\kappa_\theta} \frac{g}{T} \frac{\overline{\theta' w}}{(-\overline{uw})\, \partial U/\partial z} \qquad (4.4)$$

and the usual definition of the flux Richardson number is

$$R_{\mathrm{f}} = \frac{-g}{T} \frac{\overline{\theta' w}}{(-\overline{uw}) \, \partial U / \partial z} = \frac{\kappa_\theta}{\kappa} Ri \tag{4.5}$$

where the difference between $-\overline{uw}$ and $\tau_w / \varrho_w \equiv u_\tau^2$ is usually ignored. A third, related parameter for the surface layer is $z/L$, where $L$ is the Monin-Obukhov length, almost always defined by the dry *surface* transfer rates $Q_w$ and $\tau_w$ as

$$\frac{1}{L} = \frac{-\kappa g}{T} \frac{Q_w / (\varrho c_p)}{u_\tau^3}. \tag{4.6}$$

For small buoyancy effects, assuming that (1.50) applies and that $-\overline{uw} = u_\tau^2$, and ignoring the difference between $\kappa/\kappa_\theta$ and unity, the parameters $Ri$, $R_{\mathrm{f}}$ and $z/L$ are the same; more accurately, $Ri/R_{\mathrm{f}} \to \kappa/\kappa_\theta$. Because (1.50) and (4.3) are not valid when buoyancy effects are large, the three parameters differ in that case. In the earth's surface layer, $z/L$ is preferred; in other cases $z$ must be replaced by the appropriate length scale of the turbulence (Chapter 6), and it is usually easier to use $R_{\mathrm{f}}$.

Equations (1.50) and (4.3) can be written generally as

$$\frac{\kappa z}{(\tau/\varrho)^{1/2}} \frac{\partial U}{\partial z} = \phi_{\mathrm{m}} \left( \frac{z}{L} \right) \tag{4.7}$$

and

$$\frac{\kappa z (\tau/\varrho)^{1/2}}{Q/(\varrho c_p)} \frac{\partial \theta}{\partial z} = \phi_{\mathrm{h}} \left( \frac{z}{L} \right). \tag{4.8}$$

Here $\phi_{\mathrm{m}}$ is just the ratio of the apparent mixing length (Section 1.8) in a neutral layer, i.e., $\kappa z$, to the value at given $z/L$, $l_0/l$ say; compare (3.6). Here $\phi_{\mathrm{m}}(0) = 1$, $\phi_{\mathrm{h}}(0) = \kappa/\kappa_\theta$ and $\phi_{\mathrm{m}}$ and $\phi_{\mathrm{h}}$ are increasing functions of $z/L$. Usually, surface values of $\tau$ and $Q$ are used. The usual meteorological notation for the eddy viscosity and the analogous eddy conductivity is

$$K_{\mathrm{m}} = \frac{(\tau/\varrho)}{\partial U / \partial z} = \frac{-\overline{uw}}{\partial U / \partial z}, \tag{4.9}$$

$$K_{\mathrm{h}} = \frac{Q/(\varrho c_p)}{\partial \theta / \partial z} = \frac{-\overline{\theta' w}}{\partial \theta / \partial z}. \tag{4.10}$$

The ratio $K_{\mathrm{h}}/K_{\mathrm{m}} \equiv \alpha$, the reciprocal of the turbulent Prandtl number $\sigma_t$ used in engineering (Chapter 6) is simply $\phi_{\mathrm{m}}/\phi_{\mathrm{h}}$ or, in the neutral case, $\kappa_\theta/\kappa$. In the geophysical literature, (4.3) is often derived from by assuming a constant turbulent Prandtl number and invoking (1.50). This is conceptually equivalent to the direct application of inner-layer similarity arguments to the heat-transfer equations. The data of BUSINGER et al. [4.5] for $\phi_h$ and $\phi_m$, which show much smaller scatter than most other experimental results, gave $\kappa = 0.35$ instead of the popular laboratory value of about 0.41. TENNEKES [4.6] discusses the possibility of a real variation in $K$ with Reynolds number and with the Rossby number that represents the effect of the earth's rotation. The question is still open. It has been suggested, but by no means proved, that the low value of $K$ obtained by BUSINGER et al. was an experimental error, but the general shapes of the $\phi_m$ and $\phi_h$ curves are nevertheless trustworthy. Various analytic curve-fits to the $\phi_m$, $\phi_h$ data have been suggested; for details, see PANOFSKY [4.7]. The effects of buoyancy on other types of turbulent flow are expected to be generally similar to these. This analysis is used for particle-laden flows in Chapter 7.

Qualitatively, stable stratification (i.e., a positive potential temperature gradient) reduces the ability of the turbulence to transfer momentum (or requires a bigger value of $\partial U / \partial z$ for a given value of $\tau$). Heat transfer is reduced even more ($K_h/K_m < 1$) the reason being that "turbulence" in highly stable flows consists partly of internal waves, which can transfer momentum but not heat. For discussions of internal waves and their interaction with turbulence, see many papers in SAXTON [4.8] and OLSEN et al. [4.9], especially the flow-visualization work of PAO. Such waves may occur in any turbulence field stabilized by a body force, even the apparent "centrifugal" body force in a curved flow. The critical value of $R_f$ or $z/L$ at which turbulence disappears completely is difficult to establish, because these or other parameters based on local-equilibrium arguments are not sufficient descriptions of a flow containing propagating internal waves. Various idealized analyses for buoyancy problems lead to $R_{f,\,crit} = 1/4$. $Ri$, being equal to $R_f K_m/K_h$, becomes very large in highly stable flow.

Local-equilibrium arguments also fail in strongly *unstable* flows because large vertical transport of turbulent energy by the turbulence may occur (Chapter 6). Probably the published behavior of $\phi_h$ and $\phi_m$ includes quite large transport effects. For flows which are so highly unstable that production of turbulent energy by buoyancy greatly exceeds production by the mean shear, $(\tau/\varrho)^{1/2}$ is negligibly small compared to other quantities with the dimensions of velocity. The forms of (4.7) and (4.8) that remain finite in the limit $\tau \to 0$ (the "free convection" formulae) are

$$\phi_m\left(\frac{z}{L}\right) \equiv \frac{\kappa z}{(\tau/\varrho)^{1/2}} \frac{\partial U}{\partial z} \simeq 0.7 \left|\frac{z}{L}\right|^{-1/3} \tag{4.11}$$

$$\phi_h\left(\frac{z}{L}\right) \equiv \frac{\kappa z (\tau/\varrho)^{1/2}}{Q_w/(\varrho c_p)} \frac{\partial \theta}{\partial z} \simeq 0.32 \left|\frac{z}{L}\right|^{-1/3} \tag{4.12}$$

where the empirical constants are taken from the recent experimental results of TING and HAY [4.10]. It must be noted that $u_\tau \equiv (\tau/\varrho)^{1/2}$ still appears; these formulae still refer to shear layers, and would not be directly applicable to initially isotropic turbulence subjected to buoyancy effects. It must also be noted that the subject of free-convection scaling is still controversial; some authors (e.g., WYNGAARD and IZUMI [4.11] feel that the total height of the free-convection layer should appear in the formulae whereas in the analysis above it was implicitly assumed that the information about the upper boundary carried by the falling fluid between rising "thermal" currents is negligible. Furthermore the work of ANGELL [4.12] and LEMONE [4.13] on helical circulations about longitudinal axes even in moderately unstable flows suggests that buoyant convection may interact with the mean longitudinal vorticity in the skewed (three-dimensional) surface layer. We return to the subject of buoyant free convection in Section 4.2.

The turbulent energy equation (1.43) for buoyant flows, obeying the thin-shear layer approximation (Section 1.7) and the "Boussinesq approximation" that density changes are negligible except in the buoyancy term, can be written with the addition of the buoyancy body-force terms as

$$\frac{D^{\ddagger}\overline{q^2}}{Dt} = -\overline{uw}\frac{\partial U}{\partial z}(1-R_f) - \frac{\partial}{\partial z}\left(\frac{\overline{p'w}}{\varrho} + \frac{1}{2}\overline{q^2 w}\right) - \varepsilon \qquad (4.13)$$

because the ratio of the buoyant production terms, $-g\overline{\varrho'w}/\bar{\varrho} \equiv g\overline{\theta'w}/\bar{T}$, to the shear production term, $-\overline{uw}\partial U/\partial z$, is simply $-R_f$. Parameters called "Richardson number" are used in Chapter 3 in the study of the effect of extra strain rates on turbulent shear layers, and in Chapter 7 in the discussion of particle-laden flows. In these cases also, the parameter can be interpreted as (extra rate of energy production)/(shear production). The critical value of $R_f$ at which turbulence disappears is necessarily less than unity because dissipation will always be a large fraction of shear production. In stably stratified flow in the atmospheric inner layer, the vertical turbulent transport ("diffusion") term is expected to be small, and the mean transport term will be small if the surface roughness is uniform for a large "fetch" distance upstream of the point of observation (say a thousand times the height of the point of observation; this is sometimes difficult to achieve in field experiments). In unstably stratified flows mean transport—though not turbulent transport—should again be small, although one of the main difficulties in meteorological experiments is establishing proper mean conditions. There is inevitably a 24-hour cycle, so that averaging for more than an hour or so at a time is not realistic, and the appearance of a "spectral gap" between the diurnal cycle and turbulent fluctuations is more a matter of faith than of fact.

It is fair to say that the standards of accuracy desired or attained in meteorological experiments are significantly poorer than in test rigs under human control. Therefore assumptions which suffice in meteorology, such as the equality of $K_h$ and $K_m$ in neutral conditions, and the constancy of shear stress or the applicability of local-equilibrium arguments throughout the surface layer, should not be accepted uncritically for engineering purposes.

In recent years laboratory simulation of the atmospheric boundary layer, pioneered by CERMAK's group at Colorado State University [4.14], has been carried out by several groups. A large wind tunnel is needed to produce a significant Richardson number without an excessive fractional change in density, but simulation of buoyant plumes in a neutral boundary layer, e.g. [4.15], is less demanding. For a unique experiment on a small-scale turbulent boundary layer with a very large temperature difference, see NICHOLL [4.16].

Several large-scale meteorological and oceanographic experiments have been mounted in recent years, some as part of the Global Atmospheric Research Project. Publications are scattered; for a brief review of the Barbados Oceanographic and Meteorological Experiment (BOMEX) see FLEAGLE [4.17].

Recent reviews of the atmospheric boundary layer, which refer to and update the standard textbooks, include the useful non-mathematical introduction by TENNEKES [4.18], relevant parts of the books by PASQUILL [4.19] and TURNER [4.20], the articles by PANOFSKY [4.7] and SMITH [4.21] and the conference proceedings edited by HAUGEN [4.22]. An older set of conference proceedings including those of the upper atmosphere was edited by SAXTON [4.8]. The book by MONIN and YAGLOM [1.18] treats atmospheric turbulence in some detail, and many other books on turbulence include at least a mention of buoyancy effects. Journals devoting a large fraction of their space to micrometeorology, apart from those referenced above, include *Boundary-Layer Meteorology, Journal of Geophysical Research*, and the *Quarterly Journal of the Royal Meteorological Society*.

### 4.1.2 The Ocean Surface Layer

Very few direct measurements have been made of the distribution of turbulent kinetic energy as a function of space and time in the upper ocean, and as yet no attempt has been made to construct a quantitative theoretical model of energy distribution on the basis of hypotheses concerning the terms in the turbulent kinetic energy equation. Still less has there been any quantitative attempt to describe the spectral form of this energy distribution as a function of space, time, three wave-number components and frequency. The main difficulty has been that available instruments are not sufficiently sensitive to measure the turbulent velocity fluctuations near the surface in the presence of a far stronger field of non-turbulent motion associated with the orbital motions of surface waves. Indeed, so limited are the available methods that we have no reliable measurements of the vertical shear of horizontal mean current in the top hundred meters of the ocean and hence it is impossible at present to calculate the Reynolds stress term in the turbulent kinetic energy equation. Furthermore, experimental errors in measurements of temperature and salinity gradients are generally larger than the total differences in these quantities across the surface wind-mixed layer, so calculations of the buoyancy flux term are equally unreliable. Recent attempts to estimate the viscous energy dissipation term from measurements of motions in the vicinity of the Kolmogorov scale ($\sim 1$ cm), are also at the stage of order-of-magnitude, rather than 10%, error.

It is therefore premature to discuss the distribution and spectral properties of turbulence in the surface layers of the ocean on the basis of experimental data comparable with those available for the atmospheric boundary layer. However, a new generation of instruments now being tested in the sea (e.g., the laser current meter) promises to equip ocean turbulence investigators with the means to embark on quantitative discussion of the turbulent kinetic energy equation by the end of the present decade.

Meanwhile it is possible to offer a qualitative description of the structure and distribution of upper ocean turbulence on the basis of *in situ* flow visualization and by drawing on the results of laboratory experiments relating to particular aspects of the complex sources and sinks in Nature. Since the atmospheric boundary layer has been far better explored, it is helpful to draw parallels and to point out contrasts between it and the oceanic boundary layer.

Firstly, seawater is a thousand times denser than air at sea level, so (neglecting for a moment the effects of buoyancy) the Ekman layer thickness, $H \sim (\tau_w/\varrho)^{1/2} f$, where $f$ is the vertical component of the earth's rotation rate, will be reduced by a factor of $1000^{1/2}$ or about 30. Therefore we are dealing with a neutral oceanic boundary layer of some tens of meters thick as against an atmospheric one of about a thousand meters thick. Equally the current speeds in the upper ocean (order 0.1 m/s) are much weaker than the surface winds (order 10 m/s) and we therefore expect that the turbulent kinetic energy dissipation per unit mass of seawater will be very much less than in the air. A reasonable guess for the former in the conditions given above, assuming an inertial subrange based purely on dimensional arguments, in which $\varepsilon \sim u_\tau^3/h$ where $u_\tau \sim 3$ cm/s and $h \sim 10\text{--}100$ m, is $\varepsilon \sim 10^{-2} - 10^{-1}$ erg/gs. The lifetimes of the corresponding energy containing eddies will be of order

$$\tau = \frac{h}{u_\tau} \sim 1\text{--}10 \text{ min} .$$

If the vertical fluxes of momentum and buoyancy, $\overline{uw}$ and $-g\overline{\varrho'w}$, have spectral peaks in the same range (i.e., spectrally separated from the surface wave orbital motions at frequencies typically in the range 0.1–1 Hz), there is some hope that the corresponding terms in the turbulent kinetic energy equation may be calculated from measurements with instruments whose response time need not be less than, say, one second.

These preliminary calculations guide our attention to the appropriate length and time scales for the energy containing eddies in an Ekman layer in an idealized unstratified ocean with a surface stress of order 1 dyne/cm$^2$ ($u_\tau \sim 1$ cm/s). In practice, this analysis is appropriate only in the rather special (but by no means rare) situation of an ice-covered ocean. Nevertheless, the numbers turn out to be not very different from those estimated on the basis of quite different models more appropriate to the ice-free open ocean, where the properties of the turbulence are modified by surface waves and by buoyancy effects, which will now be considered.

A number of tentative models for the interaction of surface waves with the turbulence in the upper ocean have been proposed. These concern the following effects, which by no means exhaust the whole range of possible interactions:

1) Turbulent jets from individual wave breaking events probably dominate the turbulence field in the top one meter of the ocean (TOBA et al. [4.23], LONGUET-HIGGINS and TURNER [4.24]).

2) Interaction of a whole field of wave breaking events to produce convergence and hence vertical motion is likely to become the dominant factor below the top meter (Longuet-Higgins, 1974, unpublished).

3) Positive feedback between horizontal gradients in the surface current, due to 2) above, and surface waves to create Langmuir cells probably controls mixing most of the time in wind speeds greater than 2 m/s [4.25].

Translations of these idealized models based on simple monotonic wave trains into stochastic models including more realistic surface wave spectra have not yet been attempted (and may have to await clarification of uncertainty about the latter). But it seems likely that the wave-turbulence interactions are powerful and that the structure of the turbulence in the top few tens of meters of the ocean in windy weather will normally be controlled by these sources of turbulent kinetic energy, rather than by the direct action of the surface stress, as in the atmosphere or under ice. The reader is directed to KITAIGORODSKII's book [4.26] for a more detailed discussion of the subject.

The role of vertical density gradients is even more important in the upper ocean than in the atmosphere. An upward buoyancy flux, contributing to the turbulent kinetic energy, occurs at night, when heat is lost from the surface by long wave radiation, evaporation and, to a lesser extent, by conduction. During daylight hours over most of the ocean (i.e., except at the highest latitudes) the solar heat input into the sea is more than sufficient to supply these losses and heat is temporarily accumulated in the upper layer. Nevertheless, the sunlight is absorbed below the surface and the heat needed to supply the surface loss must be conducted upwards, for a distance $h$ which depends on the clarity of the water and the fraction of solar input needed to supply the loss. In calm weather $h$ is of order 1 cm, and molecular conduction carries the heat flux; in rough weather, depending on latitude, time of day, season, and water clarity, $h$ may range up to order 10 m. The upward heat flux within this surface layer provides a small contribution to the kinetic energy and may affect the structure of the turbulence, but no detailed analysis has yet been published. To summarize, there is a convective contribution to the turbulence in the top ten or so meters over most of the ocean both day and night, with a strong coupling between the depth $h$ and the heat balance during the day such that in calm weather there is normally no convection. At night, in calm weather when there is no contribution to the turbulence from the wind, it is possible that convection takes the form of rolls or plumes as in the atmosphere, but this conjecture ignores the role of two-dimensionally isotropic "turbulence" to be discussed at the end of this section.

Below the convective layer the density increases with depth, and the surplus solar heat being absorbed during the day is transported downwards by the turbulence, thereby increasing the potential energy of the upper ocean at the expense of the turbulent kinetic energy. In mid- and high latitudes during the winter there is no net heat gain over 24 hours and the potential energy stored in this way during the day is lost again at night when the convective layer deepens. But during the summer months in mid- and high latitudes, and throughout the year in the tropics and sub-tropics, there is a net gain of heat over each 24-hour period, so the nocturnal convection does not extract all the daytime gain of potential energy. The result is an accumulation of potential energy in the surface boundary layer, associated with the steady rise in temperature of the upper ocean during the spring and summer months.

The vertical density gradient associated with this accumulation of excess heat in the upper ocean plays a major role in controlling the structure of turbulence below the convective layer. Starting at the sea surface the density profile has a characteristic form in which there is an upper layer, usually a few tens of meters thick, with a weak density gradient, (Brunt-Väisälä frequency $N \sim 3 \times 10^{-3}$ radian/second; $2\pi/N \sim 30$ min) overlying the thermocline in which the density gradient decreases roughly exponentially from a maximum ($N \sim 0.1$ rad/s; $2\pi/N \sim 1$ min) at the top with a e-folding depth of order 10 m. The bottom of the seasonal thermocline, defined as the depth at which the seasonal temperature modulation reaches some arbitrarily small value, say $0.01°$ C, is typically 100–200 m.

The structure of the turbulence is quite different inside the thermocline, where the static stability is large compared with the overlying surface layer. Flow visualization studies by Woods (e.g. [4.27]) have shown that the turbulence occurs intermittently in patches which occupy a small fraction of the total volume at any instant, the remainder of the flow being laminar (in the sense that there is no motion with eddy lifetimes shorter than $2\pi/N$). The volume fraction occupied by turbulence events at any instant varies with location and the overlying weather, ranging from a minimum of around 5% to over 50% in rough weather [4.28]. The events result from Kelvin-Helmholtz instability of sheets of enhanced vertical density gradient, giving billows, whose mean heights range from about 20 cm in calm weather to 1 m in rough weather, with the largest normally occurring on the interface that marks the top of the thermocline. The billows occur in patches some ten meters across: individual events last about five minutes [4.27]. Kelvin-Helmholtz instability occurs when the gradient Richardson number falls below 1/4, as predicted by several workers (e.g. [4.29]), the shear across the density sheet being provided roughly equally from a) two-dimensionally isotropic turbulence and b) internal waves. The increase in billow turbulence events during rough weather is probably due to increased internal wave activity due to coupling between surface and internal waves [4.30][1].

Assuming that the billow height equals the Ozmidov length scale

$$L = \left(\frac{\varepsilon}{N^3}\right)^{1/2}$$

---

[1] It is important to remember that inside the thermocline, the vertical transport of momentum (as against scalars such as temperature, salinity and other constituents of seawater) may be effected largely by the internal wave field, quite independently of turbulence [4.31].

we estimate that the local energy dissipation rate in the billow turbulence events is $10^{-4} - 10^{-3}$ erg/gs, which gives a mean value in the range $5 \times 10^{-6} - 5 \times 10^{-5}$ erg/gs in calm weather and ten times larger in rough weather. These estimates are consistent with values of $\varepsilon$ derived from velocity variance spectra measured with hot-film flow-meters [4.28, 32]; Gibson [4.33]. The Reynolds numbers of individual billows lie in the range 100 to 1000, so the spectrum of individual events and even the ensemble spectrum of many events may not exhibit the universal Kolmogorov form expected for three-dimensionally isotropic turbulence, but currently available measurements are inadequate to test this possibility.

In shallow seas, the turbulent boundary layer formed by flow over the sea-bed (especially strong tidal currents) may penetrate through to coalesce with the surface wind-mixed layer, so that there is no thermocline and no central band of intermittent billow turbulence. In the Irish Sea, for example, the thermocline occurs only in regions where the water is particularly deep and/or the tidal currents are locally rather weak; elsewhere there is a mixed column from the surface down to the sea-bed. In extreme winter conditions (e.g., when the mistral blows over the Gulf of Lyons), surface cooling can become so intense that the convective layer penetrates down to the ocean floor at depths of several kilometers in a column some 20 km wide lasting several days (the period being controlled by the wind). Conversely, some shallow regions are strongly stabilized throughout by salinity or temperature gradients [4.34].

This brief account of turbulence in the upper ocean has been concerned with the oceanic equivalents of atmospheric boundary-layer turbulence and, in the thermocline, clear-air turbulence [4.35]. To a first approximation these resemble laboratory wind tunnel turbulence described in other chapters; they are roughly isotropic in all three dimensions and as far as one can determine from rather inadequate observations they appear to obey the similarity laws of the inertial subrange, despite the rather low Reynolds numbers and high Richardson numbers in the seasonal thermocline. Until relatively recently it was generally believed that a detailed knowledge of this turbulence based on improved observations and more detailed analysis would explain the observed bulk transport properties of the upper ocean; for example, the seasonal heating cycle, the assimilation of fresh rainwater deposited onto the surface and the acceleration of the upper layers in response to a downward transport of momentum received from the wind. But there is increasing evidence from upper ocean heat budget studies (Woods et al. unpublished) that the downward heat flux (and hence presumably the salt and momentum fluxes) in *calm* weather (when flow visualizations show that the sea is laminar) is larger than would be calculated for molecular conduction and that even in *rough* weather (when turbulence of the kind discussed above is present) there is an additional contribution from some other form of larger scale turbulence. Some tentative steps have been made towards developing a theory for this large-scale turbulence (4.36) and towards interpreting temperature microstructure patterns on scales larger than the Ozmidov length scale ($L \lesssim 1$ m in the thermocline; $L \sim 30$ m in the surface layer) in terms of the distribution of a quasi-passive scalar by it. The results to date show that the motions on scales longer than $2\pi/N$ cannot be three-dimensionally isotropic but that they are statistically two-dimensionally isotropic (i.e., $\overline{w^2}$ is much weaker than $\overline{u^2}$ and $\overline{v^2}$), at least for horizontal scales up to some hundreds of kilometers. While not *turbulence* in the sense considered elsewhere in this book, these motions demand similar statistical treatment [4.36] and, after allowing for the effects of Archimedes and Coriolis forces, they appear to have many physical similarities with small-scale turbulence. It seems likely that a complete description of turbulent transport processes in the upper ocean must include both types of turbulence.

Papers on dynamic oceanography appear in journals such as *Journal of Geophysical Research*, *Deep-Sea Research* and *Deutsche Hydrographische Zeitschrift*.

### 4.1.3 Pollutant Plume Dispersion in the Surface Layer

The classical example of this problem is the spread of smoke or hot gas from a stack in the atmospheric wind [4.37]. In practical cases the difference in density between the smoke and the air is reduced to a very small fraction after a short initial period of plume rise, and the centroid of the plume cross section thereafter follows the atmospheric motion. (Descent of a heavy plume from stack height to ground level is permitted by the laws of nature but generally forbidden by the laws of man.) The concentration at large distances downstream is therefore closely the same as would result from emission of a neutrally buoyant plume from a taller stack, and the vertical component of momentum of the emergent smoke can also be related to an increase in effective stack height, usually small. There is some controversy over the correct scaling, and formulae with dimensional coefficients are in use [4.15]. For instance, the rise $Z$ of a plume of gas chemically close to air can be empirically represented by the formula

$$Z = A Q^{1/4} x^{*3/4}/U \tag{4.14}$$

where $Q$ is the heat emission rate of the stack, $U$ is a mean wind speed and $x^*$ is an empirically defined quantity, initially equal to the distance downstream but asymptoting to a constant value so that the final rise $Z_{max}$ is

$$Z_{max} = \alpha Q^{1/4}/U \qquad (4.15)$$

where for tall stacks and "average" meteorological conditions $\alpha \approx 500 \text{ m}^2 \text{ s}^{-1} \text{ MW}^{-1/4}$. The one-quarter power has some theoretical justification but so has the alternative one-third power. These of course are simple formulae for practical use in the complicated conditions of the real atmosphere: several groups of investigators have produced more scientific but less general correlations (e.g. [4.38]), and there has also been much recent work on intermittent or continuous plumes in still air [4.37–39] as well as earlier work summarized in [4.40]. For a general review see [4.19]: frequently, research papers appear in *Atmospheric Environment*. Cumulus clouds are the result of natural, intermittent plumes: even before condensation occurs, the process is affected by the presence of water vapor in the plume and therefore falls outside the scope of this chapter. Papers on clouds appear, of course, in meteorological journals, notably *J. Atmos. Sci.*

## 4.2 Laboratory-Scale Buoyant Flows

### 4.2.1 Buoyant Free Convection from Horizontal Surfaces

The linearized equations describing laminar convection between an upper cold plate and a lower hot plate are identical with those describing flow between an inner rotating cylinder and an outer fixed one. In both cases the development of fully turbulent flow from laminar flow, as the temperature difference (i.e., the Rayleigh number[2]) or the rotational speed (i.e., the Taylor number) is increased, is extremely complicated. For the case of rotating cylinder flow, highly detailed explorations have been made by COLES [4.41]. In free convection KRISHNAMURTI [4.42] describes the evolution up to the point at which turbulence first appears. This is the first of the stages reviewed by DALY [4.43] who distinguishes "the transition from laminar to turbulent flow, the

---

[2] $Ra = g\alpha\beta d^4/k\nu$, where $\alpha$ is the coefficient of thermal expansion ($1/T$ for gases), $\beta$ is the temperature gradient $\partial T/\partial y$ (for gases, $\partial\theta/\partial y$, although the distinction is unimportant scale) and $k$ is the thermometric conductivity.

transition from low to locally high intensity turbulence, the transition to uniformly high intensity turbulence, and the transition from a buoyancy-dominated turbulence to a shear-dominated turbulence". The quantitative results presented by DALY are numerical simulations but are accurate enough for illustrative purposes at least. These two authors review much of the recent work on the simplest case of uniformly heated parallel plates (Bénard flow). Cases in which the upper boundary is not a solid surface can lead to further complications. Variations of surface tension with temperature can interact with convection patterns in liquids and there is still some doubt whether buoyant convection above a single horizontal plate in an infinite fluid is a well-posed problem. For other work with and without an upper solid surface see [4.44–47].

An interesting feature of highly unstable convection is that most of the turbulent energy is not produced directly by buoyancy but by shear near the edges of convection cells (this is Daly's "shear-dominated turbulence"). Once recognized, this is a natural consequence of a well-organized cellular structure, which is itself a natural consequence of the need to transfer more heat than molecular or unstructured turbulent diffusion allows. BUSSE and WHITEHEAD [4.48] see also [4.1], p.11) have investigated the "remarkable degree of structure" that exists even at Rayleigh numbers of order $10^5$–$10^6$ ($Ra = 10^6$ for a $10° C$ temperature difference between plates 10 cm apart in air). For a theoretical attack on stratified turbulence which pays attention to structural details, see LONG [4.49], and for a short review by the same author see [4.50].

## 4.2.2 Density Interfaces

The practical case is a stable interface (heavier fluid below). The applications are mainly geophysical (or at least hydrological); the data are mainly from the laboratory. A recent paper which reviews previous work is that of CRAPPER and LINDEN [4.51]. In some cases the fluid below (or above) the interface may be almost non-turbulent, as when a turbidity current flows below, or a river discharge above, an ocean which is itself stably stratified. Even when both layers are turbulent the interfacial region may be virtually free of turbulent velocity fluctuations. In CRAPPER and LINDEN's work the flux across the interface was less than that produced by molecular diffusion alone. The failure of the interface to transfer horizontal momentum in a vertical direction leads to the Kelvin-Helmholtz instability to internal waves, mentioned in Subsection 4.1.2. For a field which is uniformly turbulent at large positive or negative distances from the interface, the ratio of the entrainment velocity (rate of increase of interface thickness, $h$) to the velocity scale of the turbulence, $u$, is a function of the overall Richardson number $-g\Delta\varrho h/(\varrho u^2)$, documented by ELLISON and TURNER [4.52]. The complications introduced by more involved boundary conditions are almost limitless, and application of modern methods of turbulence modelling, or of complete time-dependent simulation, seems to be overdue. Theoretical treatment is hampered (but time-dependent simulation made easier) by the fact that even geophysical examples of density interfaces constitute flows at quite low Reynolds number so that viscosity affects the energy-containing range of the turbulence [4.53]. LINDEN and TURNER ([4.1], pp. 8–10) report recent interface studies, including some unpublished work on turbulence modelling by LUMLEY and SIESS. The extreme example of a density interface is that between air and water at the ocean surface. The role of air turbulence in water-wave generation is not certain and may be negligible; of the other surface phenomena, spray is discussed, in the guise of droplet suspensions, in Chapter 7, while the effect of surface roughness on a turbulent boundary layer was outlined in Section 2.3.

### 4.2.3 Convection from Inclined or Vertical Surfaces

Again the processes of transition from laminar to turbulent flow are complicated. LOYD et al. [4.54] observed longitudinal vortices on an inclined heated plate. These are analogous to the Taylor-Görtler vortices that appear in boundary layers on concave surfaces where "centrifugal force" takes the place of buoyancy. Two-dimensional convection cells can theoretically occur on a horizontal heated plate but often appear in practice only when there is a preferred horizontal direction, as near the edge of the plate [4.55]. JALURIA and GEBHART [4.56], studying transition on a vertical heated plate, found that the Grashof number $Gr \equiv (g/T)(T_w - T_\infty) x^3/v^2$—which is related to $1/\sigma$ times a Rayleigh number—was not a sufficient parameter but that transition depended upon the vertical coordinate $x$ itself. The explanation may well be the influence of unsteadiness or other nonuniformity of the air in the room (referred to again in the discussion of plane plume transition in [4.44]) but the difficulty is mentioned here to emphasize again that, as a class, free-convection problems tend to be ill posed.

A group of experimental papers [4.58–60] on the turbulent free-convection boundary layer on a vertical plate appeared in 1968–69. A review and paper by GEBHART [4.61, 62] deal mainly with laminar flow and instability but throw light on the turbulent problem. The flow is very similar to that in a wall jet (Section 2.2) except that it accelerates up the plate; the peak velocity is proportional to $[g(T_w - T_\infty) x/T]^{0.5}$. Simple entrainment arguments predict $\delta \alpha x$, but CHEESEWRIGHT [4.58] points out that different length scales may be needed for the inner and outer regions. This is the case in a wall jet too, but is even more noticeable in a convection boundary layer. As pointed out in [4.63] turbulent energy in buoyant flow up a plate is produced mainly by the mean shear and not by buoyancy. This ties up with the known behavior of convection cells (Subsection 4.2.1), and the implication that the turbulence structure may be fairly close to that of conventional wall layers is reinforced by successful predictions by CEBECI and KHATTAB (see p. 325) using an eddy-viscosity formula derived for boundary layers. However, free-convection layers, like many other buoyancy-driven turbulent flows, have rather low Reynolds numbers whose effects on the turbulence structure may complicate the picture [4.64].

The behavior of convection layers on inclined plates clearly spans the whole range between vertically oriented convection cells on a horizontal plate and the spanwise-homogeneous flow up a vertical plate. It appears that inclined plumes or roll vortices with their axes "uphill" replace convection cells (or the less well-organized patterns that appear when there is no upper boundary) at all but very small angles of inclination. TRITTON [4.65] found a well-established boundary-layer-like flow up the plate for angles as small as 10 degrees. For part of the time, however, plumes will erupt from the "boundary layer" and continue to rise into the still air above. The highly stable flow *under* an inclined plate has also been studied by TRITTON—but other data seem rare.

### 4.2.4 Turbulent Convection in Confined Spaces

In a roughly cubic container closed on all sides and uniformly heated from below (say), the multicell convection pattern of quasi-steady multiple cells found between infinitely wide horizontal plates (i.e., in a very shallow container) is replaced by a single region of hot upflow near the axis and a fairly thin descending layer of cold fluid on the walls [4.66]. This of course is an enclosed version of a single convection cell. The fluid surrounding the central column recirculates in a toroidal vortex, with mean circumferential vorticity approximately proportional to radius (Ref. [1.8], p. 508). This is the equivalent of a two-dimensional recirculation zone with *constant* vorticity, and the flow as a whole is similar to the momentum-driven recirculation in the separated flow behind a bluff body except that it is rather steadier. If the walls are cooled, the downgoing return flow may resemble an isolated free-convection boundary layer closely enough for results from the latter to be useful.

The flow in a roughly cubic closed container with a point—or line—source of buoyancy on the floor [4.66] is similar to the case above, but a different kind of flow arises if the container is wider than its height. In that case BAINES and TURNER [4.67] showed that the plume fluid settles as a stable layer on the roof of the container, displacing a roughly uniform downward flow. The reason is simply that after impinging on the roof the plume loses its (now horizontal) momentum before it manages to reach the walls. This is the customary behavior of a hot plume in the atmospheric surface layer below an inversion (simulating the container roof). The flow in a shallow container with a *uniformly* heated floor approximates to the classical parallel-plate Bénard flow discussed in Subsection 4.2.1.

When opposite *walls* of a container are heated and cooled, respectively, the flow depends strongly on the height/width ratio. Presumably—though there seem to be no data—the side-wall layer in a shallow container will spread out over the roof or floor, behaving like half the plume from a line source in the middle of the floor. Casual observation suggests that the downdraft from a (closed) window behaves in this way (height/width of the order of 0.5 for typical rooms). For height/width ratios of unity or somewhat above, a single-cell pattern is likely. For very tall narrow containers ("slots") the central region of the flow is apparently [4.68] highly turbulent but nearly stationary in the mean. Its source of energy is not certain but is probably associated with the attempt of one side-wall layer to entrain fluid from the other, leading to horizontal diffusion of turbulent energy.

There is considerable industrial interest in mixed convection in vertical pipes, where the buoyancy force and the pressure difference applied to the pipe aid or oppose each other. Detailed data for the turbulent case are scarce; two recent experimental papers are [4.69, 70], while [4.71] discusses the reverse transition from turbulent to laminar flow that can occur in "aiding" flow.

Stratified flows in *horizontal* pipes, and in the related situations of open channels and river estuaries, are discussed by TURNER [4.20]; for a review of a related confined-plume problem see [4.72].

TURNER's book [4.20] is a general review of its subject with an emphasis on laboratory experiments (frequently as idealizations of geophysical problems). The paper by THORPE (Ref. [4.73], p. 95) deals specifically with laboratory experiments in stably stratified fluids, while [4.73] as a whole is an excellent entry to current work on stable layers. A recent set of conference proceedings [4.74], nominally on geophysical turbulence, actually contains many reports of laboratory-scale work on buoyant and non-buoyant flows. As will be clear from the reference list, a large fraction of the basic research work on laboratory-scale buoyant flows appears in the *Journal of Fluid Mechanics*, and most of the rest in geophysical journals.

# References

4.1   P.F.LINDEN, J.S.TURNER: J. Fluid Mech. **67**, 1 (1975)
4.2   E.A.SPIEGEL: Ann. Rev. Astron. Astrophys. **9**, 323 (1971)
4.3   J.C.WYNGAARD, S.P.S.ARYA, O.R.COTÉ: J. Atmos. Sci. **31**, 747 (1974)
4.4   T.I.MCLAREN, A.D.PIERCE, T.FOHL, B.L.MURPHY: J. Fluid Mech. **57**, 229 (1973)
4.5   J.A.BUSINGER, J.C.WYNGAARD, Y.IZUMI, E.F.BRADLEY: J. Atmos. Sci. **28**, 181 (1971)
4.6   H.TENNEKES: J. Atmos. Sci. **30**, 234 (1973)
4.7   H.A.PANOFSKY: Ann. Rev. Fluid Mech. **6**, 147 (1974)
4.8   J.A.SAXTON (ed.): Radio Sci. **4**, 1099 (1969)
4.9   J.H.OLSEN, A.GOLDBURG, M.ROGERS (eds.): *Aircraft Wake Turbulence and Its Detection* (Plenum Press, New York 1971)
4.10  C.L.TING, D.R.HAY: J. Atmos. Sci. **32**, 637 (1975)
4.11  J.C.WYNGAARD, Y.IZUMI: unpublished paper, Air Force Cambridge Res. Lab. (1973)
4.12  J.K.ANGELL: J. Atmos. Sci. **29**, 1252 (1972)
4.13  M.A.LEMONE: J. Atmos. Sci. **30**, 1077 (1973)
4.14  J.E.CERMAK: Trans. ASME **97**I, 9 (1975)
4.15  D.J.MOORE, A.G.ROBINS: Proc. Inst. Mech. Engrs. **189**, 33 (1975)
4.16  C.I.H.NICHOLL: J. Fluid Mech. **40**, 361 (1970)
4.17  R.G.FLEAGLE: Science **176**, 1079 (1972)
4.18  H.TENNEKES Phys. Today **27**, No. 1 (1974)
4.19  F.PASQUILL: *Atmospheric Diffusion*, 2nd ed. (Ellis Horwood, Chichester 1974)
4.20  J.S.TURNER: *Buoyancy Effects in Fluids* (University Press, Cambridge 1973) pp. 157—161
4.21  F.B.SMITH: Sci. Progr. **62**, 127 (1975)
4.22  D.A.HAUGEN (ed.): Proceedings, Workshop on Meteorology (American Meteorological Soc., New York 1973)
4.23  Y.TOBA, M.TOKUDA, K.OKUDA, S.KAWAI: "Systematic Convection Accompanying Wind Waves"; Proc. Symp. on Ocean Microstructure, Grenoble, 1975. (To be published by IAPSO)
4.24  M.S.LONGUET-HIGGINS, J.S.TURNER: J. Fluid Mech. **68**, 1 (1975)
4.25  C.GARRETT: J. Marine Res. **34**, 117 (1976)
4.26  S.A.KITAIGORODSKII: *The Physics of Air-Sea Interaction* (Israel Program for Scientific Translations, Jerusalem 1970)
4.27  J.D.WOODS: J. Fluid Mech. **32**, 791 (1968)
4.28  P.NASMYTH: Ph.D. dissertation, Dept. of Oceanography, Univ. of British Columbia (1972)
4.29  P.HAZEL: J. Fluid Mech. **51**, 39 (1971)
4.30  B.J.WEST, K.M.WATSON, J.A.THOMPSON: Phys. Fluids **17**, 1059 (1974)
4.31  P.MUELLER: Hamburger Geophys. Einzelschriften **23** (1974)
4.32  H.L.GRANT, A.MOILLIET, W.M.VOGEL: J. Fluid Mech. **34**, 443 (1968)
4.33  C.H.GIBSON: Adv. Geophys. **18**A (1974)
4.34  G.K.BATCHELOR (ed.): *The Collected Works of Sir Geoffrey Ingram Taylor* (University Press, Cambridge 1960) Vol. 2, p. 240
4.35  K.A.BROWNING, G.W.BRYANT, J.R.STARR, D.N.AXFORD: Quart. J. Roy. Met. Soc. **99**, 608 (1973)
4.36  J.D.WOODS: Mem. Soc. Royale des Sciences de Liege, 6e serie 7, 171 (1975)
4.37  J.A.FAY: Ann. Rev. Fluid Mech. **5**, 151 (1973)
4.38  P.R.SLAWSON, G.T.CSANADY: J. Fluid Mech. **47**, 33 (1971)
4.39  H.J.SNECK, D.H.BROWN: Trans. ASME **96**C, 232 (1974)

4.40 E. M. Sparrow, R. B. Husar, R. J. Goldstein: J. Fluid Mech. **41**, 793 (1970)
4.41 D. Coles: J. Fluid Mech. **21**, 385 (1965)
4.42 R. Krishnamurti: J. Fluid Mech. **60**, 285 (1973)
4.43 B. J. Daly: J. Fluid Mech. **64**, 129 (1974)
4.44 D. B. Thomas, A. A. Townsend: J. Fluid Mech. **2**, 473 (1957)
4.45 A. A. Townsend: J. Fluid Mech. **5**, 209 (1959)
4.46 J. W. Deardorff, G. E. Willis: J. Fluid Mech. **28**, 673 (1967)
4.47 J. W. Deardorff, G. E. Willis, D. K. Lilly: J. Fluid Mech. **35**, 7 (1969)
4.48 F. H. Busse, J. A. Whitehead: J. Fluid Mech. **66**, 67 (1974)
4.49 R. R. Long: J. Fluid Mech. **42**, 349 (1970)
4.50 R. R. Long: Appl. Mech. Rev. **25**, 1297 (1972)
4.51 P. F. Crapper, P. F. Linden: J. Fluid Mech. **65**, 45 (1974)
4.52 T. H. Ellison, J. S. Turner: J. Fluid Mech. **6**, 423 (1959)
4.53 J. S. Turner: J. Fluid Mech. **33**, 639 (1968)
4.54 J. R. Loyd, E. M. Sparrow, E. R. G. Eckert: Intern. J. Heat Mass Transfer **15**, 457 (1972)
4.55 J. W. Deardorff, G. E. Willis: J. Fluid Mech. **23**, 337 (1965)
4.56 Y. Jaluria, B. Gebhart: J. Fluid Mech. **66**, 309 (1974)
4.57 R. G. Bill, B. Gebhart: Intern. J. Heat Mass Transfer **18**, 513 (1975)
4.58 R. Cheesewright: Trans. ASME **90**C, 1 (1968)
4.59 C. Y. Warner, V. S. Arpaci: Intern. J. Heat Mass Transfer **11**, 397 (1968)
4.60 G. C. Vliet, C. K. Liu: Trans. ASME **91**C, 519 (1969)
4.61 B. Gebhart: Appl. Mech. Rev. **22**, 691 (1969)
4.62 B. Gebhart: Trans. ASME **91**C, 293 (1969)
4.63 K. T. Yang, V. W. Nee: Rept. THEMIS-UND-1 Engrg. Dept., Univ. Notre Dame (1969)
4.64 G. S. H. Lock, F. J. deB. Trotter: Intern. J. Heat Mass Transfer **8**, 1225 (1968)
4.65 D. J. Tritton: J. Fluid Mech. **16**, 282 (1963)
4.66 K. E. Torrance, L. Orloff, J. A. Rockett: J. Fluid Mech. **36**, 21 (1969)
4.67 W. D. Baines, J. S. Turner: J. Fluid Mech. **37**, 51 (1969)
4.68 J. W. Elder: J. Fluid Mech. **23**, 99 (1965)
4.69 A. D. Carr, M. A. Connor, H. O. Buhr: Trans. ASME **95**C, 445 (1973)
4.70 H. O. Buhr, E. A. Horsten, A. D. Carr: Trans. ASME **96**C, 152 (1974)
4.71 A. Steiner: J. Fluid Mech. **47**, 503 (1971)
4.72 D. R. F. Harleman, K. D. Stolzenbach: Ann. Rev. Fluid Mech. **4**, 7 (1972)
4.73 R. Munn (ed.): Boundary-Layer Met. **5**, 1 (1973)
4.74 F. N. Frenkiel, R. E. Munn (eds.): Turbulent Diffusion in Environmental Pollution; Adv. Geophys. **18**A (1974): see also **18**B

# 5. Calculation of Turbulent Flows

W. C. Reynolds and T. Cebeci

With 3 Figures

This chapter deals only with the calculation of the velocity field. Calculation of heat or pollutant transfer, which depends on knowledge of the velocity field, is discussed in Chapter 6. The present discussion refers mainly to shear layers, being the commonest type of flow with significant Reynolds stresses, and some of the methods presented are restricted to particular types of thin shear layer, notably boundary layers.

At the time of writing, calculation methods are being developed by many different research groups, and it would be misleading to present an apparently definitive review of the most recent methods; the speculations of today may be the standard methods, or the anathemata, of tomorrow. Section 5.1 of the chapter is an outline of the strategy of the different types of calculation methods, current research being quoted mainly as an illustration of the principles involved. Subsections 5.2.1–5.2.4 discusses, in rather more detail, the older—and generally simpler—methods which have reached a fairly stable stage of development. Although some of the latest methods are intended to apply to a very wide range of flow configurations, their abilities have not yet been fully tested, primarily because of a lack of experimental data. The would-be user with an exotic flow geometry might do better to begin with one of the older methods, possibly accepting the need to adjust some of the empirical input to optimize his results. It is for this reason that the review which follows is not exclusively forward looking.

## 5.1 Strategy and Recent Developments

W. C. Reynolds[1]

### 5.1.1 Background

The computation of turbulent flows, essentially the search for a model of the Reynolds stresses appearing in (1.18), has been a problem of

---

[1] Synopsis of a paper: Computation of Turbulent Flows (Volume 8, Ann. Rev. Fluid Mech, 1976).

major concern from the time of Osborne Reynolds. Until the advent of high-speed computers, the range of turbulent flow problems that could be handled was very limited. Most boundary-layer prediction methods were highly empirical and based on ordinary differential equations, always including the momentum integral equation (2.21). Perturbed flows like those discussed in Section 2.3 could not be predicted with any degree of reliability.

Midway through this century computers began to have a major impact, and by the mid-1960's several workers were actively developing methods based on the governing partial differential equations (pde's) of Chapter 1. The first such methods used only the mean-motion equations but later methods began to incorporate turbulence pde's (transport equations) related to (1.35).

In 1968 Stanford University held a specialists' conference designed to assess the accuracy of the then current turbulent boundary-layer prediction methods [3.72]. The main impact of this conference was to legitimize pde methods, which proved to be more accurate and more general than the best integral methods. Vigorous development of pde turbulence models then followed. The ability of these more complicated models to produce predictions for the detailed features of turbulent flows has outstripped the available storehouse of data against which the predictions can be compared; moreover the predictions now include quantities which are difficult or impossible to measure.

Recently, some of the largest computers have been used to solve the three-dimensional time-dependent Navier-Stokes equations for the large-scale motion in turbulent flows, with a simple semi-empirical model for the statistics of the motion at scales smaller than the finite-difference grid size. It seems quite likely that this type of calculation will eventually be useful at the engineering level, but it is already serving as a way of performing "experiments" to yield some of the unmeasurable quantities mentioned above, in some cases producing numerical values for the empirical constants used in Reynolds-stress models.

This review will outline the essential ingredients and effectiveness of several levels of turbulent flow pde models: for methods based on ordinary differential equations, which still have some uses, see Section 5.2.

1) *Zero-equation models*—models using only the pde's for the mean velocity field, and *no* turbulence pde's.

2) *One-equation models*—models involving one pde relating to a turbulence velocity scale, in addition to the mean-flow pde's.

3) *Two-equation models*—models using an additional pde related to a turbulence length scale.

4) *Stress-equation models*—models involving pde's for all components of the Reynolds stress tensor and in general for a length scale as well.

5) *Large-eddy simulations*—computations of the three-dimensional time-dependent large eddy structure and a low-level model for the small-scale turbulence.

Class 1 is also called the class of "mean-field" closures, classes 2 to 4 being "transport-equation" closures. An alternative, open-ended classification is based on the highest order of velocity product for which a transport equation is used. Zero-equation models use pde's for $U_i$ only and are therefore "first-order" models, classes 2 to 4 use pde's for $\overline{u_i u_j}$ and are "second-order" closures, while some class 4 models approach the stage of using transport equations for "third-order" products $\overline{u_i u_j u_l}$.

Zero-equation models are in common use in the more sophisticated engineering industries, and one-equation models find use there on occasion. Two-equation models, currently popular among academics, have not been used extensively for engineering applications. Stress-equation modeling is now under intensive development; it is essential for handling the more difficult flows and will probably become standard practice in ten years' time. Large-eddy simulations are in their infancy, and are serving mainly to help assess the lower-level models. However, in the long term, large-eddy simulation may be the only way to deal accurately with the difficult flows that stress-equation models are currently trying to handle.

Zero-equation pde models, mostly based on the eddy-viscosity and mixing-length concepts mentioned in Chapter 1, are discussed in detail in Subsection 5.2.3, and in Subsections 5.1.2–5.1.5 we outline the current state of levels 2 to 5 above. Other recent reviews are [5.1–5.4]. We ignore numerical problems; for an introduction see [1.25], and for discussions of particular cases see [2.208, 209] and the papers on modelling referenced below. Note that unsteady flow presents only numerical problems, providing that the turbulence model can handle sufficiently large mean transport terms. The same applies to three-dimensional flow, with slightly stronger reservations about the applicability of the turbulence model.

### 5.1.2 One-Equation Models

In typical one-equation turbulence models, the turbulent kinetic energy equation (1.36) forms the basis for a model equation for the turbulence velocity scale $q \equiv \sqrt{\overline{q^2}} \equiv \sqrt{\overline{u_i^2}}$, following PRANDTL [5.5]. The eddy viscosity (assumed to be a scalar) is *defined* by

$$\overline{u_i u_j} \equiv \tfrac{1}{3}\overline{q^2}\delta_{ij} - v_{\mathrm{T}}(U_{i,j} + U_{j,i}) \equiv \tfrac{1}{3}\overline{q^2}\delta_{ij} - 2v_{\mathrm{T}}S_{ij} \tag{5.1}$$

following (1.3) and using the comma-suffix notation for differentiation as explained in Section 1.2. Equation (5.1) defines the mean rate of strain $S_{ij}$. The difference between zero-equation and one-equation models is that in the latter $v_{\mathrm{T}}$, instead of being related directly to the mean-flow scales, is modeled by

$$v_{\mathrm{T}} = c_2 q l . \tag{5.2}$$

An alternative, for thin shear layers in which the only important Reynolds stress gradient is that of $\overline{u_1 u_2}$, is to relate $\overline{u_1 u_2}$ directly to $q$ by

$$\overline{u_1 u_2} = a q^2 \tag{5.3}$$

where $a$ is a dimensionless constant or a function of $y/\delta$ where $\delta$ is the shear-layer thickness. A length scale is still needed, because the turbulent transport (diffusion) and dissipation terms in the $q$ equation have dimensions of (velocity)$^3$/(length) and are modeled as functions of the scales $q$ and $l$. The use of (5.1) and (5.2) unhooks $v_{\mathrm{T}}$, though not $\overline{u_i u_j}$, from the mean velocity field, which is a first step to representing the dependence of $\overline{u_i u_j}$ on flow history as expressed by (1.35). Use of (5.3) produces an equation for $\overline{u_1 u_2}$ of the same form as would be obtained by modelling (1.35) directly but uses the better-documented equation (1.36). In both approaches $l/\delta$ is taken as an empirical function of $y/\delta$.

Recalling from Chapter 1 that at high turbulence Reynolds numbers (e.g., high $ql/v$) the dissipation rate $\varepsilon$ is independent of viscosity and determined by the properties of the energy-containing eddies, we can write it as a multiple of $q^3/l$. Again at high turbulence Reynolds numbers, the difference between $\varepsilon$ and the "isotropic dissipation" $\mathscr{D} = v(\overline{u_{i,j} u_{i,j}})$ is small; we use the latter for later convenience. We write

$$\mathscr{D} = c_3 q^3 / l . \tag{5.4}$$

The turbulent transport is usually modeled by using an eddy viscosity proportional to $v_{\mathrm{T}}$, so

$$\frac{1}{\varrho}\overline{p' u_j} + \tfrac{1}{2}\overline{u_i^2 u_j} - \overline{v q_{,j}^2} = -(c_4 v_{\mathrm{T}} + v)q_{,j}^2 . \tag{5.5}$$

If $\mathscr{D}$ is substituted for $\varepsilon$ the last term on the left must be substituted for the exact viscous transport term.

NORRIS and REYNOLDS [5.6] proposed a one-equation model for use in the viscous sublayer as well as in the fully turbulent regions. Noting that at very low $ql/v$ the dissipation should scale as $vq^2/l^2$, they use

$$\mathscr{D} = c_3 \frac{q^3}{l}\left(1 + \frac{c_5}{ql/v}\right). \tag{5.6}$$

They argue that the length scale $l$ should do nothing special in the viscous region, but should behave like $l = Ky$ all the way down to the wall. However they assume that the turbulent transport is suppressed by the presence of the wall and hence take

$$v_T = c_2 ql[1 - \exp(-c_6 qy/v)] \tag{5.7}$$

in (3.1) and (3.5). A similar approach was adopted by the Imperial College group, reported by WOLFSHTEIN [5.7]. However WOLFSHTEIN allowed the length scale to depart from $Ky$ in the viscous region, but kept the same behavior (5.4) for the dissipation. When placed in comparable form, the constants used by WOLFSHTEIN and by NORRIS and REYNOLDS are quite similar.

Most workers have abandoned one-equation models in favor of two-equation or even stress-equation models. However it may be that one can do better with this sort of one-equation model in most flows of interest, for it may be easier to specify the length-scale distribution than to compute it with a pde. This would be particularly true if the length scale really should be governed by the global features of the flow through an integro-differential equation. Hence, further study of extended one-equation models is encouraged.

### 5.1.3 Two-Equation Models

In attempts to eliminate the need for specifying the turbulence length scale $l$ as a function of position throughout the flow, several workers have explored the use of a second turbulence pde which in effect gives $l$. In most cases (5.1) is still used, so that $\overline{u_i u_j}$ responds at once to changes in the mean strain field. Exact equations for turbulence length scales related to the integral scale (Section 1.4) have been derived by ROTTA [5.8]. The groups at Imperial College and at Stanford both experimented with *ad hoc* transport equations for $l$ with no real success. However these and other groups have achieved greater success using an equation for the isotropic dissipation $\mathscr{D}$, based on the exact transport equation for $\mathscr{D}$ which follows from differentiating the $x_i$-component Navier-Stokes equation with respect to $x_j$, multiplying by $\partial u_i/\partial x_j$ and averaging. Of course, this equation implies a transport equation for $l$ through (5.6), using the transport equation for $q^2$ ($c_3$ being disposable and usually taken as unity). Since neither the terms in Rotta's length-scale equation nor those in the $\mathcal{D}$ equation have ever been measured, the reason for the greater fruitfulness of the $\mathcal{D}$ equation is not entirely clear. However it can be shown that modelling turbulent transport of $l$ by an eddy viscosity gives impossible results in the inner layer of a wall flow, and most workers have responded by abandoning the $l$ equation rather than the eddy viscosity concept. The alternative to eddy viscosity is the bulk-convection hypothesis introduced by Townsend [1.22] in which turbulent energy and other quantities are supposed to be transported by the turbulence at a velocity that depends on the energy distribution but not on the local gradient (see also Subsect. 6.3.2). This is a more plausible model than gradient transport (eddy viscosity) for transport by the large eddies. However there is as yet no quantitative model for the behavior of the bulk-convection velocity. For a general (and adversely critical) discussion of gradient-transport concepts in turbulence see Corrsin (Ref. [4.74], p. 25).

Using a superscript dot to denote time differentiation, the $\mathcal{D}$ equation is

$$\dot{\mathcal{D}} + U_j \mathcal{D}_{,j} = -W - H_{j,j}. \tag{5.8}$$

Here

$$W = 2\nu \overline{u_{i,j} u_{j,l} u_{l,i}} + 2\nu^2 \overline{u_{i,jj} u_{i,ll}}$$
$$+ 2\nu (\overline{u_{i,j} u_{i,l}} U_{j,l} + \overline{u_{i,l} u_{j,l}} U_{i,j}) + 2\nu \overline{u_j u_{i,l}} U_{i,jl} \tag{5.9}$$

$$H_j = \nu \overline{u_{i,l} u_{i,l} u_j} + 2\nu \overline{u_{j,i} p'_{,i}} - \nu \mathcal{D}_{,j}. \tag{5.10}$$

$H_j$ represents the diffusive flux of $\mathcal{D}$ in the $j$ direction.

The systematic workers have insisted that their two-equation models first describe properly the decay of isotropic turbulence, and then have worried about the behavior of their models in homogeneous shear flows where the transport terms vanish. The reader will note that it is thereby assumed that the same empirical input suffices for both types of flow. For the isotropic decay problem,

$$\dot{q}^2 = -2\mathcal{D} \qquad \dot{\mathcal{D}} = -W. \tag{5.11a, b}$$

$W$ is a scalar for which a closure assumption is needed. In this problem $W$ must be a function of the only other variables available, $q^2$ and

$\mathscr{D}$, and from dimensional arguments must be (at high Reynolds number)

$$W = c_7 \mathscr{D}^2/q^2 .\qquad (5.12)$$

The exact solution for the decay is

$$q^2 = q_0^2 (1+t/a)^{-n} \qquad \mathscr{D} = \mathscr{D}_0 (1+t/a)^{-(n+1)} \qquad (5.13\text{a, b})$$

$$a = nq_0^2/(2\mathscr{D}_0) \qquad n = 2/(c_7 - 2) . \qquad (5.14\text{a, b})$$

Here $q_0^2$ and $\mathscr{D}_0$ are the initial values. Early experiments suggested $n=1$, which gives $c_7 = 4$. COMTE-BELLOT and CORRSIN [5.9] took special care to obtain better isotropy, and their data reveal $n$ values in the range 1.1–1.3. LUMLEY and KHAJEH-NOURI [5.10] suggested that slight anisotropies are responsible for these differences, and proposed a higher-order model to take this into account. But this theory does not explain the different values observed in truly isotropic decay, as revealed in Table 3 of [5.9]. It can be shown (REYNOLDS [5.4]) that the details of the low wave-number portion of the spectrum are responsible for these differences. Since these details are in no way represented by the scales $q^2$ and $\mathscr{D}$, there is no way that a model using only these scales can exactly predict the decay of laboratory grid turbulence. However, it˙is possible to make a fairly rational choice of $c_7$. We really should expect the model to work only when the large-scale structure is devoid of any scales, i.e., when the large-scale energy is uniformly distributed over all wave **vectors.** This occurs only when $\Phi_{ii}(\mathbf{k})$ is the same at all $\mathbf{k}$ not too near zero. The three-dimensional energy spectrum function (see p. 22) is $E(k) = 2\pi k^2 \Phi_{ii}(\mathbf{k})$, and represents the energy associated with a shell of wave vector space. Hence, in "equi-partitioned" large-scale turbulence, $E(k) \sim k^2$. On this basis we recommend $n=6/5$, which gives $c_7 = 11/3$. This is close to the value used by [5.10] and [3.21].

When strain is applied to the flow, there is every reason to expect an alteration in $W$; something must provide a "source" of $\mathscr{D}$, and this must depend in some way on the mean flow. Lumley has argued that this cannot come from the terms in $W$ explicitly containing the mean velocity, but must come from the first two terms in $W$ [see (5.9)], which are very large but of opposite sign. Lumley feels that the alteration of $W$ by strain should be modeled in terms of the anisotropy of the Reynolds stress tensor.

It seems most desirable to model the source of $\mathscr{D}$ by reference to experiments in nearly homogeneous flow, where the transport would not confuse the issue. There are two types of such flows, those involving pure strain and those involving pure shear. TUCKER and REYNOLDS [5.11] and Marechal [5.12] studied the pure strain case; Champagne et al. [5.13] and Rose [5.14] studied homogeneous shearing flows. For their use in modeling, see Reynolds [5.4].

The assumption that $v_T$ in (5.1) is a scalar forces the principal axes of $\overline{u_i u_j}$ and the mean strain rate $S_{ij}$ to be aligned. This is true in pure strain, but not true in any flow with mean vorticity. One is tempted to try a modified constitutive equation (see Saffman [5.15])

$$\overline{u_i u_j} = \frac{q^2}{3} \delta_{ij} - 2v_T S_{ij} - c_{11} l^2 (S_{il}\Omega_{lj} + S_{jl}\Omega_{li}) \tag{5.15}$$

where $\Omega_{ij} = 1/2(U_{i,j} - U_{j,i})$ is the rotation tensor. In a two-equation model $l$ could be expressed in terms of $q^2$ and $\mathcal{D}$. Equation (5.15) does produce the right sort of normal-stress anisotropy in shear flows, but the new terms do not alter the shear stress, and hence (5.15) works no better than (5.1) in practice. Two-equation models also fail to predict the return to isotropy after the removal of strain, or the isotropizing of grid-generated turbulence ([5.9]). This failure arises because of the need for a constitutive equation for the $\overline{u_i u_j}$. Thus, one should not really expect two-equation models to be very general, although they might be made to work well with specific constants in specific cases, such as boundary layers.

In spite of these difficulties with models based on constitutive equations, their simplicity makes them attractive. Two-equation models have been studied by a number of groups, and it is significant that these workers inevitably turn to stress-equation models because of the difficulties outlined above. Stress-equation models have their own problems, and so there probably is still considerable room for development of two-equation models. Of particular interest is turbulent boundary layer separation, where anisotropy of the normal stresses is known to be important. Since (5.1) cannot give this properly in a shear layer, but (5.15) can, the use of (5.15) in conjunction with two-equation models should be explored further.

To use the two-equation model outlined above in an inhomogeneous flow, one needs to assess (or neglect) the effects of inhomogeneity on $W$, and also to model the transport term $H_j$. Jones and Launder [5.16, 17] assume that $W$ is not modified by inhomogeneity and use a gradient-transport model for $H_j$,

$$H_j = -(v + c_{12} v_T)\mathcal{D}_{,j} \tag{5.16}$$

with $c_{12} = 0.77$. Lumley (see Ref. [4.73], p. 169) argues on formal grounds that the diffusive flux of dissipation should depend as well on the gradients in turbulence energy, and vice versa, in the manner of coupled flows such as thermoelectricity and thermodiffusion studied by the methods of irreversible thermodynamics. If this is true, one really should use models of the form

$$J_j = -A_{11}q^2_{,j} - A_{12}\mathcal{D}_{,j}, \tag{5.17a}$$

$$H_j = -A_{21}q^2_{,j} - A_{22}\mathcal{D}_{,j}. \tag{5.17b}$$

One difficulty with using the $\mathcal{D}$ equation as the basis for a second model equation has escaped the model developers. This arises from the second term in (5.10), the pressure gradient-velocity gradient term in the transport $H_j$. Since the pressure field depends explicitly upon the mean velocity field (Sect. 5.4), mean velocity gradients can explicitly give rise to $\mathcal{D}$ transport. This could be an extremely important effect, especially near a wall. The omission of this consideration would seem to be a serious deficiency in all $\mathcal{D}$ equation models that have been studied to date.

Other two-equation models have been heuristically conceived. Of these the most well developed is the SAFFMAN-WILCOX [5.19] model. Instead of a $\mathcal{D}$ equation they use an equation for a "pseudo-vorticity" $\Omega$, a typical inverse time-scale of the energy-containing eddies (see also [5.20]),

$$\dot{\Omega}^2 + U_j\Omega^2_{,j} = [\alpha\sqrt{U_{i,j}U_{i,j}} - \beta\Omega]\Omega^2$$
$$+ [(\nu + \sigma\nu_T)\Omega^2_{,j}]_{,j}. \tag{5.18}$$

In conjunction with this they use the $q^2$ equation (1.36) with the production and "isotropic dissipation" terms modeled as $\mathcal{P} = \alpha^*\sqrt{2S^2}\,q^2/2$ and $\mathcal{D} = \beta^*q^2\Omega/2$, respectively. The production term $\mathcal{P}$ as modeled in [5.19] is inconsistent with the $\overline{u_iu_j}$ transport equation; this seems to be an internal inconsistency in the model, but it may in fact be a strength. The model of the production is based on the experimental fact that the *structure* of the turbulence in the wall region of a boundary layer is essentially independent of the strain rate, and hence the production should be proportional to $q^2$. Hence, the SAFFMAN-WILCOX ([5.19]) model is a curious blend of the "Newtonian" and "structural" alternatives, (5.1) and (5.3); see also [5.2].

### 5.1.4 Stress-Equation Models

In turbulent shear flows, the energy is usually first produced in one component and then transferred to the others by turbulent processes.

Exact equations for $\overline{u_i u_j}$ were derived from the Navier-Stokes equations in Chapter 1; for an incompressible fluid, (1.35) can be written formally as

$$\dot{\overline{u_i u_j}} + U_l \overline{u_i u_j}_{,l} = P_{ij} + \phi_{ij} - J_{ijl,l} - D_{ij} \tag{5.19}$$

where $P_{ij}$ is the "generation tensor", $\phi_{ij}$ is the pressure-strain "redistribution tensor", $D_{ij}$ is the "isotropic dissipation tensor", and $J_{ijl}$ is the "transport" of $\overline{u_i u_j}$.

$P_{ij}$ is explicit, but models are needed for $\phi_{ij}$, $D_{ij}$, and $I_{ijl}$. In addition, one must specify at least one length scale, explicitly or by a $\mathcal{D}$ equation. An eddy viscosity is not needed to derive $\overline{u_i u_j}$, but may be used in modeling $J_{ijl}$.

The one fact that seems very clear from experiments is that, at high Reynolds number, the small-scale dissipative structures are isotropic. Hence all workers now use

$$D_{ij} = \tfrac{2}{3} \mathcal{D} \delta_{ij} . \tag{5.20}$$

The redistribution term $\phi_{ij}$ has been the subject of most controversy and experimentation. In a flow without any mean strain, this term is responsible for the return to isotropy. However, in deforming flows the situation is much more complicated. Guidance is provided by the exact equation for the fluctuating pressure (1.11) which can be written, after separating mean and fluctuating parts, as

$$p'_{,ii} = -2u_{i,j} U_{j,i} - u_{i,j} u_{j,i} . \tag{5.21}$$

The source term in this Poisson equation contains two parts, each of which will be responsible for a part of the pressure field. One part involves the mean deformation explicitly, and its contribution to $\phi_{ij}$ can be obtained for homogeneous fields in terms of the Fourier transform of the velocity field, as $\phi_{ij2}$ ($T_{1ij}$ in the notation of [5.4])

$$\phi_{ij2} = 2U_{q,p} G_{ijpq} \tag{5.22a}$$

where

$$G_{ijpq} = \int \left[ \frac{k_j k_q}{k^2} \Phi_{ip}(k) + \frac{k_i k_q}{k^2} \Phi_{jp}(k) \right] dk \tag{5.22b}$$

and $\Phi(k)$ is the wave number spectrum (1.23).

Models for $G_{ijpq}$ have been proposed; see HANJALIĆ and LAUNDER [3.56], LAUNDER et al. [3.21] and LUMLEY [5.21].

The remaining part of $\phi_{ij}$, which we denote by $\phi_{ij1}$, should not change instantly when the mean deformation is changed, and hence should not depend explicitly on the mean deformation. Reference [5.10] ignored this requirement, and allowed $\phi_{ij1}$ to depend on the rotation tensor. LUMLEY [5.21] has now abandoned this position. LAUNDER and his coworkers, and others, have followed ROTTA in assuming

$$\phi_{ij1} = -A_0 \mathscr{D} b_{ij}. \qquad (5.23)$$

The constant $A_0$ determines the rate of return to isotropy. Its value has been the subject of much uncertainty. The flow ([5.11]) implies a value $A_0 = 6$, while the data of [5.9] suggest a much lower value is appropriate. It does not seem that the data justify the inclusion of higher-order terms, and so (5.23) is recommended, at least for homogeneous flows away from boundaries at high turbulence Reynolds numbers.

KWAK and REYNOLDS [5.22] studied the flow ([5.11]) in a numerical simulation, and found a much slower return to isotropy than indicated in the experiments of [5.11]. However, different components return at decidedly different rates. SHAANAN et al. [5.23] carried out a similar calculation for a homogeneous shear flow. In a computation the shearing can be removed, which cannot be done experimentally. These calculations also showed a marked difference in the return rate for different components, probably because of great difference in the length scales in the three directions. We conclude that current stress equation models will not do a very good job in handling the return to isotropy; however, the models may work well in flows dominated by other effects.

Inhomogeneities greatly complicate the $\phi_{ij}$ modelling, especially $\phi_{ij2}$. In a wall region one might well expect $\phi_{ij2}$ to be determined by a region at least as wide as the distance to the wall, and hence a complex integral model is really needed for such flows. This is a very unsatisfactory aspect of present stress-equation modelling, and an area that should receive considerable attention in the future.

In addition to modifications in $\phi_{ij}$, inhomogeneities require modeling of $J_{ijl}$. The gradient diffusion model is usually employed; [3.56] and [3.21] set

$$J_{ijl} = -A_2 \frac{q^2}{\mathscr{D}} (R_{in} R_{jl,n} + R_{jn} R_{il,n} + R_{ln} R_{ij,n}) \qquad (5.24)$$

where $R_{ij} = \overline{u_i u_j}$. HANJALIĆ and LAUNDER [3.56] gave some justification for this form by consideration of the dynamical equation for $\overline{u_i u_j u_l}$.

Noting that $J_{ijl}$ contains one pressure-velocity term, and since $p'$ will have a part that depends explicitly on the mean velocity gradients, it does seem that $J_{ijl}$ also should be explicitly linear in the mean gradients, though this need has escaped notice.

Two approaches have been used in stress-equation modelling. The earlier work (Donaldson [5.24]) involved specification of the length scale and use of (5.6) to determine $\mathscr{D}$. Hanjalić and Launder [3.5b] used the $\mathscr{D}$ equation model outlined above in conjunction with the $\overline{u_i u_j}$ equations. At this writing this work is in a state of rapid development, and undoubtedly improvements will be made by the time this book is published. Interested persons should follow most carefully the work of Launder and Lumley. It will be some time before these models are sufficiently well developed to be better than simpler models for use in engineering analysis.

There is a basic difficulty in this general approach to turbulence models. One would like to model only terms that respond on time scales short compared to that of the computed quantities. It is well known that the small scales respond to change much faster than the large scales, and hence it is reasonable to express a quantity dominated by small scales, such as $D_{ij}$, as a function of quantities dominated by large scales, such as $\overline{u_i u_j}$. However, terms like $J_{ijl}$ have time scales comparable with that of $\overline{u_i u_j}$, and thus one really should not expect an equilibrium constitutive relationship to exist between $J_{ijl}$ and $\overline{u_i u_j}$. In general, it seems that higher order statistical quantities take longer to reach steady state than lower order statistics. Any model obtained by truncation at some statistical order would suffer from this difficulty. What one really needs to do is truncate at some level of *scale*, and thereby take advantage of the fact that the smaller scales do adjust faster to local conditions. Then, by truncating at smaller scales, one has at least some hope of convergence, a hope that is at best dim when one truncates at higher and higher orders of statistical quantities that have comparable time scales. The large-eddy simulation described in the next section provides one avenue to a scale-truncation approach.

### 5.1.5 Large-Eddy Simulations

This line of approach is just beginning to bear fruit. The idea is to do a three-dimensional time-dependent numerical computation of the large scale turbulence. It will always be impossible to compute the smallest scales in any real flow at high turbulence Reynolds number, so they must be modeled. Care must be taken to define what it is that is being computed, and the early work was not done with sufficient care.

In 1973 the group at Stanford began a systematic program of development and exploration of this method, in close cooperation with NASA-Ames Laboratory. The first contribution was made by LEONARD (Ref. [4.73], p. 237), who clarified the need for spatial filtering. We now define the large-scale variables (which are still functions of space and time) by (see KWAK and REYNOLDS [5.22])

$$\bar{f}(x) = \int G(x - x') f(x') dx' \tag{5.25a}$$

where the filter function is

$$G(x - x') = \left( \sqrt{\frac{6}{\pi} \frac{1}{\Delta_a}} \right)^3 \exp[-6(x - x')^2 / \Delta_a^2]. \tag{5.25b}$$

Here $\Delta_a$ is the averaging scale, which need not and should not be the same as the grid mesh width. We use this particular filter because of its advantages in Fourier transformation. When this operation is applied to the Navier-Stokes equation, and an expansion is carried out, one finds (neglecting molecular viscosity)

$$\dot{\bar{U}}_i + \bar{U}_j \bar{U}_{i,j} = -\frac{1}{\varrho} \bar{p}_{,i} + \left[ -\frac{\Delta_a^2}{24} (\bar{U}_i \bar{U}_j)_{,u} - R_{ij} \right]_{,j} + O(\Delta_a^4) \tag{5.25c}$$

where $-\varrho R_{ij}$ are the "sub-grid-scale Reynolds stresses". The unusual term appearing before $R_{ij}$ is an additional stress-like term resulting from the filtering of the nonlinear terms; we now call these the "Leonard terms", and view $-\varrho \Delta_a^2 (\bar{U}_i \bar{U}_j)_{,le}/24$ as the "Leonard stresses".

KWAK and REYNOLDS [5.22] solved the isotropic decay problem, adjusting the sub-grid scale eddy viscosity to obtain the proper rate of energy decay. The calculations were started using an isotropic field with zero skewness, but the proper skewness develops in only a few time steps. Next they simulated the flow of [5.11], first with an initial distribution. One has problems in setting anisotropic initial conditions that are free of shearing stresses, and so the isotropic starting is probably a better approach. It is remarkable that the salient features of the experiments ([5.11]) were captured quite well in a computation using only $16^3$ points! The calculation was executed on a CDC 7600, using 120 time steps, in approximately 5 minutes.

SHAANAN et al. [5.23] experimented with a staggered grid approach that is second-order accurate and does not require explicit inclusion of the Leonard stresses. There are some difficulties in providing suitable initial conditions, so comparison with experiments is not easy. Nevertheless, the salient features of the flow ([5.13]) can also be produced with $16^3$ points.

One objective of this work is to test the turbulence models, particularly the stress equation model. We can compute the pressure strain terms directly (both $\phi_{ij1}$ and $\phi_{ij2}$), and are doing this presently. We had hoped that the calculations would serve as a basis for evaluating constants in the stress-equation models; instead they seem to be high-lighting the weaknesses of these models, as discussed in Subsection 5.1.4. However, the fact that a very coarse grid produces such remarkably good results leads us to believe that large eddy simulations might, after considerable development, eventually be useful for actual engineering analysis. See [5.25, 26], and also the work of ORSZAG and ISRAELI [5.27], who are carrying out similar calculations using Fourier rather than grid methods.

The gap in computing time between Reynolds-stress models and large-eddy simulations is a large one, and it is tempting to try to bridge it by using a form of averaging which preserves more of the turbulence structure than conventional Reynolds averaging (Section 1.3) does. This would be the computational equivalent of the conditional sampling used by experimenters (Section 1.10); so far the only significant step in this direction seems to be the work of LIBBY [5.28].

## 5.2 Simpler Methods

T. CEBECI

### 5.2.1 Discussion

The numerical procedures which are associated with turbulence models to make complete calculation methods can be divided into integral and differential types.

Differential methods involve direct assumptions for the Reynolds stresses at a point and seek the solution of the governing equations in their partial differential equation (pde) form; for examples see Section 5.1.

Integral methods involve the integral parameters of the shear layer (momentum thickness, shape parameter, skin-friction coefficient, etc.). They avoid the complexity of solving the shear-layer equations in full pde form. Instead, in two-dimensional flows they solve a system of ordinary differential equations, whose independent variable is $x$ and whose dependent variables are the profile parameters: in three-dimensional flows the equations are pde's in the plane of the shear layer.

It is most important to note that any of the turbulence models could be incorporated into either a differential or an integral procedure, with the sole exception that a turbulence model whose assumptions are formulated only in terms of integral parameters, like the entrainment model discussed in Subsection 5.2.2, would not be used in a differential method. The method of integral relations (generalized Galerkin method) can be used to convert pde's in two independent variables into ordinary differential equations. Therefore the important distinction between calculation methods is the type of turbulence model (Section 5.1) rather than the type of numerical procedure; for a further discussion see [5.37]. However most integral methods use zero-equation turbulence models, generally formulated in terms of integral parameters.

As pointed out in Section 5.1, the turbulence models in most common use are those of zero-equation or one-equation level. Most of the others are still in the development stage, although some two-equation models have been used in industry. Clearly, the more advanced methods are likely to come into more common use quite soon, and the reader who comes to this book some years after publication should review the literature to find out the current position. Calculation methods using well-established types of turbulence model are usually published in the engineering journals, such as ASME J. Fluids Engrg. or AIAA Journal, while advances in turbulence modeling are more often reported to J. Fluid Mech. or Phys. Fluids.

In the rest of this chapter we shall describe some of the simpler practical calculation methods, concentrating on the turbulence models but briefly commenting on the numerical procedures. The methods were all originally developed for thin shear layers. The numerical problems of merging them with a method for calculating the external inviscid flow will not be discussed here. In Subsection 5.2.2 two integral methods which can be satisfactorily used to compute two-dimensional incompressible boundary layers will be discussed. One uses a zero-equation turbulence model and the other a one-equation model. In that subsection we shall also discuss the extension of the zero-equation model to three-dimensional incompressible or compressible boundary layers. In Subsection 5.2.3 a discussion of differential methods will be given. Zero-equation differential methods, sometimes called mean-velocity methods, can be used to compute two-dimensional or three-dimensional shear layers for incompressible, compressible, steady or unsteady flows for a wide range of boundary conditions. Their advantage over integral methods is that the restrictions and inaccuracy that arise from the need to parameterize velocity profiles are avoided, although computing times are longer. One-equation differential methods have the further advantage of including history effects. The methods in Subsections 5.2.2 and 5.2.3

normally require the use of computers, although Head's method was developed by hand. Sometimes it is useful to have simpler methods or formulae that do not require the use of computers. Although the accuracy of such methods may not be as high and although they are invariably restricted to simple boundary conditions, they are very useful in preliminary engineering calculations. These "short-cut" methods for external boundary layers and for free shear layers are discussed in Subsection 5.2.4; the discussion of free shear layers supplements that of mixing layers in Chapter 3, where simple formulae for duct and pipe flows were also given. For highly complex and poorly documented flows, intelligent use of short-cut formulae may be more cost-effective than detailed calculations.

## 5.2.2 Integral Methods

There is no shortage of integral methods for two- and three-dimensional turbulent boundary layers [3.72, 5.29]. We shall discuss two methods for two-dimensional flow, those due to Head [2.60] and to Green et al. [5.30], in some detail. Integral methods for three-dimensional boundary layers are not as accurate as those for two-dimensional flow, even if the accuracy of the turbulence model is unimpaired, because of difficulties in parameterizing the "crossflow" profile, that of the velocity component normal to the external stream. However, the method due to Myring [5.31], which is an extension of Head's method to three-dimensional flows (see also [2.319]), is a very general, useful one and will be discussed briefly. In this method, the governing equations are written for a non-orthogonal curvilinear coordinate system. Thus, the user has some flexibility in selecting the proper coordinate system for his three-dimensional flow problem. The method has been used for compressible adiabatic flows, with slight modifications, by Smith [5.32].

*Head's Method*

Head [2.60] assumes that the dimensionless entrainment velocity $V_E/U_e$ (Subsection 2.3.1) is a function of shape factor $H_1$,

$$\frac{V_E}{U_e} = \frac{1}{U_e}\frac{d}{dx}[U_e(\delta - \delta^*)] = F(H_1). \tag{5.26}$$

Here $H_1$ is defined by

$$H_1 = \frac{\delta - \delta^*}{\theta}. \tag{5.27}$$

Equation (5.26) is a typical integral equivalent of a zero-equation pde model (Subsection 5.1.1). Using (5.27), we can write (5.26) as

$$\frac{d}{dx}(U_e \theta H_1) = U_e F . \tag{5.28}$$

Head also assumes that $H_1$ is related to the shape factor $H \equiv \delta^*/\theta$ by

$$H_1 = G(H) . \tag{5.29}$$

($G$ is not related to the "Clauser parameter" $G$ of Chapter 3).

The functions $F$ and $G$ are determined from experiment. A best fit to several sets of experimental data shows that those functions can be approximated by

$$F = 0.0306(H_1 - 3.0)^{-0.6169} \tag{5.30}$$

$$G = \begin{cases} 0.8234(H-1.1)^{-1.287} + 3.3 & H \leq 1.6 \\ 0.5501(H-0.6778)^{-3.064} + 3.3 & H \geq 1.6 . \end{cases} \tag{5.31}$$

The momentum integral equation (Subsection 2.3.5),

$$\frac{d\theta}{dx} + (H+2)\frac{\theta}{U_e}\frac{dU_e}{dx} = \frac{c_f}{2} , \tag{5.32}$$

has three unknowns $\theta$, $H$ and $c_f$ for a given external velocity distribution $U_e(x)$. Eq. (5.28), with $F$, $H$, and $G$ defined by (5.29), (5.30), and (5.31), respectively, provides a relationship between $\theta$ and $H$ that can be written as

$$\theta \frac{dG}{dH}\frac{dH}{dx} + G\left(\frac{\theta}{U_e}\frac{dU_e}{dx} + \frac{d\theta}{dx}\right) = F . \tag{5.33}$$

Substitution for $d\theta/dx$ from (5.32) leads to

$$\theta \frac{dG}{dH}\frac{dH}{dx} + G\left(\frac{c_f}{2} - (H+1)\frac{\theta}{U_e}\frac{dU_e}{dx}\right) = F , \tag{5.34}$$

which is a linear relation between $dH/dx$ and $dU_e/dx$ if $F$ and $G$ are known functions of $H$.

Another equation relating $c_f$ to $\theta$ and/or $H$ is needed. It has been shown, for example by Ludwieg and Tillmann [2.13], that one shape parameter, such as $H$, is sufficient to define a boundary layer profile quite accurately except very near the surface. Therefore unique relations such as (5.29) between *any* two parameters can be expected. If one parameter is $H$ and the other the ratio of the velocity at, say, $y=0.1\delta$ to the external-stream velocity $U_e$, then the empirical relation between these two parameters, together with the logarithmic velocity profile law (1.49) applied at $y=0.1\delta$, yields the skin-friction coefficient $c_f$ as a function of $H$ and the Reynolds number $U_e\delta/v$. That Reynolds number can be related to $R_\theta \equiv U_e\theta/v$ by using (5.27) and (5.29). LUDWIEG and TILLMANN obtain, and HEAD uses,

$$c_f=0.246 \times 10^{-0.678H}R_\theta^{-0.268}.\tag{5.35a}$$

FERNHOLZ [5.33] gives a slightly different form which also covers strong adverse pressure gradients

$$c_f=0.0580\gamma^{1.705}R_\theta^{-0.268}\tag{5.35b}$$

where

$$\gamma=\log_{10}(8.05/H^{1.818}).\tag{5.35c}$$

An alternative method of deriving a skin-friction formula is to write (2.11) at $y=\delta$, giving $c_f=f(u_\tau y/\delta, \Pi)$, while (2.11) itself implies $\theta/\delta = f(u_\tau\delta/v, \Pi)$, etc., leading to $c_f=f(H, R_\theta)$ again. The more refined profile family of THOMPSON [2.66] could be used similarly. The system (5.28) to (5.35) can be solved numerically for a specified external velocity distribution to obtain the boundary-layer development. We note that to start the calculations, say at $x=x_0$, values of two of the parameters $\theta$, $H$, and $c_f$ must be specified.

   This method, like most integral methods, uses a chosen value of the shape factor $H$ as the criterion for separation [(5.35) predicts $c_f=0$ only if $H$ tends to infinity]. Although it is not possible to give an exact value of $H$ corresponding to separation, separation is assumed to occur when $H$ is between 1.8 and 2.4. The difference between the lower and upper limits of $H$ makes very little difference in locating the separation point, since close to separation the shape factor increases very rapidly.

*Green's "Lag-Entrainment" Method* [5.30]

This method is an extension of Head's method in that the momentum integral equation and the entrainment equation are supplemented by an equation for the streamwise rate of change of entrainment coefficient $F$. This additional equation, derived from an equation for shear stress which BRADSHAW et al. [5.34] derived from the turbulent kinetic energy equation and used in their one-equation differential method, explicitly represents the balance between the advection, production, diffusion and dissipation of turbulent kinetic energy. It allows for more realistic calculations in rapidly changing flows and is a significant improvement upon Head's method. In effect this is an "integral" version of the "differential" method of BRADSHAW et al.

This method employs Eqs. (5.28) and (5.32) as before. It also considers the "rate of change of entrainment coefficient" equation given by

$$\theta(H_1 + H)\frac{dF}{dx} = \frac{F(F+0.02)+0.2667c_{f_0}}{F+0.01}$$

$$\times \left\{ 2.8[(0.32c_{f_0}+0.024F_{eq}+1.2F_{eq}^2)^{1/2} \right.$$

$$-(0.32c_{f_0}+0.024F+1.2F^2)^{1/2}]$$

$$\left. + \left(\frac{\delta}{U_e}\frac{dU_e}{dx}\right)_{eq} - \frac{\delta}{U_e}\frac{dU_e}{dx} \right\}, \tag{5.36}$$

where the numerical coefficients arise from curve fits to experimental data and the empirical functions of BRADSHAW et al.

Here $c_{f_0}$ is the flat plate skin-friction coefficient calculated from the following empirical formula

$$c_{f_0} = \frac{0.01013}{\log_{10} R_\theta - 1.02} - 0.00075. \tag{5.37}$$

The subscript "eq" refers to equilibrium (self-preserving; see Subsection 2.3.2) flows. The functional forms of the equilibrium values of $F_{eq}$ and $(\delta/U_e dU_e/dx)_{eq}$ are given by

$$F_{eq} = H_1 \left[ \frac{c_f}{2} - (H+1)\left(\frac{\theta}{U_e}\frac{dU_e}{dx}\right)_{eq} \right], \tag{5.38}$$

$$\left(\frac{\theta}{U_e}\frac{dU_e}{dx}\right)_{eq} = \frac{1.25}{H}\left[\frac{c_f}{2} - \left(\frac{H-1}{6.432H}\right)^2\right], \tag{5.39}$$

$$\left(\frac{\delta}{U_e}\frac{dU_e}{dx}\right)_{eq} = (H+H_1)\left(\frac{\theta}{U_e}\frac{dU_e}{dx}\right)_{eq}. \tag{5.40}$$

The skin-friction formula and the relationship between the shape factors $H$ and $H_1$ complete the number of equations needed to solve the system of ordinary differential Eqs. (5.28), (5.32), and (5.36) For $c_f$

$$\left(\frac{c_f}{c_{f_0}}+0.5\right)\left(\frac{H}{H_0}-0.4\right)=0.9 \tag{5.41}$$

where

$$1-\frac{1}{H_0}=6.55\left(\frac{c_{f_0}}{2}\right)^{1/2} \tag{5.42}$$

so that (5.37), (5.41) and (5.42) give $c_f$ as a function of $H$ and $R_\theta$ with values fairly close to (5.35).

For $H_1$,

$$H_1=3.15+\frac{1.72}{H-1}-0.01(H-1)^2 \tag{5.43}$$

giving values fairly close to (5.31).

GREEN et al. discuss the extension of their method to axisymmetric flows and wakes. Also, they take account of the effect on the turbulence structure of extra strain rates such as longitudinal curvature, lateral divergence and—in compressible flow—bulk compression, by the methods outlined in Subsection 3.1.1. Comparisons with experiment show that the method is quite accurate in incompressible flows. In compressible flows, the accuracy of the method in zero pressure gradient is assured by its derivation, but the available experimental data do not enable its accuracy in flows with strong pressure gradient to be assessed with any finality.

HEAD and PATEL [2.61] have produced an improved version of Head's method which also incorporates a form of lag into the entrainment equation.

*Myring's Method*

A detailed description of this method would be rather lengthy, but an outline is presented here as an example of the algebraic and numerical complication introduced by three-dimensional flow. The turbulence model is essentially Head's. Full details are given by MYRING [5.31] and by SMITH [5.32].

MYRING uses a coordinate system in which the $z$-axis is normal to the surface and $x$ and $y$ form a non-orthogonal curvilinear mesh on the body surface. The two momentum integral equations given by MYRING are

$$\frac{1}{h_1}\frac{\partial}{\partial x}\theta_{11}^* + \theta_{11}^*\left[\frac{(2-M_e^2)}{h_1}\frac{1}{U_s}\frac{\partial U_s}{\partial x} + \frac{1}{q}\frac{\partial}{\partial x}\left(\frac{q}{h_1}\right) + k_1\right] + \frac{1}{h_2}\frac{\partial}{\partial y}\theta_{12}^*$$

$$+ \theta_{12}^*\left[\frac{(2-M_e^2)}{h_2}\frac{1}{U_s}\frac{\partial U_s}{\partial y} + \frac{1}{q}\frac{\partial}{\partial y}\left(\frac{q}{h_2}\right) + k_3\right]$$

$$+ \Delta_x\left(\frac{1}{h_1}\frac{1}{U_s}\frac{\partial U_e}{\partial x} + k_1\frac{U_e}{U_s}\right)$$

$$+ \Delta_y\left(\frac{1}{h_2}\frac{1}{U_s}\frac{\partial U_e}{\partial y} + k_2\frac{U_e}{U_s} + k_3\frac{U_e}{U_s}\right) + \theta_{22}^*k_2 = \frac{c_{f_x}}{2}, \qquad (5.44)$$

$$\frac{1}{h_1}\frac{\partial\theta_{21}^*}{\partial x} + \theta_{21}^*\left[\frac{(2-M_e^2)}{h_1}\frac{1}{U_s}\frac{\partial U_s}{\partial x} + \frac{1}{q}\frac{\partial}{\partial x}\left(\frac{q}{h_1}\right) + l_3\right] + \frac{1}{h_2}\frac{\partial\theta_{22}^*}{\partial y}$$

$$+ \theta_{22}^*\left[\frac{(2-M_e^2)}{h_2}\frac{1}{U_s}\frac{\partial U_s}{\partial y} + \frac{1}{q}\frac{\partial}{\partial y}\left(\frac{q}{h_2}\right) + l_2\right]$$

$$+ \Delta_x\left(\frac{1}{h_1}\frac{1}{U_s}\frac{\partial V_e}{\partial x} + l_1\frac{U_e}{U_s} + l_3\frac{V_e}{U_s}\right)$$

$$+ \Delta_y\left(\frac{1}{h_2}\frac{1}{U_s}\frac{\partial V_e}{\partial y} + l_2\frac{V_e}{U_s}\right) + \theta_{11}^*l_1 = \frac{c_{f_y}}{2}. \qquad (5.45)$$

Here $M_e$ is the Mach number at the boundary-layer edge and $h_1$, $h_2$, and $g$ are the metric coefficients of the coordinate system in which a length element in the body surface is given by

$$ds^2 = h_1^2 dx^2 + h_2^2 dy^2 + 2g dx\,dy. \qquad (5.46)$$

The various integral thicknesses appearing in (5.44) and (5.45) together with the quantities $k_1$, $k_2$, $l_1$, $l_2$, $l_3$, and $q$ are defined in MYRING [5.31]. The velocity components in the $x$, $y$ directions at the edge of the boundary layer are denoted by $U_e$, $V_e$, and the resultant velocity at the boundary-layer edge, denoted by $U_s$, is defined by

$$U_s^2 = U_e^2 + V_e^2 + \frac{2g}{h_1 h_2}U_e V_e. \qquad (5.47)$$

Myring writes the three-dimensional version of the entrainment Eq. (5.26) as

$$\frac{1}{\varrho_e U_s q}\left\{\frac{\partial}{\partial x}\left[\frac{\varrho_e q}{h_1}(U_e\delta - U_s\Delta_x)\right] + \frac{\partial}{\partial y}\left[\frac{\varrho_e q}{h_2}(V_e\delta - U_s\Delta_y)\right]\right\}$$

$$= \frac{1}{U_s}\left(\frac{U_e}{h_1}\frac{\partial\delta}{\partial x} + \frac{V_e}{h_2}\frac{\partial\delta}{\partial y} - W_e\right) = F. \qquad (5.48)$$

Although Eqs. (5.44), (5.45), and (5.48) can be solved in the form presented, they are better expressed in terms of the more familiar thicknesses associated with streamline coordinates. Thus the solution is obtained in streamline-coordinate variables while the calculations are made in the non-orthogonal coordinate system. The conversion is rather length and the reader is again referred to Myring's report for details. As an example, the crossflow momentum thickness $\theta_{22}^*$ can be expressed as

$$\theta_{22}^* = (\sin^2 \lambda)^{-1}[\theta_{11} \sin^2 \alpha + (\theta_{12} + \theta_{21}) \sin \alpha \cos \alpha + \theta_{22} \cos^2 \alpha]. \quad (5.49)$$

Here $\alpha$ is the angle between the $x$-axis and an external streamline and $\lambda$ is the angle between the $x$ and $y$ axes. In a similar way, $\theta_{11}^*$, $\theta_{12}^*$, $\theta_{21}^*$, $\Delta_x$, and $\Delta_y$ can be expressed in terms of $\theta_{11}$, $\theta_{12}$, $\theta_{21}$, $\theta_{22}$, $\Delta_s$, $\Delta_n$, $\lambda$, and $\alpha$.

In the above conversions, the following relationships between the velocity components hold

$$U = U_s \frac{\sin(\lambda - \alpha)}{\sin \lambda} - V_n \frac{\cos(\lambda - \alpha)}{\sin \lambda}, \quad (5.50)$$

$$V = U_s \frac{\sin \alpha}{\sin \lambda} + V_n \frac{\cos \alpha}{\sin \lambda}. \quad (5.51)$$

If the indicated conversion is carried through, we still have far more unknowns than equations. However, now we may apply empirical relationships applicable to streamline coordinate variables and reduce the number of unknowns further. Following Head, a unique relationship[2] between the shape parameters $H$ and $H_1$ is postulated:

$$H_1 = \frac{2H}{H-1}. \quad (5.52)$$

This is exact for power-law velocity profiles of the streamwise velocity component and is a good approximation in practice for $R_{\theta 11} > 10^4$. For the crossflow profile, Myring uses the relationship proposed by Mager [2.345].

$$\frac{V_n}{U_s} = \left(1 - \frac{z}{\delta}\right)^2 \tan \beta \quad (5.53)$$

---

[2] It is noteworthy that the functions $H_1(H)$ used by Head, Green, and Myring differ by as much as 20% in the working range $1.3 < H < 1.9$; see Fig. 5.1.

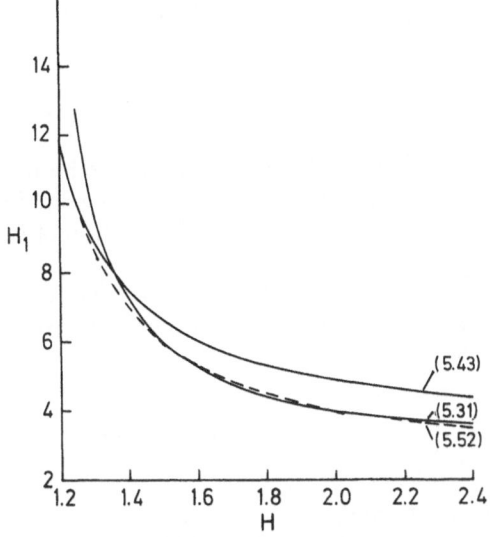

Fig. 5.1. Empirical relations between boundary-layer shape parameters: numbers refer to equations in the text

where $\beta$ denotes the angle between the external and the limiting (surface) streamline.

Equations (5.27) and (5.28) allow the thicknesses $\theta_{12}$, $\theta_{21}$, $\theta_{22}$, and $\delta_n^*$ to be expressed in terms of $\theta_{11}$, $H_1$, and $\beta$:

$$\theta_{12} = \theta_{11} m_{12} \tan \beta$$
$$\theta_{21} = \theta_{11} m_{21} \tan \beta$$
$$\theta_{22} = \theta_{11} m_{22} \tan \beta$$
$$\delta_n^* = \theta_{11} m_2 \tan \beta . \tag{5.54}$$

Here the functions $m$ depend on $H$ only. They can be found in Myring's report, which also contains equivalent expressions for Johnston's cross-flow profiles [2.346].

For the streamwise $c_f$ given by (5.35), the components of $c_f$ in the $x$ and $y$ directions, namely, $c_{f_x}$ and $c_{f_y}$, can be obtained from

$$c_{f_x} = c_f \left[ \frac{\sin(\lambda - \alpha) - \cos(\lambda - \alpha) \tan \beta}{\sin \lambda} \right], \tag{5.55a}$$

$$c_{f_y} = c_f \left[ \frac{\sin \alpha + \cos \alpha \tan \beta}{\sin \lambda} \right]. \tag{5.55b}$$

The non-dimensional rate of change of mass flow in the boundary layer, namely $F$ in (5.48), is assumed to follow the relationship derived by

Green [5.35] in his earlier (unlagged) version of Head's method for two-dimensional flow

$$F = 0.025H - 0.022 . \tag{5.56}$$

With expressions similar to (5.49) and with all the empirical relationships, various thicknesses $\theta_{11}^*$, $\theta_{21}^*$, $\theta_{12}^*$, $\Delta_x$, $\Delta_y$, can be expressed in terms of $\theta_{11}$ and $\tilde{F}_j$ where $\tilde{F}_j$ is a new parameter which is a function of $H$, $\alpha$, $\lambda$, and $\gamma$. Here $\beta$ has been replaced by $\gamma$ in order to allow for the use of Johnston's crossflow profiles. For Mager profiles $\gamma$ is identical to $\beta$, but for Johnston's profiles the following empirical relationship between $\gamma$ and $\beta$ holds

$$\tan \beta = \tan \gamma \left( \frac{0.10}{\sqrt{c_f} \cos \beta} - 1 \right) . \tag{5.57}$$

If these new relations together with those given by (5.55) are substituted into (5.44), (5.45), and (5.48), there result three unknowns $\theta_{11}$, $H$, and $\beta$ defined by three pde's in $x$ and $y$. The $x$-axis is chosen as the direction of forward integration, and the $y$-derivatives are approximated by finite differences in a standard technique for solving pde's in two dimensions. Thus

$$F_{11} \frac{\partial \theta_{11}}{\partial x} + \theta_{11} F_{11H} \frac{\partial H}{\partial x} + \theta_{11} F_{11\beta} \frac{\partial y}{\partial x} = s_1$$

$$F_{21} \frac{\partial \theta_{11}}{\partial x} + \theta_{11} F_{21H} \frac{\partial H}{\partial x} + \theta_{11} F_{21\beta} \frac{\partial y}{\partial x} = s_2$$

$$J_1 \frac{\partial \theta_{11}}{\partial x} + \theta_{11} J_{1H} \frac{\partial H}{\partial x} + \theta_{11} J_{1\beta} \frac{\partial \gamma}{\partial x} = s_3 . \tag{5.58}$$

Here the functions $F$, $J$, and $s$ depend on $x$ and $y$ and are tabulated in Myring's report. Equation (5.58) can now be solved for $\partial \theta_{11}/\partial x$, $\partial H/\partial x$, and $\partial y/\partial x$ and the three unknowns $\theta_{11}$, $H$, and $\gamma$ can be calculated. The solution procedure is rather tricky and the reader is referred to Smith [5.32].

### 5.2.3 Differential Methods

*Zero-Equation Methods*

Most of these methods are based on the "eddy viscosity" concept (Section 1.8) in some form. The two most popular versions are the "mixing-length" formula (e.g. [2.208, 212])

$$-\overline{uv} = l^2 \left| \frac{\partial U}{\partial y} \right| \frac{\partial U}{\partial y} \qquad (5.59)$$

and the original eddy viscosity formula (e.g. [1.4, 5.2]) which for $\overline{uv}$ in a thin shear layer becomes

$$-\overline{uv} = \nu_T \frac{\partial U}{\partial y} . \qquad (5.60)$$

These two formulae are based on erroneous physical arguments, but can be regarded as definitions of quantities $\nu_T$ and $l$ which, in simple flows, are simpler to correlate empirically than $\overline{uv}$ itself. In most boundary-layer calculation methods the viewpoint of Subsection 2.3.1 is adopted, the turbulent boundary layer is regarded as a composite layer consisting of inner and outer regions, and the distributions of $l$ and $\nu_T$ are described by two separate empirical expressions in each region. For example, if the viscous sublayer close to the wall is excluded, $l$ is proportional to $y$ in the inner region, and it is proportional to $\delta$ in the outer region. Therefore,

$$l = \begin{cases} \kappa y & y_0 \leqq y \leqq y_c , & (5.61a) \\ \alpha_1 \delta & y_c \leqq y \leqq \delta , \quad \text{where} \quad \delta = \delta_{995}, \text{ say.} & (5.61b) \end{cases}$$

Here $y_0$ is a small distance from the wall, the viscous sublayer thickness $40\nu/u_\tau$, and $y_c$ is another distance obtained from the continuity of mixing length. The empirical parameters $\kappa$ and $\alpha_1$ vary slightly according to experimental data. For flows at high Reynolds numbers ($R_\theta > 5000$), they are generally taken to be about $\kappa = 0.4$ and $\alpha_1 = 0.075$. According to recent studies [2.40, 5.36] at lower Reynolds numbers ($R_\theta < 5000$) $\alpha_1$ is a function of Reynolds number (Subsection 2.3.1).

Similarly, according to various studies, $\nu_T$ varies linearly with $y$ in the inner region where $-\varrho \overline{uv}$ is nearly equal to $\tau_w$, and is nearly constant in the outer region. Its variation across the boundary layer can conveniently be described by the following formulas:

$$\nu_T = \begin{cases} l^2 \left| \dfrac{\partial U}{\partial y} \right| & y_0 \leqq y \leqq y_c , & (5.62a) \\ \alpha \left| \int_0^\infty (U_e - U) dy \right| & y_c \leqq y \leqq \delta & (5.26b) \end{cases}$$

with $l$ given by (5.61a). The parameter $\alpha$ is generally assumed to be a universal constant equal to 0.0168 for $R_\theta > 5000$. Like $\alpha_1$, it also varies with

Reynolds number when $R_\theta$ is less than 5000. According to Cebeci [5.37], it is given by

$$\alpha = 0.0168 \frac{1+\Pi_0}{1+\Pi} \qquad (5.63)$$

where $\Pi$ is the "strength of the wake" defined in Subsection 2.3.1, its high-$R_\theta$ value $\Pi_0$ being about 0.55 and its variation with $R_\theta$ being quite well fitted by

$$\Pi = 0.55[1 - \exp(-0.243 z_1^{1/2} - 0.298 z_1)] \qquad (5.64a)$$

$$z_1 = R_\theta/425 - 1 . \qquad (5.64b)$$

There have been numerous attempts to extent (5.61a) to include the viscous sublayer, by multiplying the mixing length by the function $f(u_\tau y/\nu)$ of (1.51) or otherwise (e.g. [5.6]). Van Driest [5.38], using an analogy with the laminar flow on an oscillating flat plate, suggested

$$l = \kappa y[1 - \exp(-y/A)] \qquad (5.65)$$

for a smooth flat-plate flow. Here $A$ is a damping-length constant for which the best dimensionally correct empirical choice is $A^+ \nu(\tau_w/\varrho)^{-1/2}$ with $A^+$ denoting an empirical constant equal to about 26. $A^+$ is expected to vary somewhat with pressure gradient, transpiration, etc. (see below).

Abbott et al. [5.39], using an intermittent model of the sublayer, obtain an expression similar to [5.65] by considering the unsteady one-dimensional vorticity equation rather than the unsteady one-dimensional momentum equation considered by Van Driest. For the additive constant in the logarithmic velocity profile (1.49), they get

$$C = \frac{1}{6} S^2 p^+ + \frac{\sqrt{\pi}}{2} S + \frac{1}{\kappa}\left(1 + \frac{\gamma_0}{2} - \ln 2S\right) \qquad (5.66)$$

where

$$y^+ = \frac{u_\tau y}{\nu}, \qquad p^+ = -\frac{\nu}{\varrho u_\tau^3} \frac{dp}{dx},$$

$$S = S_0[1 + 10 \operatorname{sgn}(p^+) p^{+2/3}]^{-1}, \qquad S_0 = 10.618 \qquad (5.67)$$

with $\gamma_0$ denoting the Euler constant. This model, which is developed for incompressible flows, has not yet been used and tested so thoroughly as the Van Driest model. Note that although $A^+$ as recommended above does not depend on $p^+$, the Van Driest prediction of $C$ would do so because $\tau/\tau_w = 1 + p^+ y^+$ approximately. For a pde model of the sublayer see [5.6] and Section 5.1.2.

Needless to say, the eddy viscosity and mixing-length formulas, like most (if not all) expressions for turbulent flows, are empirical. Over the years, several empirical corrections to these formulas have been made, to account for the effects of low Reynolds number, transitional region, compressibility, mass transfer, pressure gradient, and transverse curvature. Here we shall present an eddy-viscosity formulation developed by CEBECI and SMITH [1.4] that accounts for all of those effects and appears to give satisfactory results. According to that formulation, $\nu_T$ is defined by the following expressions:

$$\nu_T = \begin{cases} (\nu_T)_i = L^2 \left(\frac{r}{r_0}\right) \left|\frac{\partial U}{\partial y}\right| \gamma_{tr} & \text{(5.68a)} \\ (\nu_T)_0 = \alpha |\int_0^\infty (U_e - U) dy| \gamma_{tr} \, . & \text{(5.68b)} \end{cases}$$

Here $r_0$ is body radius and

$$L = 0.4 r_0 \ln \left(\frac{r}{r_0}\right) \left\{ 1 - \exp\left[ -\frac{r_0}{A} \ln \left(\frac{r}{r_0}\right) \right] \right\} \tag{5.69a}$$

$$A = 26 \frac{\nu}{N} \left(\frac{\tau_w}{\varrho_w}\right)^{-1/2} \left(\frac{\varrho}{\varrho_w}\right)^{1/2} \tag{5.69b}$$

$$N = \left\{ \frac{\mu}{\mu_e} \left(\frac{\varrho_e}{\varrho_w}\right)^2 \frac{p^+}{v_w^+} \left[ 1 - \exp\left( 11.8 \frac{\mu_w}{\mu} v_w^+ \right) \right] + \exp\left( 11.8 \frac{\mu_w}{\mu} v_w^+ \right) \right\}^{1/2} . \tag{5.69c}$$

$$v_w^+ = \frac{v_w}{u_\tau} . \tag{5.69d}$$

The expression $\gamma_{tr}$, which accounts for the transitional region between a laminar and a turbulent flow, is given by

$$\gamma_{tr} = 1 - \exp\left[ -Gr_0(x_{tr}) \left(\int_{x_{tr}}^x \frac{dx}{r_0}\right) \left(\int_{x_{tr}}^x \frac{dx}{U_e}\right) \right]. \tag{5.69e}$$

Here

$$G = \frac{3}{C^2} \left(\frac{U_e^2}{\nu^2}\right) R_{x_{tr}}^{-1.34} , \qquad R_{x_{tr}} = \frac{U_e x_{tr}}{\nu}$$

$$C = 60 + 4.86 M_e^{1.92} \qquad 0 < M_e < 5 \tag{5.70}$$

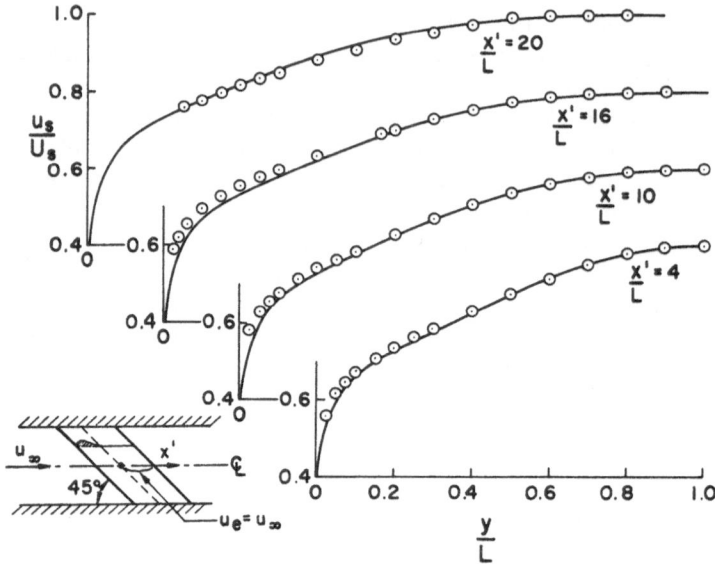

Fig. 5.2. Predictions of a swept-wing boundary layer using (5.72). ○ experiments [2.343]; ────── calculations

and $x_{tr}$ is the location of the start of transition. For two-dimensional incompressible flows, the transitional Reynolds number, $R_{x_{tr}}$, can be calculated from the following empirical formula given by CEBECI and SMITH [1.4]:

$$R_{\theta_{tr}} = 1.174 \left[ 1 + (22400/R_{x_{tr}}) \right] R_{x_{tr}}^{0.46} . \tag{5.71}$$

The eddy viscosity formulation (5.36) has also been generalized to three-dimensional incompressible and compressible flows by the following expressions (see CEBECI [5.40], CEBECI and ABBOTT [5.41], CEBECI et al. [2.293] who, unlike MYRING (Subsection 5.2.2), follow the usual convention that $y$ is normal to the surface):

$$v_T = \begin{cases} (v_T)_i = L^2 |S(y)| & (v_T)_i \leqq (v_T)_0 \\ (v_T)_0 = 0.0168 \int_0^\infty (|q_e| - |q(y)|) dy & (v_T)_i \geqq (v_T)_0 \end{cases} \tag{5.72}$$

where

$$L = 0.4y[1 - \exp(-y/A)]$$
$$A = 26(v/|S_w|)^{1/2} .$$

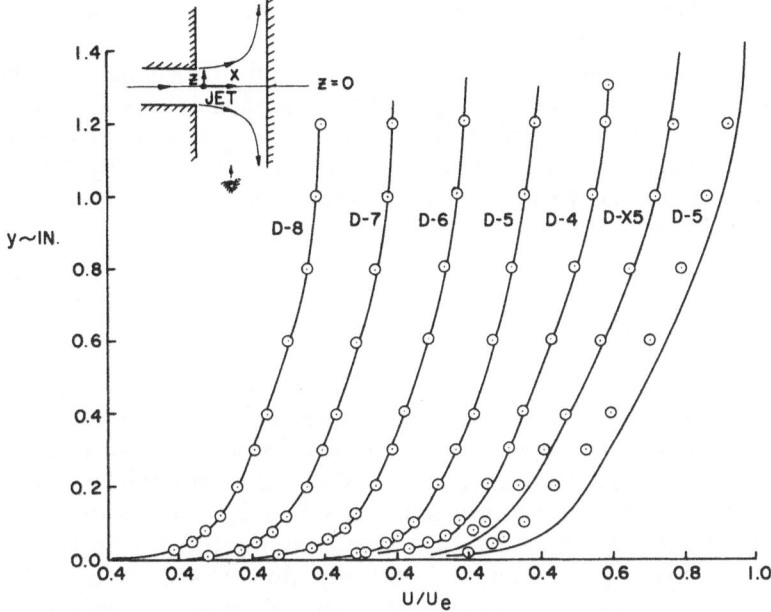

Fig. 5.3. Predictions of a laterally diverging boundary layer using (5.72). ○ experiments [2.327]; ——— calculations

Here $q$ denotes the velocity vector parallel to the wall, $q=(U, W)$ and $S$ denotes the shear or strain vector. $S = \partial q/\partial y = (\partial U/\partial y, \partial W/\partial y)$. The subscripts e and w denote the edge and the wall, respectively.

The expressions in (5.72) have been used to compute a number of three-dimensional flows ranging from swept infinite cylinders to arbitrary wings for incompressible and compressible flows. The governing boundary-layer equations in orthogonal and non-orthogonal coordinate systems were solved by a very efficient numerical method for several practical problems, and comparisons were made with experiment and with the predictions of Bradshaw's method [2.352]. Figure 5.2 shows the results for a 45° infinite swept wing and Fig. 5.3 shows the results for a laterally diverging boundary layer. The experimental data in Fig. 5.2 were obtained by BRADSHAW and TERRELL [2.343] on the flat rear of the wing in a region of nominally zero pressure gradient and decaying cross flow. Figure 5.2 shows the calculated results of Cebeci compared with the experimental data. The experimental data in Fig. 5.3 were obtained by JOHNSTON [2.327] in an apparatus consisting of a rectangular inlet duct from which an issuing jet impinged on an end wall from the outlet of the channel. The jet was confined on the top and bottom by flat surfaces and the boundary layer which developed on the floor of the test section was probed.

Other zero-equation methods based on mixing length or eddy viscosity have also been extended to three-dimensional flows. One such model, which leads to satisfactory results, is due to HUNT et al. [5.42]. Their expression has been adopted and used by ADAMS [2.320] and by HARRIS and MORRIS [5.43] to several three-dimensional compressible boundary layers.

*One-Equation Methods*

Few entirely new one-equation models have been developed since the Stanford meeting in 1968 [3.72]. Of those then current, probably the one most thoroughly developed since is Bradshaw and Ferriss' method [5.34, 44] which has been applied to a wide range of boundary-layer problems including three-dimensional [2.335, 352] compressible flow with heat transfer [5.45]. It has also been used in internal flows [3.59], free shear layers [3.60], and in unsteady flow [5.46]. It uses the "structural" assumption (5.3), rather than the "Newtonian" assumption (5.1), to derive a shear-stress transport equation from the turbulent energy equation, and uses an empirically specified bulk convention velocity (Subsection 5.1.3) for the turbulent transport terms. The more conventional PRANDTL one-equation model has also been extended beyond simple two-dimensional boundary layers (e.g., LAUNDER et al. (Ref. [2.2], p. 361), but neither of these models can be used outside thin shear layers without the addition of a length-scale equation, which removes them from the category of "simple" methods to which this part of the present chapter is devoted.

### 5.2.4 Short-Cut Methods

*Incompressible Flow on a Smooth Flat Plate*

A useful expression for the skin-friction coefficient in zero pressure gradient, obtained by substituting the overlap-law $c_f$ formula (Subsection 2.3.1) into the momentum integral equation, is von Karman's formula

$$\frac{1}{\sqrt{c_f}} = 1.7 + 4.15 \log(c_f R_x) . \tag{5.73}$$

A formula for the average skin friction, $\bar{c}_f = (1/x) \int_0^x c_f dx'$, was derived by Schoenherr from (5.73) as

$$\frac{1}{\sqrt{\bar{c}_f}} = 4.13 \log(\bar{c}_f R_x). \tag{5.74}$$

He also related $\bar{c}_f$ to $R_\theta$ by the following expression

$$\frac{1}{\sqrt{\bar{c}_f}} = 1.24 + 4.13 \log R_\theta. \tag{5.75}$$

These expressions assume that the flow is turbulent from the leading edge and are valid only if $R_x$ is much greater than its value at transition. The Clauser shape parameter $G$ is constant for a flat-plate boundary layer and equal to about 6.8. This gives the relation, valid for $R_\theta > 5000$ approx.,

$$H = 1/(1 - 4.8 \sqrt{c_f}). \tag{5.76}$$

Much simpler but less accurate relations between $\delta$, $c_f$, $\delta^*$, $\theta$, and $H$ can also be obtained if one uses the "power-law" assumption for the velocity profiles, that is, $U/U_e = (y/\delta)^{1/n}$. They are given by

$$\frac{\delta^*}{\delta} = \frac{1}{1+n} \tag{5.77a}$$

$$\frac{\theta}{\delta} = \frac{n}{(1+n)(2+n)}, \tag{5.77b}$$

$$H = \frac{2+n}{n}. \tag{5.77c}$$

Here $n$ varies slightly with Reynolds number; a value $n = 7$ is often used. Equations (5.77) yield Myring's Eq. (5.52).

Other formulae obtained from the one-seventh power law assumption, given in SCHLICHTING [1.3], are the following:

$$c_f = \frac{0.045}{R_\theta^{1/4}}, \tag{5.78a}$$

$$\bar{c}_f = \frac{0.074}{R_x^{1/5}}, \tag{5.78b}$$

$$\frac{\delta}{x} = \frac{0.37}{R_x^{1/5}}, \tag{5.78c}$$

$$\frac{\theta}{x} = \frac{0.036}{R_x^{1/5}}. \tag{5.78d}$$

These equations are valid only for Reynolds numbers $R_x$ between $5 \times 10^5$ and $10^7$, the power laws being only approximations to the logarithmic results above.

Equation (5.73) was obtained on the assumption that the boundary layer is turbulent from the leading edge onwards. If the Reynolds number is not very large, then the portion of the laminar flow which preceeds the turbulent flow should be considered. There are several empirical formulas which account for this effect. One is the formula obtained by Schlichting [1.3],

$$\bar{c}_f = \frac{0.455}{(\log_{10} R_x)^{2.58}} - \frac{C}{R_x}. \tag{5.79}$$

Another, obtained from (5.78b), is

$$\bar{c}_f = \frac{0.074}{R_x^{1/5}} - \frac{C}{R_x}, \quad 5 \times 10^5 < R_x < 10^7. \tag{5.80}$$

Here $C$ is a constant which depends on the transition Reynolds number $R_{x_{tr}}$. It is given by

$$C = R_{x_{tr}}(\bar{c}_{f_t} - \bar{c}_{f_l}) \tag{5.81}$$

where $\bar{c}_{f_t}$ and $\bar{c}_{f_l}$ correspond to the values of average skin-friction coefficient for turbulent and laminar flow at $R_{x_{tr}}$. While (5.80) is restricted to the indicated $R_x$-range, (5.79) is valid for a wide range of $R_x$ and has given good results up to $R_x = 10^9$.

*Prediction of Flow Separation in Two-Dimensional Incompressible Flows*

For two-dimensional steady flows, the separation point is defined as the point where the wall shear stress is equal to zero, that is

$$\left(\frac{\partial U}{\partial y}\right)_w = 0. \tag{5.82}$$

With the use of high-speed computers, the prediction methods discussed in Subsection 5.2.3, as well as transport equation methods, can be solved numerically and the separation point can be determined. The accuracy of several current methods for predicting separation without a full boundary-layer calculation was studied by Cebeci et al. [5.47]. The study showed that a simple formula due to Stratford [5.48] predicts flow separation quite satisfactorily. For a flow in which the pressure is constant for $0 \leq x \leq x_0$ and has a fairly sharp rise when $x$ exceeds $x_0$, this formula is

$$C_p \left( x \frac{dC_p}{dx} \right)^{1/2} (10^{-6} R_x)^{-1/10} = f(x).  \tag{5.83}$$

Here $c_p$ is defined as

$$C_p = 1 - \left( \frac{U_e}{U_0} \right)^2  \tag{5.84}$$

with $U_0$ denoting the velocity at the beginning of the adverse pressure gradient. The function $f(x)$ increases as separation is approached and decreases after separation; its value must be found empirically. For that reason, after applying his method to several flows with turbulent separation, Stratford observed that if the maximum value of $f(x)$ is a) greater than 0.40, separation is predicted when $f(x) = 0.40$; b) between 0.35 and 0.40, separation occurs at the maximum value; c) less than 0.35, separation does not occur. On the other hand, in the study conducted by CEBECI et al. [5.47], STRATFORD'S method gave better agreement with experiment, provided that the range of $f(x)$ was slightly changed from that given above, namely, if the maximum value of $f(x)$ is a) greater than 0.50, separation is predicted when $f(x) = 0.50$; b) between 0.30 and 0.50, separation occurs at the maximum value; c) less than 0.30, separation does not occur.

The analysis leading to (5.83) assumes an adverse pressure gradient starting from the leading edge, as well as fully turbulent flow everywhere. When there is a region of laminar flow or a region of turbulent flow with a favorable pressure gradient, STRATFORD makes the assumption that at the minimum pressure point, $x = x_m$, the velocity profile is approximately that of a flat-plate turbulent boundary layer starting from a false origin $x = x'$. Thus we replace $x$ by $(x - x')$ in (5.83) and take the value of $R_x$ as $U_m(x - x')/v$, with $x'$ given by

$$x = x_m - 58 \frac{v}{U_m} \left[ \frac{U_{tr}}{v} \int_0^{x_{tr}} \left( \frac{U_e}{U_m} \right)^5 dx \right]^{3/5} - \int_{x_{tr}}^{x_m} \left( \frac{U_e}{U_m} \right)^4 dx .  \tag{5.85}$$

*Similarity Solutions for Free Shear Flows*

As in laminar flows, under certain restrictions the turbulent shear layers admit similarity solutions for two-dimensional and axisymmetric jets, wakes and mixing layers. The solutions become valid only at large distances from the origin, say 20 nozzle diameters in jets or 100 body diameters in wakes. Here we shall give a brief summary of similarity solutions for free shear flows. For details the reader is referred to Schlichting [1.3] and Cebeci and Bradshaw [1.24].

For a two-dimensional jet coming out from a slot and mixing with the surrounding fluid at rest, the dimensionless velocity profile predicted by assuming an eddy viscosity independent of $r$ is

$$\frac{U(x, y)}{U_c(x)} = \text{sech}^2 \sqrt{B/2}\,\eta .\tag{5.86}$$

Here $B$ is an empirical constant equal to 1.55, and $\eta$ is defined by

$$\eta = \frac{y}{\delta(x)} \tag{5.87}$$

where $\delta(x)$ is the shear layer thickness usually chosen to scale the $y$-coordinate. For a two-dimensional jet it is the $y$-distance where $U/U_c = 0.5$; it is given by

$$\delta = 0.115x . \tag{5.88}$$

The centerline velocity $U_c$ and the eddy viscosity $\nu_T$ are given by

$$U_c = 2.40 \left( \frac{J}{\varrho} \frac{1}{x} \right)^{1/2} , \tag{5.89}$$

$$\nu_T = 0.037 \delta U_c . \tag{5.90}$$

Here $J$ represents the total momentum in the $x$-direction, $\int_{-\infty}^{\infty} \varrho U^2 dy$.

For an axisymmetric jet coming out from a circular hole and mixing with the surrounding fluid at rest, the dimensionless velocity profile predicted by assuming an eddy viscosity independent of $r$ is

$$\frac{U(x, r)}{U_c(x)} = \frac{1}{(1 + \frac{1}{8} A_1 \eta^2)^2} . \tag{5.91}$$

Here $A_1$ is again an empirical constant equal to 3.31 and $\eta$ is defined by

$$\eta = \frac{r}{\delta(x)} \tag{5.92}$$

where $\delta(x)$ is given by

$$\delta = 0.085x \,. \tag{5.93}$$

The centerline velocity $U_c$ and the eddy viscosity $v_T$ are given by

$$U_c = 7.415 \frac{\sqrt{J/\varrho}}{x}, \tag{5.94}$$

$$v_T = 0.0256 \delta U_c \,. \tag{5.95}$$

The similarity solutions for two-dimensional and axisymmetric wakes are at best valid only far enough downstream (100 diameters) for the pressure disturbances introduced by the body to be negligible. In addition the solutions are subject to the restriction that the velocity defect in the wake

$$U_1(x, y) = U_e - U(x, y) \tag{5.96}$$

is small compared with the velocity of the free stream $U_e$ so that higher order terms in the momentum equation can be dropped.

For a two-dimensional wake, the velocity defect is given by

$$U_1 = -\frac{F}{\varrho b} (2v_T U_e d)^{-1/2} (2\pi)^{-1/2} \exp\left(-c_2 \frac{\eta^2}{2}\right). \tag{5.97}$$

Here $F$ denotes the flux of momentum defect, $b$ the span; $c_2$ and $v_T$ are constants that must be determined from experiment. According to measurements behind circular cylinders of diameter $d$,

$$\delta = \tfrac{1}{4}(x C_D d)^{1/2} \tag{5.98}$$

and

$$\frac{v_T}{U_e C_D d} = 0.0222 \,. \tag{5.99}$$

Here $C_D$ is a drag coefficient defined by

$$C_D = \frac{F}{\tfrac{1}{2}\varrho U_e^2 b d} \,. \tag{5.100}$$

The shear layer thickness $\delta$, defined as half width at half depth, is given by

$$\delta = 1.675 \left(\frac{\nu_T x}{U_e}\right)^{1/2} .$$    (5.101)

Solving (5.100) for $F$ and inserting the resulting expression into (5.97), we get the mean velocity distribution behind a *circular cylinder* as

$$\frac{u_1}{U_e} = 0.141 \left(\frac{U_e C_D d}{\nu_T}\right)^{1/2} \left(\frac{x}{C_D d}\right)^{-1/2} \exp(-0.70\eta^2)$$    (5.102)

where $\eta$ is given by (5.87).

*Acknowledgements.* The preparation of Section 5.1 was supported by the National Science Foundation, the NASA Ames Research Center and the Air Force Office of Scientific Research. Section 5.2 is published by permission of McDonnell Douglas Corp.

# References

5.1   W. C. Reynolds: Adv. Chem. Engrg. **9**, 193 (1974)
5.2   G. L. Mellor, H. J. Herring: AIAA J. **11**, 590 (1973)
5.3   P. Bradshaw: Aeronaut. J. **76**, 403 (1972)
5.4   W. C. Reynolds: Ann. Rev. Fluid Mech. **8**, 183 (1976)
5.5   L. Prandtl: Nachr. Akad. Wiss., Göttingen, Math-Phys. Klasse (1945) p. 6
5.6   L. H. Norris, W. C. Reynolds: Rept. FM-10, Mech. Engrg. Dept., Stanford Univ. (1975)
5.7   M. Wolfshtein: Intern. J. Heat Mass Transfer **12**, 301 (1969)
5.8   J. C. Rotta: Z. für Phys. **129**, 547 (1951) and **131**, 51 (1951)
5.9   G. Comte-Bellot, S. Corrsin: J. Fluid Mech. **25**, 657 (1966)
5.10  J. L. Lumley, B. Khajeh-Nouri: Phys. Fluids (in press)
5.11  H. J. Tucker, A. J. Reynolds: J. Fluid Mech. **32**, 657 (1968)
5.12  J. Marechal: J. de Mécanique **11**, 263 (1972)
5.13  F. H. Champagne, V. G. Harris, S. Corrsin: J. Fluid Mech. **41**, 81 (1970)
5.14  W. G. Rose: J. Fluid Mech. **25**, 97 (1966)
5.15  P. G. Saffman: Studies Appl. Math. **53**, 17 (1974)
5.16  W. P. Jones, B. E. Launder: Intern. J. Heat Mass Transfer **15**, 301 (1972)
5.17  W. P. Jones, B. E. Launder: Intern. J. Heat Mass Transfer **16**, 1119 (1973)
5.19  P. G. Saffman, D. C. Wilcox: AIAA J. **12**, 541 (1974)
5.20  D. B. Spalding: Progr. Heat Mass Transfer **2**, 255 (1969) and Rept. TM/TN/A/16, Mech. Engrg. Dept., Imperial College, London (1971)
5.21  J. L. Lumley: Lecture Series No. 76, von Karman Inst. Belgium (1975)
5.22  D. Kwak, W. C. Reynolds: Rept. TF-5, Mech. Engrg. Dept., Stanford Univ. (1975)
5.23  S. Shaanan, J. H. Ferziger, W. C. Reynolds: Rept. TF-6, Mech. Engrg. Dept., Stanford Univ. (1975)
5.24  C. du P. Donaldson: AIAA J. **10**, 4 (1972)

5.25  J. W. DEARDORFF: J. Fluid Mech. **41**, 453 (1970)

5.26  J. W. DEARDORFF: Trans. ASME **95**I, 429 (1973)

5.27  S. A. ORSZAG, M. ISRAELI: Ann. Rev. Fluid Mech. **6**, 281 (1974)

5.28  P. A. LIBBY: J. Fluid Mech. **68**, 273 (1975)

5.29  F. M. WHITE: *Viscous Fluid Flow* (McGraw Hill, New York 1974)

5.30  J. E. GREEN, D. J. WEEKS, J. W. F. BROOMAN: ARC R. & M. 3791 (1973)

5.31  D. F. MYRING: RAE Tech. Rept. 70147 (1970)

5.32  P. D. SMITH: ARC R & M 3739 (1974)

5.33  H. H. FERNHOLZ: Z. Flugwiss. **16**, 401 (1968)

5.34  P. BRADSHAW, D. H. FERRISS, N. P. ATWELL: J. Fluid Mech. **28**, 593 (1967)

5.35  J. E. GREEN: RAE Tech. Rept. 72079 (1972)

5.36  D. M. BUSHNELL, A. M. CARY, B. B. HOLLEY: AIAA J. **13**, 1119 (1975)

5.37  T. CEBECI: AIAA J. **11**, 102 (1973)

5.38  E. R. VAN DRIEST: J. Aero. Sci. **23**, 1007 (1955)

5.39  D. E. ABBOTT, J. D. A. WALKER, R. E. YORK: Lecture Notes in Physics, Vol. 35, ed. by R. D. RICHTMYER (Springer, Berlin, Heidelberg, New York 1975) p. 34

5.40  T. CEBECI: AIAA J. **12**, 779 (1974)

5.41  T. CEBECI, D. E. ABBOTT: AIAA J. **13**, 829 (1975)

5.42  J. L. HUNT, D. M. BUSHNELL, I. E. BECKWITH: NASA TN D-6203 (1971)

5.43  J. E. HARRIS, D. J. MORRIS: Lecture Notes in Physics, Vol. 35, ed. by R. D. RICHTMYER (Springer, Berlin, Heidelberg, New York 1975) p. 204

5.44  P. BRADSHAW, D. H. FERRISS: Trans. ASME **94**D, 345 (1972)

5.45  P. BRADSHAW, G. A. MIZNER, K. UNSWORTH: Aero Rept. 75—04, Imperial College, London (1975): AIAA J. **14**, 399 (1976)

5.46  E. A. EICHELBRENNER (ed.): *Recent Research on Unsteady Boundary Layers* (Laval Univ. Press, Quebec 1972)

5.47  T. CEBECI, G. J. MOSINSKIS, A. M. O. SMITH: J. Aircraft **9**, 618 (1972)

5.48  B. S. STRATFORD: J. Fluid Mech. **5**, 1 (1959)

# 6. Heat and Mass Transport

B. E. LAUNDER

With 7 Figures

As mentioned in Chapter 1 the equations governing heat transfer with small temperature differences also govern transfer of small concentrations of other passive scalar contaminants. Particle-laden flows (Chapter 7) are excluded unless the particles are so small that they move with the flow and are negligibly affected by gravity. For convenience the discussion below is phrased in terms of heat transfer but translation to pollutant-transfer problems is straightforward. We use $C, c$ for the mean and fluctuating parts of the concentration or temperature, for generality.

The flow configurations of interest are the same as in Chapters 2 and 3 (see also Chapter 4) so that no further discussion of geometry is necessary. This chapter is a discussion of calculation methods for heat transfer, a sequel to the treatment of calculation methods for the velocity field in Chapter 5. Again it is convenient to start with a discussion of transport-equation modeling, and specialize to eddy-diffusivity models, in this case for the "eddy conductivity" or the turbulent Prandtl number; this occupies Section 6.2. Section 6.3 deals with modeling of the turbulent transport terms, and also discusses recent work on the temperature-field equivalent of the eddy length scale of Chapter 5. There is no discussion comparable with that of the simple zero-equation turbulence models or simple formulae in Sections 5.6 to 5.9 because their equivalent is the assumption of a simply behaved turbulent Prandtl number already discussed in Section 6.2 (see also Subsection 2.3.9).

Because heat transfer on the environmental scale usually involves buoyancy effects, while buoyant flows always include heat or contaminant transfer, a treatment of buoyancy effects has been fully integrated into the main discussion of heat transfer. The discussion also includes a review of heat-transfer data for a wide range of buoyant and non-buoyant flows.

# 6.1 Background

### 6.1.1 Areas of Importance

A large proportion of engineering heat- or mass-transfer problems in turbulent flow are reducible in their essentials to one of two forms:

the problem of achieving *very high* rates of heat or mass transfer at low cost

or

the problems of achieving *very low* rates of transfer—again at low cost.

Because their aims are opposite, the two basic problem types tend to be associated with different types of flow and heat-transfer configurations. When the aim is to diminish heat exchange the flow is usually of thin-shear-layer type over the majority of the region of interest. As an example, one may cite the external cooling of gas turbine blades by the ejection of relatively cool by-pass air through slots or holes in the blade inclined at a small angle with the surface. Similar film-cooling arrangements have been used for many years in gas-turbine combustion chambers, not to mention the de-mister of the automobile windscreen. The majority of engineering problems of heat transfer in turbulent flow, however, may be linked with the aim of *promoting* heat transfer—"heat exchangers" in the widest sense. In these cases there is a variety of complicating features that may be present. Particularly on the gas-stream side of a heat exchanger, it may be desirable to create flow separation to raise the level of effective thermal conductivity of the gas near the wall. Alternatively, swirl motion may be imparted to a ducted flow with the twofold aim of raising the overall level of heat-transfer coefficient and of producing a more uniform distribution of heat transfer. The insertion of twisted tapes and the coiling or bending of the flow passage are methods that are commonly used for thus inducing strong secondary motions in the fluid stream. Even where the flow is predominantly parallel to the duct axis, weak secondary motion will nearly always be present generated by the anisotropic turbulent stress field (Subsection 3.1.2). Although these velocities will rarely exceed one or two percent of the axial velocity, they nevertheless significantly affect the pattern of heat transport [6.1, 2].

Because of complexities such as those noted above, there are today relatively few problems of heat-exchanger performance that can be reliably predicted from a knowledge of heat transport in two-dimensional thin shear flows. Even after judicious simplification of the actual flow geometry, a meaningful analysis will usually need to account for simultaneous turbulent transport in at least two directions. Thus, whereas for many practical problems in aerodynamics a two-dimensional thin shear flow model is adequate, in the area of heat-exchanger design at least a two-dimensional "elliptical flow" treatment is usually needed. This is the main reason why heat-exchange equipment has usually been designed from results of experiments on models of prototype components.

In the last few years, however, our ability to obtain numerical solutions of coupled partial differential equations for elliptic and three-dimensional flows has advanced considerably [2.209, 6.3, 4]. Already the gas-turbine and nuclear-power industries in Western Europe and the USA make use of such numerical approaches in solving heat convection problems in turbulent flow. In the next few years, we may foresee a considerable extension of this pattern. In this chapter, as elsewhere in the book, we will therefore assume that purely numerical problems can be solved.

The value of such computational methods largely depends on how well the various scalar transport processes are characterized in the computer program. The mathematical model of turbulence must now account for both the turbulent stress *and* heat-flux fields. Thus, on sheer weight of numbers, the chances of getting the answer wrong has to be greater than if the velocity field alone were under study. The prospect is not quite as daunting as it may seem, however, for while an error of 15% in skin friction coefficient may be an unacceptably large error in many aerodynamic problems, to predict the level of heat-transfer coefficient to that degree of accuracy would suffice for most applications.

### 6.1.2 Models for Heat and Mass Transport

Models of heat transport have been proposed at various levels, corresponding with the different types of stress closure discussed in Chapter 5. When the flow field is found from a prescribed or calculated value of the effective turbulent ("eddy") viscosity, it is consistent to base the heat-transport model on an effective turbulent thermal diffusivity, $\gamma_T$, such that $-\overline{u_i c} = \gamma_T \partial C / \partial x_i$, at least for $i = 2$. The turbulent diffusivity must of course be prescribed or calculated in some way. In fact the magnitude of $\gamma_T$ is always of the same order of magnitude as the kinematic eddy viscosity $\nu_T$; attention is therefore focussed on their ratio, the turbulent Prandtl number, $\sigma_t \equiv \nu_T / \gamma_T$ (Subsection 2.3.9). Numerous simple models of heat transport, including an early proposal by PRANDTL [6.5] himself, have taken $\sigma_t$ to be unity, thus implying equal diffusivities of heat and momentum. The assumption is sometimes spoken of as "Reynolds' analogy" though purists would argue that this label should be applied only to the formula $St = c_f / 2$, where the

Stanton number, $St$, and skin friction coefficient, $c_f$, have been defined in Chapter 2. In fact a flow in which $\sigma_t = 1$ will satisfy the Stanton-number/skin-friction relation only for the same restrictive conditions as would a laminar flow; that is, for a molecular Prandtl number of unity, a uniform wall temperature and negligible pressure gradients and heat sources. These conditions are so rare that it is hard to understand the prominence still given to the analogy in many textbooks.

Knowledge of $\sigma_t$ is always used as a means of finding $\gamma_T$ from an already computed value of $v_T$ (rather than vice versa). There is thus, inevitably, the tendency to look to changes in the heat-transfer mechanism to explain variations in $\sigma_t$ that may actually occur from one point in the flow to another. This trend of thought needs revising, for the transport of a scalar such as heat or chemical species is a more appropriate quantity than momentum to take as the basic gradient-transport process. It would also be helpful to dispel the idea that a turbulent Prandtl number of unity was in any sense the "normal" value. We shall see in a later section that a value of about 0.7 has a far stronger claim to normality. The values close to unity found near a wall seem to be a consequence of wall-affected pressure fluctuations modifying the Reynolds-stress structure. At least, this is the conclusion that is suggested by models based on transport equations for $\overline{u_i c}$, analogous to the transport equations for $\overline{u_i u_j}$ discussed in Section 1.6 and Chapter 5. The problem of devising closed forms of these equations is the main theme of the present chapter; it is thus to be regarded as the heat-transfer equivalent of Chapter 5, especially Section 5.1.4. Section 6.2 reviews current attempts at devising suitable approximate forms for the generation/destruction terms in the equations. This completes the closure process for flows in which the turbulent transport terms (analogous to term 3 of (1.35)) are negligible.

In Section 6.3 attention shifts to non-homogeneous flows; Subsection 6.3.1 considers models of the turbulent flux transport while in Subsection 6.3.2 the transport equation for $\overline{c^2}$, the mean square level of scalar fluctuations, is examined. The latter correlation is important for several reasons, not least through its appearance in the characteristic time scale of the thermal turbulence field, $\overline{c^2}/\varepsilon_c$, where $\varepsilon_c$ is the rate of destruction of $\overline{c^2}/2$ by molecular diffusivity (Section 1.5).

A first approximation of this time scale is $\overline{q^2}/\varepsilon$, i.e., the time scale of the velocity fluctuations. As more and more experimental data become available, however, it begins to appear that the connection between the two is not strong enough for the latter to serve generally as an approximation of the former. The implication is that a transport equation for $\varepsilon_c$ is also needed, analogous to the transport equation for the turbulence energy dissipation rate presented in Section 5.3; the provision of such an equation is also discussed in Subsection 6.3.2. Here it might be said that second-order closures for scalar transfer are not as well developed as the second-order stress closures summarized in Chapter 5. There are good reasons for this; the turbulent stresses are a very important *input* to the heat-flux equations (both directly and through their influence on the mean velocity field). Inevitably, therefore, development of the stress closure precedes that of the heat-flux model. Moreover, there is still a great shortage of thermal turbulence measurements of the right kind. For every comprehensive heat-transfer study of a thin shear flow there are five or ten aerodynamic studies. In several cases the best available data of a simple, fundamental flow contain inconsistencies of 50% or more in the measured levels of turbulent heat flux. Thus, although the general pattern of a second-order heat-flux closure is already beginning to emerge, numerous refinements will doubtless be made over the next few years as more basic experimental studies of thermal turbulence are published.

Subsection 6.3.3 reviews the experimental data for scalar transport in thin shear layers, covering both free and near-wall flows. Subsection 6.3.4 is concerned with the flow which occurs in the immediate vicinity of a wall, including the viscous sublayer (Section 1.8). Here the problem of inhomogeneity is compounded with that of low turbulence Reynolds number. A satisfactory heat-flux transport model for this region is probably some years off. However, the region is so thin and the scales of turbulent motion so small that for many thin shear flows streamwise convection is negligible; the thermal properties of the flow (properly non-dimensionalized) should thus be functions only of $y^+$ and the molecular Prandtl/Schmidt number. Here trends of the experimental data are reviewed together with some of the simple transport models that these data have spawned. Finally, Subsection 6.3.5 considers flow in the complex geometries found in practical heat exchangers. Models for separated flows are first reviewed, following which some of the recent predictions of flow in heat exchange components are summarized.

Inevitably in a book such as this many topics have been omitted or have received only passing mention. Discussion is limited entirely to single-phase transport processes; thus nothing is said of flows where boiling or condensation is present. Natural convection, though mentioned briefly in Subsection 6.3.4, is discussed more fully in Chapter 4. With the exception of buoyant effects, we also restrict attention to flows where gradients of density and transport properties do not introduce significant additional turbulent correlations; more general forms of the exact equations for $\overline{u_i c}$ and $\overline{u_i u_j}$ are provided by RUBESIN and ROSE [6.6]. Only oblique reference is made to Lagrangian analyses for the dispersion of a scalar (here a recent article by CORRSIN (Ref. [4.74], p. 25) provides a good entry to the literature). Likewise we do not include discussion of dispersion from line or point sources in the absence of mean strain. Discussion and analysis of this topic are provided in the textbooks by HINZE [1.7] and TENNEKES and LUMLEY [1.15]. Finally, apart from the discussion of separated flow in Subsection 6.3.5, little is said about problem geometries because the less specialized configurations have already been discussed in Chapters 2 to 4.

## 6.2  Scalar Fluxes in Flows Near Local Equilibrium

### 6.2.1  Some Proposals for Closing the Scalar Flux Equations

The subject of this section is the correlation $\overline{u_i c}$ representing the rate of transport in direction $x_i$ of the scalar $C$ by turbulent velocity fluctuations. The exact transport equation for this correlation, analogous to (1.35), may be obtained according to the rules outlined in Chapter 1. We take the $u_i$ velocity component, multiply it by the equation for the instantaneous value of the scalar $(C+c)$, and add it to the $x_i$-component of the Navier-Stokes equations multiplied by $c$. Upon ensemble averaging, the result may be expressed as

$$\underbrace{\frac{D\overline{u_i c}}{Dt}}_{} = -\underbrace{\left(\overline{u_i u_l}\frac{\partial C}{\partial x_l} + \overline{u_l c}\frac{\partial U_i}{\partial x_l}\right)}_{\text{(i)}} + \underbrace{\frac{\overline{\varrho' c}}{\varrho}g_i}_{\text{(ii)}}$$

$$-\underbrace{(\gamma+\nu)\frac{\overline{\partial c}\,\overline{\partial u_i}}{\partial x_l\,\partial x_l}}_{\text{(iii)}} + \underbrace{\frac{\overline{p'}}{\varrho}\frac{\partial c}{\partial x_i}}_{\text{(iv)}} \qquad (6.1)$$

$$-\underbrace{\frac{\partial}{\partial x_l}\left(\overline{u_l u_i c} + \frac{\overline{p' c}}{\varrho}\delta_{il}\right)}_{\text{(v)}}$$

where $p'$ and $\varrho'$ are the fluctuating parts of the pressure and density, their mean parts being $p$ and $\varrho$. Equation (6.1) is valid for incompressible flows and where gradients in $C$ are small enough for $\gamma'/\gamma$ and $\nu'/\nu$ to be entirely unimportant ($\gamma'$ and $\nu'$ being the fluctuating parts of $\gamma$ and $\nu$) and for $\varrho'/\varrho$ to be significant only in the gravitational term. Even where these conditions do not pertain, (6.1) still contains the main physical processes controlling the level of the scalar-flux correlation. Equation (6.1) neglects the influence of viscous transport but includes the gravitational term in the Navier-Stokes equations. Term (ii), representing augmentation of $\overline{u_i c}$ due to gravitational forces contributing to $u_i$ via a body force per unit mass $f_i \equiv (\varrho' g_i)/\varrho$ in (1.2), is conveniently modified as follows:

$$\frac{\overline{\varrho' c}}{\varrho} g_i = -\alpha \frac{\overline{c^2}}{C} g_i \qquad (6.2)$$

where the dimensionless coefficient $\alpha$ is defined as

$$\alpha \equiv -\left. \frac{\partial \varrho}{\partial C} \right|_p \frac{C}{\varrho}. \qquad (6.3)$$

If $C$ stands for temperature and the fluid is an ideal gas, $\alpha$ is unity.

Term (i) expresses the rate of creation of $\overline{u_i c}$ due to the combined actions of mean velocity and mean scalar gradients, the former tending to increase the velocity fluctuations and the latter the magnitude of the scalar fluctuations. The dissipative correlation (iii) is zero in isotropic turbulence and will be negligible also in non-isotropic turbulence provided that, as we assume here, the turbulence Reynolds number is high. The pressure-temperature gradient correlation, (iv), is the counterpart of the pressure-strain correlation in the stress equations. With the direct dissipation negligible, this provides the mechanism which limits the growth of the fluxes. Finally, term (v) denotes the rate of spatial transport of $C$ due to velocity and pressure fluctuations. In the homogeneous flows considered in this section its effect will be negligible and we therefore defer until Section 6.3 the question of its approximation.

For convenience let us rewrite the scalar-flux transport equation including the simplifications noted above

$$\frac{D\overline{u_i c}}{Dt} = -\left( \overline{u_k u_i} \frac{\partial C}{\partial x_k} + \overline{u_k c} \frac{\partial U_i}{\partial x_k} + \frac{\alpha \overline{c^2} g_i}{C} \right) + \frac{\overline{p'}}{\varrho} \frac{\partial c}{\partial x_i}. \qquad (6.4)$$

For non-buoyant flows approximation of the last term in (6.4) would close the equations for $\overline{u_i c}$ and $C$, if the velocity and stress fields are regarded as known. The succeeding paragraphs will outline recent and current work at devising a widely valid, closed form of (6.4). To judge the adequacy of the approximation we shall draw comparisons with experimental data in some simple two-dimensional shear flows where convective transport may be considered small or nearly balanced by the diffusive flux. Equation (6.4) then merely expresses a balance between the flux-generation terms and the pressure-temperature gradient correlation. As may be expected the resultant formulae show some kinship with conventional effective-conductivity representations—but also sufficient differences to emphasize how imprudent it is to assume an isotropic thermal diffusivity.

As a first step in approximating $\overline{(p'/\varrho)\partial c/\partial x_i}$ it is convenient, as in Chapter 5, to turn to the Poisson equation for $p'$ (Eq. (1.11) with $u$ replacing $U+u$). After multiplication by $\partial c/\partial x_i$ and ensemble averaging, we obtain

$$
\phi_{ic} \equiv \frac{\overline{p'}}{\varrho} \frac{\partial c}{\partial x_i}
$$

$$
= \frac{1}{4\pi} \int \left( \underbrace{\frac{\partial^2 u'_l u'_m}{\partial x_l \partial x_m} \frac{\partial c}{\partial x_i}}_{\phi_{ic_1}} + 2 \underbrace{\frac{\partial U'_l}{\partial x_m} \frac{\overline{\partial u'_m}}{\partial x_l} \frac{\partial c}{\partial x_i}}_{\phi_{ic_2}} + \underbrace{\frac{\alpha g_i}{C'} \frac{\overline{\partial c'}}{\partial x_l} \frac{\partial c}{\partial x_l}}_{\phi_{ic_3}} \right) \frac{d\,Vol}{r} \tag{6.5}
$$

where, as in Chapter 1, prime superscripts on the right-hand side denote that the quantity is evaluated at $x+r$. The three contributions to the integral in (6.5) will be termed the turbulence, the mean-strain and the gravitational parts of $\phi_{ic}$. Note that although the mean rate of strain appears (from the Poisson equation) the mean scalar gradient does not. Most of the earlier proposals for approximating this term included only the first of these effects [6.7–10]. The importance of the mean-strain effect on pressure fluctuations is now generally recognized, at least in approximating the pressure-strain correlation in the stress equations. Recently LUMLEY [5.21] has shown that, for a homogeneous flow, the assumption that the two-point correlation functions in the mean strain and gravitational parts, $\phi_{ic_2}$ and $\phi_{ic_3}$, are symmetric in form leads to the result

$$
\phi_{ic_2} = 0.8\overline{u_m c}\,\partial U_i/\partial x_m - 0.2\overline{u_m c}\,\partial U_m/\partial x_i , \tag{6.6}
$$

$$
\phi_{ic_3} = \alpha \overline{c^2} g_i/3C . \tag{6.7}
$$

The writer has also obtained (6.6) and (6.7) in [6.11, 12] by assuming the general form of the resultant tensors and applying to them the symmetry properties that the corresponding integrands in (6.5) would possess in isotropic turbulence. The two approaches to the same result may in this example be considered equivalent; in the discussion below, (6.6) and (6.7) are referred to as the "quasi-isotropic" model.

From (6.7) we see that the effect of $\phi_{ic_3}$ is to obliterate one-third of the direct generation (or destruction) of $\overline{u_i c}$ by gravitational influences. The predominant contribution of the mean-strain term is also that of diminishing the effect of the direct generation due to velocity gradients. This result seems at least qualitatively what we should expect and is in line with the effect of the mean-strain contribution to the pressure-strain correlation discussed in Chapter 5. Equations (6.6) and (6.7) hold exactly for isotropic turbulence—but this idealized state can be achieved in practice only if $\phi_{ic}$ and the direct generation terms in (6.1) are negligible. For general shear flows they will be only approximately correct; but at least they do have the merit of not introducing fresh empirical coefficients into our system of equations.

A further suggestion for approximating $\phi_{ic_2}$ and $\phi_{ic_3}$ will be considered in some detail partly because it has been used more extensively than (6.6) and (6.7) and partly because it produces attractively simple algebraic results. The stress-closure exploration by LAUNDER et al. [3.21] showed that for free shear flows the mean-strain contribution to $\phi_{ij_2}$ could be adequately and simply represented by

$$\phi_{ij_2} = -0.6(P_{ij} - \tfrac{2}{3}P\delta_{ij}) \tag{6.8}$$

where $P_{ij}$ and $P$ denote the rate of generation of $\overline{u_iu_j}$ and $\tfrac{1}{2}q^2$ by inter-action with the mean rate of strain (term 1 of (1.35) and (1.36), respectively) and the coefficient 0.6 is chosen to satisfy the exact result for isotropic turbulence [6.13]. This form says that the *sole* effect of $\phi_{ij_2}$ is to redistribute the stress-production tensor so as to reduce its anisotropy. Equation (6.8) suggests that the corresponding terms in the scalar flux equations could be approximated as

$$\phi_{ic_2} + \phi_{ic_3} = -c_{2c}P_{ic} \tag{6.9}$$

where now $P_{ic}$ stands for the generation rate of $\overline{u_ic}$ arising from that part of the equation for $D\overline{u_ic}/Dt$ responsible for the pressure fluctuations, i.e., from $\overline{cDu_i}/Dt$. Thus, written out in full, (6.9) implies

$$\phi_{ic_2} = c_{2c}\overline{u_mc}\,\partial U_i/\partial x_m \tag{6.10}$$

$$\phi_{ic_3} = c_{2c}\frac{\overline{\alpha c^2}g_i}{C}. \tag{6.11}$$

By comparing (6.11) and (6.7) it is seen that $c_{2c}$ should be taken as 0.33 in order to satisfy the "quasi-isotropic" result. Reference [6.14] adopted a value of 0.5 for this coefficient in order to give the ratio of streamwise to cross-stream heat flux indicated by Webster's [6.15] nominally homogeneous shear flow under neutral conditions. It will be seen in Subsection 6.2.4 that the same value also gives about the correct effect of stable stratification on the turbulent heat fluxes, again compared with Webster's data. OWEN [6.16] has also adopted equation (6.9) with $c_{2c}$

equal to 0.5 for most of his calculations. MERONEY [6.17], too, adopted
(6.10) but with $P_{ic}$ interpreted as the *total* generation rate of $\overline{u_ic}$, including
that due to gradients[1] of $C$; he proposed that the coefficient should take
the value of 0.4.

It may be said that, though use of (6.9) gives the same form for $\phi_{ic_3}$
as the quasi-isotropic model, there is no such similarity between the
approximations of $\phi_{ic_2}$ given by (6.10) and (6.6). Which of these two
representations is closer to the truth in practically occurring shear flows?
Despite the clear-cut differences between the forms it does not yet seem
possible to choose with certainty. We cannot measure $\phi_{ic}$ directly, still
less obtain from experiment the relative contributions made by the
turbulence, mean-strain and gravitational parts of this process. We shall
see below that the method of approximating $\phi_{ic_1}$ can greatly affect conclusions as to the most appropriate way of approximating $\phi_{ic_2}$ and
$\phi_{ic_3}$.

The most widely used form for $\phi_{ic_1}$, perhaps first used by MONIN
[6.7], is

$$\phi_{ic_1} = -c_{1c}(2\varepsilon/\overline{q^2})\overline{u_ic} \tag{6.12}$$

which will be recognized as a direct counterpart of Rotta's [5.7] "linear
return-to-isotropy" approximation of the pressure-strain correlation. It
seems likely that the time scale taken as $(\overline{q^2}/2\varepsilon)$ above should contain
some weighting from the time scale of the thermal turbulence field,
$(\overline{c^2}/2\varepsilon_c)$. In local-equilibrium turbulence these two time scales may be
supposed to be proportional, however, so there is nothing to be gained at
this stage by introducing two further scalar properties as unknowns.

Just as there have been recent proposals for adopting a nonlinear
representation of $\phi_{ij_1}$, various elaborations of (6.12) have been suggested,
e.g., [6.10, 12], Ref. [4.74], p. 169. The most comprehensive proposals
are probably those outlined by LUMLEY [5.21]. Briefly, he suggests that
the coefficient $c_{1c}$ should separately be a linear function of a dimension-
less strain rate and of the "anisotropy" of the heat flux, $\overline{cu_i}\,\overline{cu_i}/\overline{q^2}\overline{c^2}$;
further terms involving products of heat fluxes and mean strain rates
are also added (in addition to those arising from $\phi_{ic_2}$). The only form for
which numerical values for the coefficients have been ventured, however,
seems to be

$$\phi_{ic_1} = -c_{1c}(2\varepsilon/\overline{q^2})\overline{u_ic} - c'_{1c}(2\varepsilon/\overline{q^2})a_{il}\overline{u_lc} \tag{6.13}$$

where $a_{il}$ is the dimensionless, anisotropic part of the Reynolds stress,
$(2\overline{u_iu_l}/\overline{q^2} - \frac{2}{3}\delta_{il})$.

---

[1] As we have noted, gradients of $C$ do not appear in (6.5).

Table 6.1. Proposed values for the coefficients in the first part of the pressure temperature-gradient approximation

| Workers | $c_{1c}$ | $c'_{1c}$ | $\phi_{ic_2}$ included |
|---|---|---|---|
| DONALDSON et al. [6.9] | 5 | — | No |
| WYNGAARD and COTE [6.10] | 4.4—9.7 | —16.7 | No |
| LAUNDER [6.14] | 3.2 | — | Yes |
| OWEN [6.16] | 4.1 | — | Yes |
| MERONEY [6.17] | 2.5 | — | Yes |
| LUMLEY and KHAJEH-NOURI (Ref. [4.74], p. 169) | 4.3 | no specific recommendation | No |
| Work in progress at Imperial College [6.19] | 3.8 | —2.2 | Yes |

Table 6.1 shows some of the values suggested for the coefficients in (6.12) and (6.13). There is an untidy disarray among the various proposals. This reflects the fact that most of the suggestions have been based on comparisons with only one type of experimental flow (and, in two cases, without *any* recourse to experimental data). Leaving aside the buoyant atmospheric boundary layers examined by WYNGAARD and COTE [6.10], however, the values proposed for $c_{1c}$ do not seem to depend significantly on whether $\phi_{ic_2}$ is included. This is in marked contrast with the situation in the case of the Reynolds-stress equations; there the value of the coefficient $c_1$ is roughly twice as large when the mean strain terms are omitted (see the survey in [6.12] and compare, for example, the two papers by LUMLEY and KHAJEH-NOURI (Ref. [4.74], p. 169, [6.18]) the first of which omits mean-strain effects). Also, the values proposed for $c_{1c}$ are about twice as large as the more plausible proposals for $c_1$. The implication is that pressure fluctuations are about twice as effective at destroying the heat-flux correlation as they are at destroying the shear stresses. The reason for this may lie partly in the somewhat different parts of the spectrum responsible for the respective correlations and partly in the omission of any contribution to (6.12, 13) from the scalar-turbulence time scale.

With the convective transport terms neglected, (6.4) provides a set of algebraic equations for the heat fluxes; the precise form depends, of course, on which of the various closures discussed above for $\phi_{ic}$ is adopted.

1) Use of (6.12) for $\phi_{ic_1}$ and of (6.9) for $\phi_{ic_2}$ (or its neglect altogether) gives the following expression for the cross-stream scalar flux in a two-dimensional thin shear layer:

$$\overline{vc} = -\frac{1}{c_{1c}} \frac{\overline{q^2 v^2}}{2\varepsilon} \frac{\partial C}{\partial y}. \tag{6.14}$$

Notice that there is no contribution from $\phi_{ic_2}$ to this equation. Workers have tended to use (6.14) as the basis of fixing $c_{1c}$; this perhaps explains why the values in Table 6.1 are virtually independent of whether $\phi_{ic_2}$ is included or not. Successful calculations of heat or species transport in a number of free shear flows have been obtained with the following similar equation (Ref. [2.2], p. 361):

$$\overline{vc} = -0.037 \frac{q^4}{\varepsilon} \frac{\partial C}{\partial y}. \tag{6.15}$$

Thus, on approximating $\overline{v^2}$ as $0.25\overline{q^2}$, we conclude that $c_{1c}$ should be about 3.4.

Use of (6.9) and (6.12) gives the following formula for the ratio of streamwise to cross-stream heat fluxes:

$$\overline{uc}/\overline{vc} = \overline{uv}/\overline{v^2} + \frac{(1-c_{2c})}{c_{1c}} \overline{q^2}/2\overline{uv}. \tag{6.16}$$

The homogeneous shear-flow data of CHAMPAGNE et al. [5.13] provide values of the stress ratios

$$\overline{uv}/\overline{v^2} \simeq -\tfrac{2}{3}; \quad \overline{q^2}/\overline{uv} \simeq -6.0 \quad \text{(yielding } \overline{v^2} = 0.25\overline{q^2} \text{ again)}$$

while the data of WEBSTER suggest a value of the scalar flux ratio of approximately $-1.1$; using the above argument [6.14] concluded that $c_{2c}$ should be about 0.5. The value of $\overline{uc}/\overline{vc}$ obtained from $c_{2c} = \tfrac{1}{3}$ (which is the value indicated by (6.7)) is $-1.25$. Webster's measured values of $\overline{u^2}/\overline{v^2}$ are some 25% less than those usually reported for the homogeneous shear layer; it is thus likely that his ratio of $\overline{uc}/\overline{vc}$ is also too low. It seems, therefore, that the choice of $\tfrac{1}{3}$ for $c_{2c}$ is as acceptable as that adopted in [6.14].

By definition, the turbulent Prandtl/Schmidt number is

$$\sigma_t \equiv \frac{\overline{uv}/(\partial U/\partial y)}{\overline{vc}/(\partial C/\partial y)}. \tag{6.17}$$

Then using the fact that, in equilibrium, turbulence-energy production and dissipation rates are equal (to eliminate $\partial U/\partial y$), and eliminating $\partial C/\partial y$ with the help of (6.14), we obtain

$$\sigma_t = 2c_{1c}\overline{uv}^2/(\overline{q^2}\overline{v^2}). \tag{6.18}$$

On inserting the values of the stress ratios given above we obtain $\sigma_t \simeq 0.7$, which is indeed in line with the average levels of turbulent Prandtl number reported for a number of inhomogeneous *free* shear flows.

2) When the quasi-isotropic form is used for $\phi_{ic_2}$ (Eq. (6.6)) the flux down the scalar gradient is given by

$$\overline{vc} = -\frac{\overline{q^2 v^2}}{2\varepsilon c_{1c}} \frac{\partial C}{\partial y} - \frac{0.2\,\overline{q^2}}{c_{1c}\,2\varepsilon} \overline{uc}\,\frac{\partial U}{\partial y}. \tag{6.19}$$

The extra term, containing mean velocity gradients, arises, of course, from the mean-strain contribution to $\phi_{ic}$. With the values noted above for $\overline{uc}/\overline{vc}$ and $\overline{q^2}/\overline{uv}$, we may recast (6.19) as

$$\overline{vc} = -(c_{1c}-0.66)^{-1}\frac{\overline{q^2 v^2}}{2\varepsilon}\frac{\partial C}{\partial y}. \tag{6.20}$$

Comparing this with (6.14) we see that the effect of mean strain on the implied value of $c_{1c}$ is significant but not major. When the corresponding expression for $\overline{uc}$ is worked out, however, it emerges that then the value of $c_{1c}$ needs to be only about 0.5 to produce the correct flux level— evidently an incompatible value! That is why, in adopting the quasi-isotropic approximation for $\phi_{ic_2}$ and $\phi_{ic_3}$, [6.19] had to introduce the nonlinear form for $\phi_{ic_1}$ (Eq. (6.13)). With the values shown in Table 6.1 for $c_{1c}$ and $c'_{1c}$ the correct level of streamwise and cross-stream scalar fluxes is secured.

Which of the two combinations for the pressure-temperature-gradient term $\phi_{ic}$ considered above is the better? Both forms require two empirically chosen coefficients; (6.9) has the advantage that it is exact in isotropic flows. Equation (6.6) has the merit of leading to algebraically simple expressions and for this reason it is the form chosen to illustrate the discussion of succeeding sections. Neither has been tested in a sufficiently wide range of flows for experimental evidence to provide the undisputed answer.

The following simple flux formula has been proposed by a number of workers, for example, DALY and HARLOW [6.20] and DALY [6.21]:

$$\overline{u_i c} = -\text{constant} \times \overline{u_l u_i}\,\frac{\partial C}{\partial x_l}. \tag{6.21}$$

This may be interpreted as implying the use of (6.9) and (6.12) to approximate $\phi_{ic}$ with the coefficient $c_{2c}$ taking the value of unity; that is, *all* mean strain (and buoyant) generation of $\overline{u_i c}$ is supposed to be obliterated by pressure fluctuations. For the case of scalar-flux transport across a thin shear layer, (6.21) reduces to (6.14) (though the implied value of $c_{1c}$ adopted by DALY is only 0.6) while the ratio of stream-wise to cross-stream fluxes is:

$$\overline{uc}/\overline{vc} = -\overline{uv}/\overline{v^2} .\qquad(6.22)$$

From the discussion following (6.16), it is evident that (6.22) seriously underestimates the magnitude of this flux ratio; in general mean-strain contributions to the heat fluxes will be substantial.

### 6.2.2 Near-Wall Turbulence and the Turbulent Prandtl Number

We now consider another kind of equilibrium turbulent flow: that in the near-wall region of a turbulent boundary layer or duct flow (Section 1.8). We exclude, for the present, the region immediately adjacent to a smooth wall extending from the wall to $y^+ \simeq 40$ where viscous effects are important. In any case this sublayer region could certainly not be termed a "nearly homogeneous" region; it is thus more appropriately discussed in Section 6.3.

The proximity of the wall exerts two principal effects on the turbulence structure in the fully turbulent region: it limits the length scale of the fluctuating motion, thus raising the level of the energy dissipation rate ($\varepsilon$ varying inversely with distance from the wall); and it modifies the fluctuating pressure field. The former modification does not concern us explicitly since we regard $\varepsilon$ as a known or calculable quantity. The second effect does, however. The near-wall modification of the pressure field has important consequences in the Reynolds stress equations [3.21]. How important are the corresponding effects in the heat-flux equation?

First let us note that, according to (6.10), there is no contribution from $\phi_{ic_2}$ to $\overline{vc}$ in a free shear flow. If it is assumed that the contribution is negligible also in near-wall flows, (6.18) for the turbulent Prandtl number still applies—though we would expect that $c_{1c}$ would be altered somewhat. In near-wall turbulence the coefficient $-\overline{uv}/q^2$ takes a mean value within a few percent of 0.135 and $\overline{v^2}$ is very nearly equal to $-\overline{uv}$. We therefore deduce that

$$\sigma_t \simeq 0.27 c_{1cw}$$

where the subscript $w$ has been added to indicate a near-wall value.

What is the evidence of experiment? KESTIN and RICHARDSON [6.22] reviewed measurements of $\sigma_t$ in pipe flows and boundary layers published up to the early 1960's. The values reported displayed much scatter, some experiments showing a far from uniform level in the near-wall region. BLOM [6.23] extended the review in his PhD thesis to include most of the experimental data reported up to 1968. The intervening years had done little to establish a conclusive value of $\sigma_t$ from experiment. It could be said that for air and fluids with molecular Prandtl numbers greater than 1, the turbulent Prandtl number appeared to be approximately unity within an uncertainty of about $\pm 0.2$. By this time, however, computational work had succeeded in narrowing $\sigma_t$ within finer limits than the reported measurements indicated. The appearance, from 1967, of numerical solving procedures for the partial differential equations governing boundary layer flow [2.208] facilitated extensive predictions of heat transfer in turbulent boundary layers (see, for example, [6.24, 2.210]). Now these procedures required the level of turbulent Prandtl number as an *input*; workers accordingly adjusted the prescribed value of $\sigma_t$ to obtain best agreement with measured temperature profiles and wall heat fluxes. These studies confirmed what earlier analyses, based upon enthalpy-integral approaches, had suggested: that the turbulent Prandtl number was approximately 0.9 under a wide range of flow conditions and could for most purposes be taken as uniform across the near-wall region.

More recently many further experimental investigations into the distribution of turbulent Prandtl number have been reported, among the most comprehensive being those of SIMPSON et al. [6.25], CHEN [6.26], BAKER and LAUNDER [2.40] and QUARMBY and QUIRK [6.27]. All except CHEN determined $\sigma_t$ by deducing the stress and heat- (or species-) flux distributions across the flow by integral momentum and enthalpy balances. The measurements of [6.23 and 2.40], obtained in a flat-plate boundary layer with and without transpiration, display similar behavior. They indicate a fairly uniform mean value for $\sigma_t$ of approximately 0.95 in the near-wall region (with scatter of $\pm 15\%$ about this value) though with a tendency for the turbulent Prandtl number to *exceed* unity as the wall is approached. Quarmby and Quirk's [6.27] heat- and mass-transfer data for pipe flow display a similar amount of scatter as the studies of [6.23 and 2.40] with $\sigma_t$ decreasing near the wall from 0.85 at $y/R = 0.2$ to only 0.65 at $y/R = 0.05$. CHEN did not attempt to obtain point derivatives of his mean velocity and temperature profiles. Instead, he carefully fitted straight lines through the near-wall semi-logarithmic region of his profiles, thus obtaining values for the coefficients $\kappa$ and $\kappa_c$ in the equations

$$U^+ = \frac{1}{\kappa} \ln y^+ + B$$

$$C^+ = \frac{1}{\kappa_c} \ln y^+ + D .$$

$C^+$ is the dimensionless temperature $\varrho U_\tau c_p (C_w - C)/Q_w$ and the subscript $w$ denotes wall values. When $\overline{uv}$ and $\overline{vc}$ are treated as constant over the region where these profiles apply, the turbulent Prandtl number is readily shown to equal $\kappa/\kappa_c$ (see Subsection 4.1.1). CHEN [6.26] reported a value of 0.885 for this ratio with no mention of scatter.

More direct measurements of $\sigma_t$ have recently been reported by members of the Heat and Mass Transfer Group at Stanford University [6.28, 29], the turbulent heat and momentum fluxes being obtained by hot-wire anemometry. The work by PIMENTA et al. [6.29] is particularly interesting since it was conducted on a rough test plate (consisting of a uniform bed of 1 mm nickel-plated copper balls brazed together). Air velocities were sufficiently high that the turbulent boundary layer exhibits a "fully rough" behavior (Subsection 2.3.7) i.e., there is no observable effect of viscosity on any of the macro-properties of the boundary layer. These measurements of $\sigma_t$ exhibit hardly any scatter and lack the near-wall "peaking" characteristic of many of the smooth-wall explorations noted above. From over thirty data points taken within the region of near-constant stress and heat flux close to the wall (in six separate profile traverses) the extreme values reported for $\sigma_t$ were 1.02 and 0.93 with only one run providing values differing by more than 3% from 0.955.

Viewed overall, the weight of experimental evidence seems to suggest that the near-wall value of $\sigma_t$ is a little above 0.9. Thus, from the earlier deduction that $\sigma_t \simeq 0.27 c_{1cw}$ we conclude that $c_{1cw}$ is approximately 3.4; that is, *the same value as deduced for free shear flows*. This is an interesting and initially surprising result, for there is a marked effect of the wall on the pressure-strain correlation, $\phi_{ij}$. It is not immediately clear why the pressure temperature-gradient correlation would be unaffected since the same fluctuating pressure field appears in $\phi_{ic}$ as in $\phi_{ij}$. Note, however, that the main effect of the wall on $\phi_{ij}$ is to modify the levels of the normal stresses $\overline{u^2}$ and $\overline{v^2}$; the direct effect on $\overline{uv}$ (which is the nearest counterpart to $\overline{vc}$ in the stress tensor) is minor[2]. It appears therefore that the difference in turbulent Prandtl number between wall-bounded and free shear flows is largely due to the different magnitudes of the normalized Reynolds stresses $\overline{v^2}/q^2$ and $\overline{uv}/q^2$ in the two types of flow, a variation in relative stress levels which stems mainly from near-wall effects in the pressure-strain correlation. Thus, as suggested in Section 6.1, the turbulent Prandtl number might be better regarded as an indicator of the state of *momentum*-transport rather than of heat-transport processes.

---

[2] There is of course an indirect effect since the generation terms in both the stress and heat flux equations contain $\overline{v^2}$.

Equation (6.4) retains no molecular transport terms; the assumption is therefore implicit that the closure formulae introduced in Subsection 6.2.1 should be independent of molecular Prandtl or Schmidt number, $\sigma \equiv \nu/\gamma$. (Of course for liquid metals, where the molecular Prandtl number is of the order of $10^{-2}$, the turbulent Peclet number ($\sigma q^4/\nu\varepsilon$) will very often be low enough for the effects of molecular conduction to be important throughout the flow; these are conditions discussed in Subsection 6.3.4. The pipe-flow experiments of GOWEN and SMITH [6.30] suggested a weak variation of $\sigma_t$ with molecular Prandtl number. Viewed in retrospect however, it appears that the values emerging from their air-flow experiments gave too high values of turbulent Prandtl number; certainly the experiments using water and aqueous ethylene glycol as fluids indicated values of $\sigma_t$ of approximately 0.9. More recently, MIZUSHINA et al. [6.31] examined the wall region of flow in a plane channel by means of a Mach-Zehnder interferometer. Their study spanned a range of molecular Prandtl number from 6 to 40. No dependence of the turbulent Prandtl number on the level of $\sigma$ could be detected, in accordance with (6.4). In confirmation of this result, QUARMBY and QUIRK [6.32] have recently repeated some of their earlier pipe-flow diffusion studies for Schmidt numbers as high as 1200. They concluded that there was no detectable difference between the turbulent diffusivities measured in this study and those obtained [6.27] where the Schmidt number was close to unity.

### 6.2.3 Streamwise and Lateral Heat Transport in Near-Wall Turbulence

In the preceding section the mechanism of heat transport normal to the wall was considered. Here we examine what is known or implied about transport in the plane normal to the direction of mean velocity gradient (the $x, z$ plane).

The early measurements of JOHNSON [6.33] and virtually all subsequent studies suggest that the ratio of streamwise to vertical heat flux is substantially larger in near-wall turbulence than in free shear flows. The ratio measured by Johnson was approximately 3.0 which is in accord with the atmospheric turbulence data of WYNGAARD et al. [6.34]. The rough-surface boundary layer experiments of PIMENTA et al. [6.29] however, indicate

$$\overline{uc}/\overline{vc} = -2.05 \pm 0.15 \tag{6.23}$$

a value which is consistent with those found in the pipe-flow measurements of BREMHORST and BULLOCK [6.35] and LAWN and WHITE [6.36] and with the atmospheric measurements of ZUBKOVSKY and TSVANG [6.37].

What heat-flux ratio is indicated by the approximation of $\phi_{ic}$ by (6.9) and (6.12)? With experimental near-wall stress levels inserted in (6.16) there follows

$$\overline{uc}/\overline{vc} = -1.0 - 4.2(1 - c_{2c})/c_{1c}. \tag{6.24}$$

We concluded above that $c_{1c}$ appeared virtually unaffected by the wall, a value of approximately 3.4 giving the correct level of turbulent Prandtl number in near-wall turbulence. With $c_{2c}$ taken as $\frac{1}{3}$, (6.24) indicates a value of $-1.8$ for the heat flux ratio; though this falls within the uncertainty of $\pm 20\%$ reported in [6.29], the magnitude is probably somewhat too small. Now $\overline{uc}$ is somewhat similar in character to $\overline{u^2}$ whose level, we know, is substantially raised by a wall [$(\overline{u^2} - \overline{q^2}/3)$ is twice as large in near-wall turbulence as in a free shear flow] so it is to be expected that there should be a wall effect on $\phi_{ic_2}$. A reduction in the value of $c_{2c}$ to 0.2 would bring the heat flux ratio given by (6.24) fully into accord with (6.23). In physical terms one is thus saying that the presence of the wall diminishes the ability of pressure fluctuations to obliterate the streamwise heat flux. Indeed it seems probable that the high values of the correlation coefficient $\overline{uc}/\sqrt{u^2}\sqrt{c^2}$ near the wall (PIMENTA et al. [6.29] consistently obtained values around 0.75 in their experiments) is due to the diminishing effectiveness of the mean-strain destruction of $\overline{uc}$ as the wall is approached.

Several experiments have been reported in which, though the velocity field was that of a two-dimensional thin shear flow, asymmetrical heating or release of chemical species yielded significant scalar fluxes in both directions normal to the mean velocity vector. The question that these experiments have aimed to answer is whether the effective diffusivity for fluxes parallel to the surface (and normal to the mean flow) is effectively the same as that for fluxes normal to the wall. Most practical problems of heat convection do involve three-dimensional heat transport so the question is of some importance.

The first reported measurements appear to be those of SPARROW and BLACK [6.38] who studied the development of the thermal field for flow in an unsymmetrically heated pipe. Although no attempt was made to extract values of circumferential diffusivities, calculations showed that the measured temperature distributions could only be explained if the circumferential diffusivity was several times larger than the radial one close to the pipe wall. Rather similar experiments have been undertaken by QUARMBY and his students over several years using, for the most part, an unsymmetrical distribution of tracer gas (rather than of temperature) to generate the three-dimensional scalar-flux field. The use of a tracer gas has some advantages over a heated tube since problems of circum-

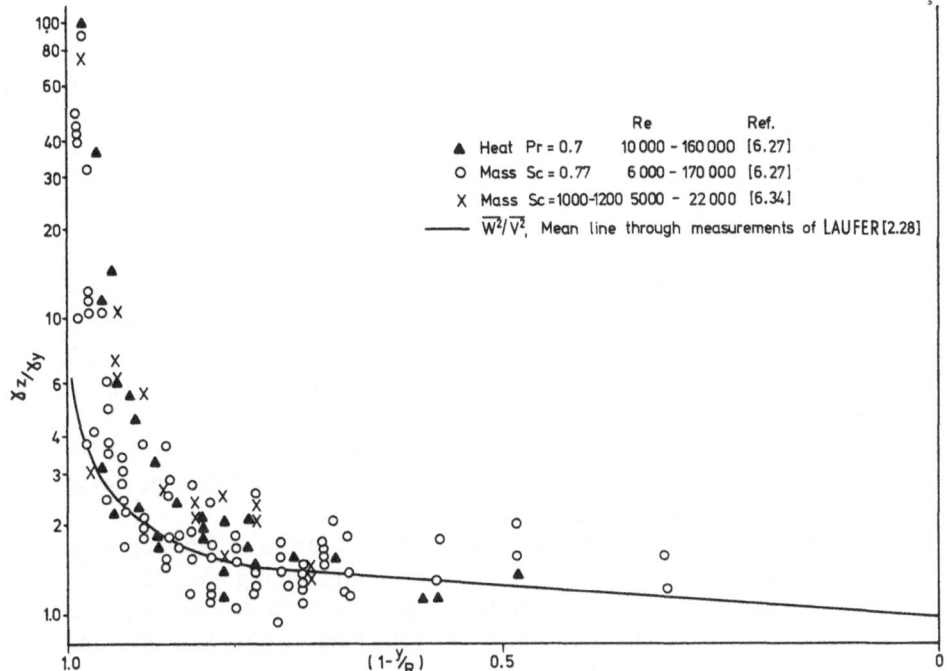

Fig. 6.1. Ratio of tangential: radial diffusivities in pipe flow; comparison with distribution of $\overline{w}^2/\overline{v}^2$

ferential conduction in the pipe wall and of properly accounting for heat losses are thereby eliminated. If the pipe wall is impervious, however, radial gradients of the species are small near the wall so it is difficult to estimate the radial diffusivity. QUARMBY and ANAND [6.39] raised a number of criticisms of the Sparrow and Black work and concluded, from predictions of their own data with an isotropic diffusion coefficient that "... the good agreement between theory and experiment suggests ... that the tangential and radial diffusivities of mass are equal". Later [6.40], however, it was acknowledged that the experiments in question did not provide a sensitive test of the isotropy of the diffusivity. Indeed, the more recent results of QUARMBY and QUIRK [6.27, 6.32], from which Fig. 6.1 has been drawn, support and quantify more precisely Sparrow and Black's conclusion that near the wall the tangential diffusivity is several times larger than the radial one.

Under local-equilibrium conditions (6.9) gives the following expression for $z$-direction heat flux in a two-dimensional velocity field:

$$-\overline{wc} = \frac{1}{c_{1c}} \frac{\overline{q^2 w^2}}{2\varepsilon} \frac{\partial C}{\partial z}. \tag{6.25}$$

Thus the effective diffusivity in the $z$-direction is

$$\gamma_z \equiv -\overline{wc}/(\partial C/\partial z) = \overline{q^2 w^2}/2c_{1c}\varepsilon .\tag{6.26}$$

Likewise from (6.14) we may write down the corresponding formula for the effective $y$-direction diffusivity. On taking their quotient we find

$$\gamma_z/\gamma_y = \overline{w^2}/\overline{v^2} .\tag{6.27}$$

There is considerable disagreement about the relative magnitudes of $\overline{v^2}$ and $\overline{w^2}$ in the literature of wall shear flows. The line on Fig. 6.1 represents the variation in the normal stress ratio indicated by LAUFER's [2.28] pipe-flow data. These fall between the extremes of measured values and may, perhaps, thus be taken as representative. The curve displays much the same behavior as the measurements of $\gamma_z/\gamma_y$ although the centroid of the data lies above the line. In view of the scatter, however, the level of agreement is probably satisfactory.

At least two other proposals exist for the variation of the tangential effective diffusivity across a pipe based on simple mixing ideas (BOBKOV et al. [6.41] and RAMM and JOHANNSEN [6.42]). Both sets of proposals give approximately the same level of agreement with Quarmby and Quirk's data as achieved by (6.27).

### 6.2.4 Buoyant Influences on Local-Equilibrium Flows

Equation (6.1) showed in symbolic form the direct mechanism by which a gravitational field can affect scalar fluxes in a turbulent shear flow. A fluid element whose instantaneous temperature exceeds by an amount $c$ the average temperature at that horizontal plane receives an upthrust proportional to $\alpha cg$. In a small time $\delta t$ the velocity induced by this source alone is $\alpha cg\delta t$ producing an extra enthalpy flux proportional to $\alpha gc(C+c)\delta t$. Now the mean value of $c$ is, by definition, zero so the average buoyant enthalpy flux induced in $\delta t$ is thus proportional to $\alpha g\overline{c^2}\delta t$ and its rate of increase to $\alpha g\overline{c^2}$, in accordance with (6.1). The gravitational source evidently does not change sign with the temperature gradient so if we take the positive vertical direction $x_2$ as upwards (i.e., $g_2 = -g$) temperature fluctuations always tend to increase the algebraic value of $\overline{u_2 c}$ [3]. The sign of $\overline{u_2 c}$, however, depends on the type of stratification so, in a stable flow, where temperature gradients are positive, $\overline{u_2 c}$ is negative. Thus the gravitational term, as we should expect, diminishes the magnitude of the heat flux in a stable flow.

---

[3] Provided that $\alpha$ is positive. Note that for consistency with the rest of the chapter we use $x_2$ as the vertical coordinate rather than $x_3$ or $z$ preferred by geophysicists (see Chapter 4).

Besides the direct effect of buoyant generation of $\overline{u_2 c}$ there is, as mentioned in Subsection 6.2.1, "feedback" due to the pressure-temperature-gradient correlation. Equation (6.7) indicated that in isotropic turbulence one-third of the buoyant generation is absorbed by the pressure fluctuations. Moreover, just as the mean velocity becomes directly coupled to the thermal field when buoyant effects are significant, so too do the corresponding turbulence quantities. The generation terms in the stress-transport equations are:

$$- \left( \overline{u_i u_k} \frac{\partial U_j}{\partial x_k} + \overline{u_j u_k} \frac{\partial U_i}{\partial x_k} \right) - \frac{\alpha}{C} (g_j \overline{u_i c} + g_i \overline{u_j c}) .$$

The additional generation terms containing the gravitational vector contribute both to the vertical normal-stress component $\overline{u_2^2}$ and to the shear stress $\overline{u_1 u_2}$[4]. In the former case this extra source term expresses the rate at which the vertical velocity fluctuations are amplified by losing potential energy. The transfer process is sustained by the heat flux continuously acting to augment the potential energy of the fluctuations.

Besides the direct gravitational input to the stress equations we must include the indirect contribution of the pressure-strain correlation. On eliminating the fluctuating pressure by means of the Poisson equation for $p'$ (1.11) we may obtain

$$\overline{p' \left( \frac{\partial u_i}{\partial x_j} + \frac{\partial u_j}{\partial x_i} \right)} = \phi_{ij_1} + \phi_{ij_2} + \underbrace{\frac{1}{4\pi} \int \frac{\alpha g_l}{C'} \overline{\frac{\partial c'}{\partial x_l} \left( \frac{\partial u_i}{\partial x_j} + \frac{\partial u_j}{\partial x_i} \right)} \frac{d \text{Vol}}{r}}_{\phi_{ij_3}} \qquad (6.28)$$

where $\phi_{ij_1}$ and $\phi_{ij_2}$ denote the turbulence and mean-strain parts of the pressure-strain term discussed in Chapter 5. We note that the two-point correlation functions appearing in (6.28) are of the same form as that in $\phi_{ic_2}$ [see (6.5)]. Thus, by treating the mean temperature $C$ as uniform over the integral and the two-point correlation function as entirely symmetric, a little algebra produces the result [5.21, 6.12] that

$$\phi_{ij_3} = -0.3(G_{ij} - \tfrac{2}{3} \delta_{ij} G) \qquad (6.29)$$

where $G_{ij}$ and $G$ are shorthand notation for the rates of generation of $\overline{u_i u_j}$ and $\tfrac{1}{2} q^2$ by buoyant action:

$$G_{ij} \equiv -\frac{\alpha}{C} (\overline{u_i c} g_j + \overline{u_j c} g_i)$$

$$G \equiv -\frac{\alpha}{C} \overline{u_i c} g_i . \qquad (6.30)$$

---

[4] And, of course, to $\overline{u_3 u_2}$ in a three-dimensional flow.

Here again we see that the action of the pressure fluctuations is to "redistribute" part of the buoyant production among the normal stresses and to absorb the corresponding proportion of the shear-stress generation. However, the magnitude of the coefficient in (6.29) is only one-half that needed in the corresponding approximation of the mean-strain part of the pressure strain hypothesis [see (6.8)]. Thus, from purely kinematic arguments it seems that pressure fluctuations should distribute the effects of gravitational sources in the stress equations only half as effectively as those arising from mean strain.

In practice the two-point correlations in (6.28) will not be symmetric so the question arises as to the adequacy of (6.29). The homogeneous buoyant shear flow, in principle, provides an unambiguous test. Approximation of the three parts of the pressure-strain correlation by (5.23), (6.8) and (6.29) enables us to write the Reynolds stresses under local-equilibrium conditions[5] as

$$(\overline{u_i u_j} - \tfrac{1}{3}\delta_{ij}q^2)/\tfrac{1}{2}q^2 = \frac{0.4}{c_1}(P_{ij} - \tfrac{2}{3}\delta_{ij}P)/\varepsilon$$

$$+ \frac{0.7}{c_1}(G_{ij} - \tfrac{2}{3}\delta_{ij}G)/\varepsilon \tag{6.31}$$

or in the rather more informative non-tensor form:

$$(\overline{u^2} - \overline{q^2}/3)/\tfrac{1}{2}\overline{q^2} = [0.53 + R_f/(1 - R_f)]c_1^{-1} \tag{6.32}$$

$$(\overline{v^2} - \overline{q^2}/3)/\tfrac{1}{2}\overline{q^2} = [-0.27 - 1.2R_f/(1 - R_f)]c_1^{-1} \tag{6.33}$$

$$(\overline{w^2} - \overline{q^2}/3)/\tfrac{1}{2}\overline{q^2} = [-0.27 + 0.2R_f/(1 - R_f)]c_1^{-1} \tag{6.34}$$

$$-\overline{uv}/\tfrac{1}{2}\overline{q^2} = \frac{0.4\overline{v^2}}{c_1\varepsilon}\frac{\partial U}{\partial y}(1 - 1.75R_s) \tag{6.35}$$

where $R_f \equiv (g\alpha\overline{vc}/C)/(\overline{uv}\partial U/\partial y)$ $(= -G/P)$, is the flux Richardson number and $R_s \equiv (g\overline{uc}/C)/(\overline{v^2}\partial U/\partial y)$ $(= -G_{uv}/P_{uv})$, is the corresponding shear Richardson number proposed by Bradshaw [2.193].

With $c_1$ taken as 2.0 (which gives approximately the correct stress levels under non-buoyant conditions when (6.8) is used for $\phi_{ij_2}$) the normal stresses vary with increasingly stable stratification as shown in Fig. 6.2. Comparison is drawn with Webster's [6.15] experimental data and with predictions of [6.14] which assumed the coefficient of the gravitational term in (6.29) to be the *same* as that of the mean-strain term (i.e., 0.6).

---

[5] i.e., $P + G = \varepsilon$.

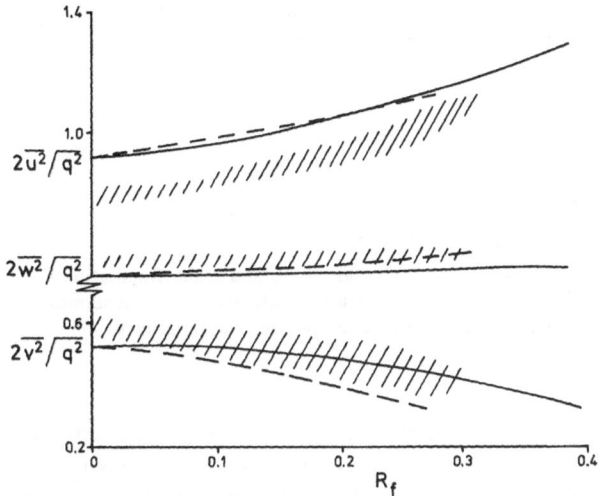

Fig. 6.2. Effect of stable stratification on normal-stress levels in stably stratified equilibrium free shear flows: ——— Eq. (6.32–34); – – – Ref. [6, 14]

Evidently, the predicted variation displays the main features of the experiments: the decrease of the proportion of energy in vertical fluctuations, the relative increase in streamwise fluctuations and the insensitivity of $\overline{w^2}/q^2$ to $R_f$. The agreement with experiment obtained by [6.14] is slightly superior to that given by (6.31). Just how closely the experiment came to achieving an equilibrium shear flow is, however, a matter that WEBSTER himself questioned. It would therefore be unwise to draw conclusions on the satisfactoriness of (6.31) based on these data alone.

The net effect of buoyancy on the shear stress and heat fluxes cannot be inferred so easily because of the intricate interlinkage of the equations. If we adopt the approximation of $\phi_{ic_2}$ and $\phi_{ic_3}$ given by (6.9) with $c_{2c} = \frac{1}{3}$ the heat flux equations may be written

$$-\overline{uc} = \frac{q^2}{2c_{1c}\varepsilon}\left(\overline{uv}\,\frac{\partial C}{\partial y} + 0.67\overline{vc}\,\frac{\partial U}{\partial y}\right), \tag{6.36}$$

$$-\overline{vc} = \frac{q^2}{2c_{1c}\varepsilon}\left(\overline{v^2}\,\frac{\partial C}{\partial y} + 0.67\,\frac{\alpha g\overline{c^2}}{C}\right). \tag{6.37}$$

From these equations and (6.32–35) it may be deduced that the stress and heat flux equations are coupled as shown in Fig. 6.3. The arrows indicate that the quantity in the box from which an arrow leaves appears in the transport equation for the correlation to which the arrow points.

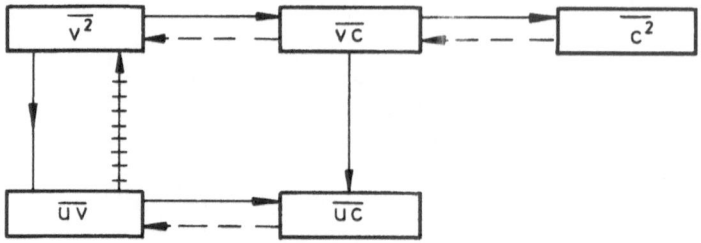

Fig. 6.3. Coupling of $\overline{u_i c}$ equations with the Reynolds stress and scalar fluctuation fields. ⟶ Generation by mean velocity or temperature gradients, – – –→ Generation by gravitational action, ⊩⊩⊩→ Indirect generation through pressure-strain effects

In a stable flow the shear stress $\overline{uv}$ is damped partly through the reduction in $\overline{v^2}$ and partly through the increase of $R_s$; in free shear flows the two contributions appear to be of about equal importance. The vertical flux of heat is also inhibited by two actions: reduction in $\overline{v^2}$ and the working of the fluctuations against the gravitational field.

Let us now eliminate $\overline{c^2}$ in favor of the time-scale-ratio, $R$, for the decay of the temperature and velocity fluctuations, i.e.

$$R \equiv \frac{\overline{c^2}}{\overline{q^2}} \frac{\varepsilon}{\varepsilon_c}. \qquad (6.38)$$

Hence on noting that, in local equilibrium, production and dissipation rates of $\overline{c^2}$ will be in balance

$$\overline{c^2} = -\frac{R\overline{q^2}\,\overline{vc}}{\varepsilon} \frac{\partial C}{\partial y}. \qquad (6.39)$$

The vertical heat flux can thus be re-expressed as

$$-\overline{vc} = \frac{\tfrac{1}{2}\overline{q^2}\,\overline{v^2}}{c_{1c}\varepsilon} \frac{\partial C}{\partial y} \bigg/ (1+0.1B) \qquad (6.40)$$

where $B$ stands for the buoyancy parameter $(R\alpha g q^4/C\varepsilon^2)\,\partial C/\partial y$. In the limit of negligible stratification (6.40) of course reduces to (6.14). It is possible to obtain explicit expressions for both the horizontal heat flux and the shear stress by patient manipulation of the above equations. The resultant expressions, however, are on the whole too cumbersome to be illuminating and are not presented here; they are given in [6.14] for the algebraically simpler case of $c_{2c} = 0.5$.

Table 6.2. Scalar flux correlation coefficient in shear flows in local equilibrium

| Workers | $(\overline{vc})^2/\overline{v^2}\overline{c^2}$ | Geometry |
|---|---|---|
| WEBSTER [6.15] | 0.15—0.30 | Homogeneous shear flow |
| BREMHORST and BULLOCK [6.35] | 0.20 | Pipe flow |
| BOURKE and PULLING [6.43] | 0.20 | Pipe flow |
| LAWN and WHITE [6.36] | 0.16 | Pipe flow |
| IBRAGIMOV et al. [6.44] | 0.6 | Pipe flow |
| ARYA and PLATE [6.45] | 0.11 | Smooth flat plate |
| JOHNSON [6.33] | 0.20 | Smooth flat plate |
| PIMENTA et al. [6.29] | 0.3—0.4 | Rough flat plate |

What is the appropriate value of the time scale ratio $R$? Under non-buoyant conditions, eqs. (6.39) and (6.40) may be combined to give

$$(\overline{vc})^2/\overline{v^2}\overline{c^2} = 1/(2c_{1c}R) .\qquad(6.41)$$

Experimental data of this correlation coefficient are summarized in Table 6.2. Only WEBSTER's measurements relate to a free shear flow. Judging by the near constancy of $\overline{uv}/\sqrt{\overline{u^2}}\sqrt{\overline{v^2}}$ in thin shear flows, however, it is unlikely that the scalar flux correlation coefficient is significantly affected by the presence of a wall. There is such a large variation among the measurements that it is not possible to make very definite statements. It seems possible, however, that the values reported by [6.36] are rather too low (since their directly measured radial heat fluxes are some 15 % less than those indicated from an enthalpy balance of the mean flow) while the magnitude of the correlation coefficients reported in [6.44–45] lies well outside those usually associated with a well-organized turbulent flow.

With $c_{1c}$ taken as approximately 3.4, (6.41) gives a value of $R$ of 0.5 according to Pimenta's data or approximately 0.7 for the Webster/Bremhorst/Bourke/Johnson experiments. There is some indirect support for the former value. SPALDING [6.46] has proposed a closed form of the $\overline{c^2}$ transport equation in which he used (6.38) to eliminate $\varepsilon_c$. He found that $R$ needed to be taken as 0.5 for the level of concentration fluctuations in a round jet to be predicted correctly[6].

Some numerical solutions of (6.36) and (6.40) are shown in Fig. 6.4 for $R$ equal to 0.5, 0.7, and 0.8, the last being the value adopted in [6.14]. The abscissa on this figure is the "gradient" Richardson number $Ri(\equiv R_f\sigma_t)$ which is the parameter used when the data were published. Webster's measurements suggest that the turbulent

---

[6] This is admittedly not a local-equilibrium flow, but the average levels of turbulence energy production and dissipation at any section are roughly the same.

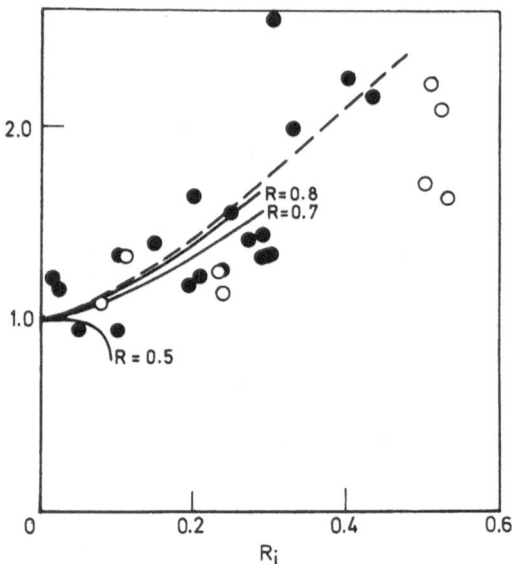

Fig. 6.4. Variation of turbulent Prandtl number in stably stratified free shear flows

Prandtl number rises fairly steeply with $Ri$, a behavior in general agreement with other less detailed investigations of stratified shear flows (see the monograph by Turner [4.20] for a survey). This means simply that heat fluxes are damped more rapidly than momentum transport in stably stratified flows. The predicted behavior for $R=0.5$, however, displays a slight *decrease* in $\sigma_t$ with $Ri$. Much better agreement is obtained for $R=0.8$, there being no substantial differences between the present results and those of [6.14], in view of the scatter in the experimental data. The result appears to support a value of the correlation coefficient in (6.41), of around 0.2. It is of course possible that $R$ itself varies when buoyancy effects are significant since gravitational terms enter the $\overline{q^2}$ equation but, as will be seen in Subsection 6.3.2, not that for $\overline{c^2}$. It seems unlikely, however, that such influences will have more than a secondary effect on $R$.

We now examine briefly some effects of buoyancy on near-wall turbulence. The most important practical case of such a flow is the earth's boundary layer (Subsection 4.1.1); understandably, therefore, meteorologists have provided most of the available measurements. The presence of the "wall" ensures that the macro-length scales of turbulence increase with height above the surface; consequently, so does the importance of gravitational effects. The Richardson number, the buoyancy parameter $B$ or any other stability parameter will therefore be height dependent. The flow is thus not homogeneous and we cannot presume that turbulent transport terms in the stress and scalar flux budgets will be negligible. One of the most complete sets of atmospheric turbulence data currently available is that obtained in Kansas in 1968 by staff of the Air Force Cambridge Research Laboratories [6.34, 47–49, 4.5]. WYNGAARD and COTÉ [6.47] analyzed the turbulent energy budget for these experiments. They found that turbulent transport by velocity fluctuations, though negligible under stable conditions, was as large as buoyant generation for unstable stratification, the sign implying a net diffusion upwards[7]. The result is not really conclusive, however, because the energy budget, based on measurements of all quantities except pressure transport, failed to close by an amount somewhat *larger* than the measured velocity transport. Possibly, therefore, pressure fluctuations may cause a diffusive transport of turbulence energy of about the same magnitude as and of the opposite sign from the velocity transport—which would leave the flow in nearly local equilibrium since (horizontal) advection is probably small.

Traditionally, experimental data are expressed in terms of $y/L$ where $y$ is the height above the ground and $L$ is the Monin-Obukhov length scale (Chapter 4), $-Cu_\tau^3/(\kappa g \overline{vc} \alpha)$. While this scaling helps to emphasize the importance of height on the flow structure, it has the undesirable effect of intermixing influences of buoyancy on the turbulence length scale with those on the Reynolds stresses and heat fluxes. A better length ratio in this respect would be $l_\varepsilon/L$ where $l_\varepsilon$ is the local dissipation length scale $(\frac{1}{2}\overline{q^2})^{3/2}/\varepsilon$. Let us assume that the net diffusive transport of $\overline{q^2}$ *is* negligible; then we may easily convert the turbulent kinetic energy equation $(P+G=\varepsilon)$ into

$$\frac{l_\varepsilon}{L} = \frac{1}{2\sqrt{2}\kappa}\left[\frac{q}{u_\tau}\right]^3\left[\frac{R_f}{1-R_f}\right].$$

(6.42)

The extensive laboratory measurements of ARYA and PLATE [6.45] in a stably stratified wind-tunnel boundary layer indicate that the ratio $q/u_\tau$ is virtually independent of $R_f$. For values of $R_f$ up to 0.05 their variation of the Richardson number is well fitted by

$$\left[\frac{y}{L}\right] = R_f/(1-15R_f).$$

(6.43)

Thus from (6.42) and (6.43) it appears that

$$(l_\varepsilon/y)\propto(1-15R_f)/(1-R_f) ;$$

(6.44)

[7] As [6.47] points out, the total diffusive contribution averaged over the earth's boundary layer thickness must be zero, so that at levels higher than explored in [6.47] the transport terms must change sign.

that is, the turbulence length scale characteristic of the energy-containing motions is substantially damped as the stratification becomes progressively more stable. Consequently, the assumption $l_\varepsilon \propto y$, which is frequently made in neutral shear flows (and which has formed the basis of several buoyant flow analyses, for example [6.9] and [6.50]), does not appear to be tenable.

A striking feature to emerge from the Kansas (and earlier) atmospheric measurements is the entirely different kind of behavior under stable and unstable stratification. The ratio of horizontal to vertical heat flux falls progressively in an unstable flow as $(-Ri)$ increases. Yet the data of (6.43) suggest that the ratio also falls as the stratification shifts from neutral to weakly stable; for larger Richardson numbers the ratio is virtually constant. This behavior in stable flows is not easy to explain since, of course, the direct buoyant damping occurs in the $\overline{vc}$ equation, not the $\overline{uc}$ equation; moreover, it is entirely contrary to the behavior in free shear flows. The turbulent Prandtl number falls steeply in unstable flows; [6.34] suggests that for $Ri = -1$ the value is about half that for a neutral flow, while the laboratory data of SCHON [6.51] indicate even greater sensitivity to the Richardson number. The level of $\sigma_t$ rises under stable stratification, though according to [6.34] never exceeds unity; the laboratory data of [6.45] suggest a steady rise in $\sigma_t$ with $Ri$, the turbulent Prandtl number reaching about 2.0 for $Ri \simeq 0.08$.

The most serious attempts at predicting the structure of the atmospheric surface layer with second-order closures seem to be those of MELLOR [6.50] and WYNGAARD [6.52]. As remarked above, MELLOR adopts a linear length-scale profile; he also includes only $\phi_{ij_1}$, $\phi_{ic_1}$ and a rather bare version of $\phi_{ij_2}$ among the various pressure-interaction terms. Agreement with experiment seems nevertheless impressive. WYNGAARD [6.52] adopts (6.9) for $\phi_{ic_2}$ and $\phi_{ic_3}$[8] and (6.29) for $\phi_{ij_3}$; the coefficients chosen are close to those adopted in [6.14] (for which free shear flow predictions have appeared in Fig. 6.4). Although Wyngaard reports encouraging agreement with experiment, his model gives $\overline{v^2}/\overline{u^2} = 0.44$ under neutral conditions, which is some 70% larger than is usually reported for the near wall region of laboratory boundary layers. The writer's experience is that, when a near-wall correction to $\phi_{ij}$ is introduced to give the more usual ratio for $\overline{v^2}/\overline{u^2}$ of about 0.20, the calculated behavior for buoyant flows by no means agrees with the actual behavior. For example, with such a low level of $\overline{v^2}$ buoyant damping of $\overline{uv}$ causes $\sigma_t$ to *decrease* as $Ri$ increases. The following seems a possible explanation of this paradox. The magnitude of $\overline{v^2}/\overline{u^2}$ is very dependent on the effect of the near-wall correction on $\phi_{ij}$. LAUNDER

---

[8] Under strongly stable conditions, however, the coefficient of $\phi_{ic_3}$ is increased by making it dependent on $Ri$; the modification is not significant for $Ri < 0.20$.

et al. [3.21] and IRWIN [6.53] concluded independently that the strength of the near-wall correction varied (roughly) linearly with $(l_\varepsilon/y)$. Now, from (6.44) $(l_\varepsilon/y)$ diminishes with the Richardson number and consequently $\overline{v^2}/\overline{u^2}$ will tend to rise—or, at least, its rate of diminution as $Ri$ increases will be reduced (which could help explain why $\overline{uc}/\overline{vc}$ does not rise under increasingly stable stratification). It seems possible, therefore, that by using a larger than "normal" value of the stress ratio, WYNGAARD was partially accounting for the diminished influence of the near-wall correction under stable stratification. Whether this *is* the case or whether major changes are needed in the model of the fluctuating pressure interactions, time will have to resolve.

## 6.3 Inhomogeneous Flows

### 6.3.1 Closure of the Scalar Flux Equations in Inhomogeneous Flows

To calculate inhomogeneous flows, the diffusive transport terms, neglected in the discussion of Section 6.2, must be retained. They require approximation of course and here we summarize briefly some of the suggestions for closure to be found in the literature.

An exact transport equation for the correlation $\overline{u_i u_j c}$ may be readily obtained by performing, on the Navier-Stokes and enthalpy or species equations, operations similar to those used to produce the $\overline{u_i c}$ equation in Section 6.2. LUMLEY [5.21], ANDRÉ et al., and KOLOVANDIN (see "Additional References") have made preliminary proposals for retaining all or nearly all the terms in these equations. Others have been content merely to note the exact $\overline{u_i u_j c}$ equation as a guide in devising much simpler models. OWEN [6.16] and LAUNDER [6.11] take

$$-\overline{u_i u_j c} \propto \frac{\overline{q^2}}{\varepsilon}\left(\overline{u_i u_l}\,\frac{\partial \overline{u_j c}}{\partial x_l} + \overline{u_j u_l}\,\frac{\partial \overline{u_i c}}{\partial x_l}\right) \qquad (6.45)$$

with the proportionality constant approximately equal to 0.1. As [6.11] notes, it looks as though a term like $\overline{u_l c}\,\partial\overline{u_i u_j}/\partial x_l$ ought really to be added on the right-hand side of (6.45) as well. WYNGAARD and his colleagues [6.10] retain only the second of the terms in (6.45), in the $\overline{u_i c}$ equation, with a coefficient of 0.15, while DONALDSON et al. [6.9] suppose that

$$-\overline{u_i u_j c} \propto \frac{q^4}{\varepsilon}\left(\frac{\partial \overline{u_i c}}{\partial x_j} + \frac{\partial \overline{u_j c}}{\partial x_i}\right). \qquad (6.46)$$

LUMLEY and KHAJEH-NOURI (Ref. [4.74], p. 169) arrive at a form similar to (6.46) purely on dimensional arguments; there is, however, an additional term included which is proportional to $q^4/\varepsilon(\partial\overline{u_m c}/\partial x_m)\delta_{ij}$. The same workers suggest that in a stratified flow, effects of buoyant transport would outweigh contributions from heat-flux gradients. Note that (6.46) suggests that, in a thin shear layer, $\overline{u^2 c}$ is zero; in an equilibrium layer (6.45) gives the same result if the layer is thick enough for $y$-gradients of $\overline{uc}$ to be neglected. This behavior arises from the neglect of the generation terms (in the *exact* $\overline{u_i u_j c}$ equation) containing mean temperature gradients. Implicit in (6.45) is the notion that these terms are negligible compared with generation due to heat-flux gradients. This *may* be an acceptable approximation when flux gradients are large but will generally not be so in local-equilibrium conditions. Fortunately, it is only in non-homogeneous flows that $\overline{u_i u_j c}$ contributes significantly to (6.1).

LUMLEY [6.54] argues that in addition to the algebraic approximations such as (6.45) and (6.46) a further term proportional to $\overline{u_i c}\,\partial(q^4/\varepsilon)/\partial x_j$ needs to be added to $\overline{u_i u_j c}$ to account for spatial variations in the transport coefficient; this would have an effect similar to that of the suggestion following (6.45). These forms have at least a little in common with the practice of representing turbulent fluxes as the product of the local value of the diffused quantity times a representative turbulent convection velocity advocated by TOWNSEND [1.22], BRADSHAW and FERRISS [6.55] and others.

No specific proposals seem to have been made for modelling the pressure transport terms in (6.1), the correlation being explicitly or implicitly neglected. An approximation for $\overline{pc}/\varrho$ could be developed in much the same way as for $\overline{p\,\partial c}/\partial x_i$ (LUMLEY [5.21] has already provided the parallel analysis for $\overline{pu_i}/\varrho$); the relative unimportance of transport in thin shear layers is presumably the main reason that the task has not been undertaken earlier.

At present none of the second-order heat-flux closures has been tested over as wide a range of flows as some of the Reynolds stress models discussed in Chapter 5. Apart from the work of WYNGAARD and OWEN noted earlier and the demonstration calculations of DONALDSON et al. [6.9] the only computations of inhomogeneous flows known to the writer are the currently unpublished predictions of free shear flows of BRYANT and SAMARAWEERA at Imperial College and the work, again unpublished, of LUMLEY and his associates at the Pennsylvania State University.

There are of course numerous simpler schemes available in the literature. The two with the most direct connection to the heat-flux transport equations are those of GIBSON and LAUNDER [6.56] and

BRADSHAW and FERRISS [6.55]. The former eliminated the flux gradient term from the $\overline{u_i c}$ equation by assuming that

$$\frac{D\overline{u_i c}}{Dt} - \mathscr{D}(\overline{u_i c}) = \frac{\overline{u_i c}}{q\sqrt{c^2}} \left[ \frac{D}{Dt}(q\sqrt{c^2}) - \mathscr{D}(q\sqrt{c^2}) \right] \tag{6.47}$$

where the operator $\mathscr{D}(\ )$ stands for "diffusion rate of ...". The right-hand side of (6.47) is then manipulated to the form

$$\overline{u_i c}[(P_c - \varepsilon_c)/\overline{c^2} + (P - \varepsilon)/\overline{q^2}] \tag{6.48}$$

where $P_c$ and $\varepsilon_c$ are the rates of generation and destruction of $\tfrac{1}{2}\overline{c^2}$ (see (6.51) below). By this means the heat-flux-transport equations (using (6.9) and (6.12) for $\phi_{ic}$) were reduced to the algebraic equation

$$-\overline{u_i c} = f_1 \frac{\tfrac{1}{2}\overline{q^2}}{\varepsilon}\,\overline{u_i u_l}\,\frac{\partial C}{\partial x_l} - f_2 \frac{\tfrac{1}{2}\overline{q^2}}{\varepsilon}\,P_{ic} \tag{6.49}$$

where

$$f_1 \equiv [c_{1c} + \tfrac{1}{2}(P/\varepsilon - 1) + \tfrac{1}{2}R(P_c/\varepsilon_c - 1)]^{-1},$$

$$f_2 \equiv (1 - c_{2c})f_1,$$

and $R$ is the time-scale ratio defined by (6.38). Reference [6.56], following RODI [6.57], made a corresponding simplification of the Reynolds stress equation. An interesting outcome of the resultant formulae is that, in a thin shear flow, the turbulent Prandtl number rises in regions where $P/\varepsilon$ and $P_c/\varepsilon_c$ are low. It will be seen in Subsection 6.3.3 that this behavior is generally in agreement with experiment.

A simplification of a different kind has been made by BRADSHAW and FERRISS [6.55] in extending the boundary-layer calculation method of BRADSHAW et al. [5.35] to heat-transfer problems. The transport equation for $\overline{vc}$ is replaced by one for the mean square temperature fluctuations $\overline{c^2}$ together with one for $\overline{q^2}$ (which is needed in any case to calculate the flow field development). The heat flux is then calculated from

$$-\overline{vc} = a_c q\sqrt{\overline{c^2}} \tag{6.50}$$

where the quantity $a_c$ is some dimensionless function of position in the shear flow. In the computations by BRADSHAW and FERRISS [6.55] $a_c$ was taken as a constant approximately equal to 0.18.

### 6.3.2 The Transport and Dissipation of Scalar Fluctuations

The preceding paragraphs have shown how two simplification schemes have brought into prominence the mean square, $\overline{c^2}$, of scalar fluctuations. In stratified flows, moreover, $\overline{c^2}$ appears directly in the buoyant source term in the vertical heat flux equation. Here, therefore, we consider the equation governing the transport of $\overline{c^2}$ and the question of its closure. The exact $\overline{c^2}$ transport equation seems to have been presented first by CORRSIN [6.58]. It is obtained by multiplying the equation for the instantaneous value of the scalar $(C+c)$ by $c$ and ensemble averaging. The result may be written as

$$\frac{D\frac{1}{2}\overline{c^2}}{Dt} = \underbrace{-\overline{u_j c}\frac{\partial C}{\partial x_j}}_{P_c} \underbrace{-\gamma \overline{\frac{\partial c}{\partial x_j}\frac{\partial c}{\partial x_j}}}_{\varepsilon_c} - \frac{\partial}{\partial x_j}\left(\overline{\frac{u_j c^2}{2}} - \gamma\frac{\partial \overline{c^2}/2}{\partial x_j}\right). \tag{6.51}$$

The equation expresses the fact that the level of $\overline{c^2}$ following a mean streamline will change through an imbalance of the generation rate of scalar fluctuations by gradients in $C$ (denoted hereafter as $P_c$), the dissipation of fluctuations due to molecular diffusion in the fine scale motions ($\varepsilon_c$) and through diffusive transport produced by turbulent velocity fluctuations and molecular dispersion. Equation (6.51) is the most straightforward of all turbulence transport equations; as CORRSIN [6.58] has commented, it resembles the turbulent kinetic energy Eq. (1.36) except that pressure transport is absent from (6.51).

The relative importance of the different processes in the $\overline{c^2}$ budget is similar to that of the corresponding terms in the turbulent energy equation; that is, in wall boundary layers, the production and dissipation terms far outweigh transport terms while in free shear flows advection and diffusion are a good deal more influential. The detailed budgets in free shear flows depend very much on the particular flow. Mixing layers developing between two streams of unequal temperatures display an overall excess of production over dissipation. At the other extreme is the axisymmetric wake studied in detail by FREYMUTH and UBEROI [6.59], whose scalar-fluctuation budget is reproduced in Fig. 6.5; here the average level of $P_c$ is barely one-third of $\varepsilon_c$, there being a correspondingly large loss by advection.

Most proposals for closing the $\overline{c^2}$ equation have adopted a gradient-type representation of $\overline{u_i c^2}$ either in the form (e.g., WYNGAARD [6.52])

$$-\overline{u_j c^2} \propto \frac{\overline{q^2}}{\varepsilon}\overline{u_j u_l}\frac{\partial \overline{c^2}}{\partial x_l} \tag{6.52}$$

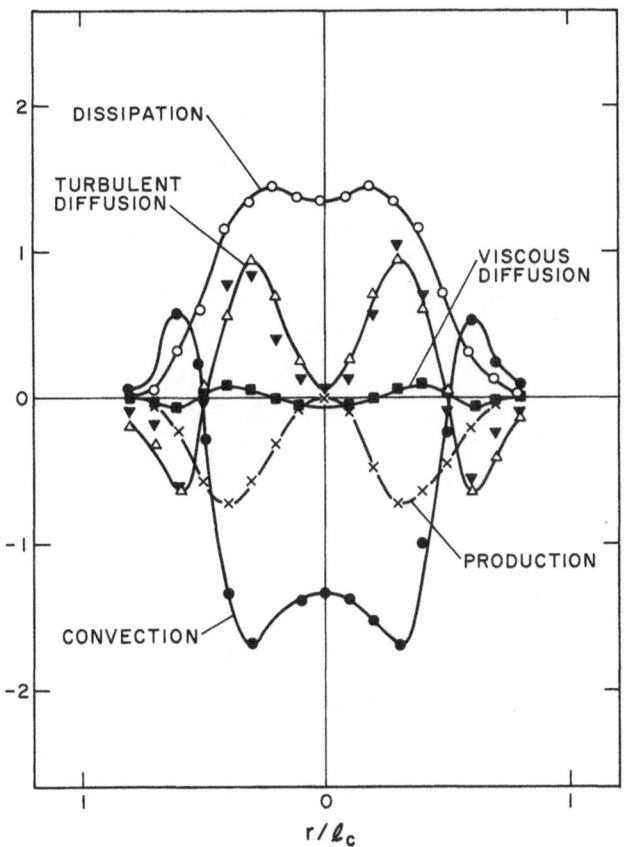

Fig. 6.5. Scalar-fluctuation balance in the wake of a sphere. (FREYMUTH and UBEROI [6.59]). Reproduced by kind permission of the American Physical Society

or (SPALDING [6.46])

$$-\overline{u_j c^2} \propto \frac{\overline{q^4}}{\varepsilon} \frac{\partial \overline{c^2}}{\partial x_j}.$$  (6.53)

Only BRADSHAW and FERRISS [6.55] adopt

$$-\overline{u_j c^2} \propto \overline{c^2} V_t'$$  (6.54)

where the turbulent bulk-convection velocity $V_t'$ (Section 5.1) is a prescribed function of position in the flow. Although the gradient transport model seems adequate in many flows, it does not match experiments for

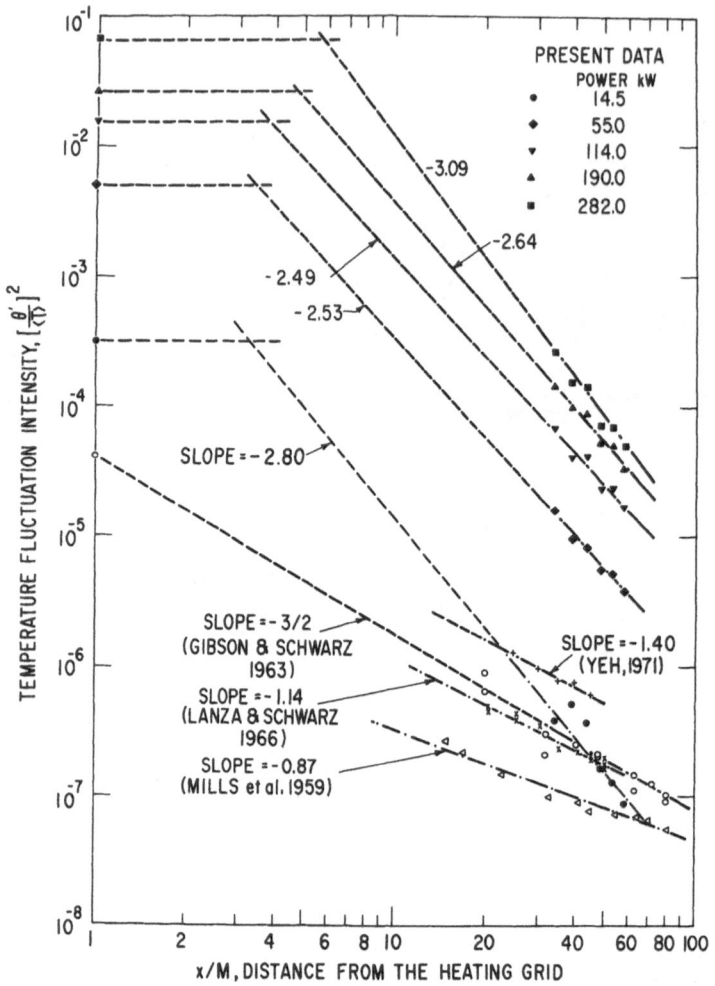

Fig. 6.6. Decay of scalar fluctuations behind a grid. (LIN and LIN [6.60]). Reproduced by kind permission of the American Physical Society

the axisymmetric wake where FREYMUTH and UBEROI'S [6.59] measurements shown in Fig. 6.5 indicate that $\overline{vc^2}$ does not change sign across the flow even though $\overline{c^2}$ displays a maximum at $r/r_{1/2} \simeq 0.6$.

The more important term needing approximation in Eq. (6.51) is the dissipation rate of $\overline{c^2}$. So far we have expressed this, by definition of the time-scale ratio $R$, as

$$\varepsilon_c = \frac{\varepsilon \overline{c^2}}{\overline{q^2} R}. \tag{6.55}$$

As remarked above, SPALDING [6.46] takes $R$ as 0.5 while in Section 6.2 it was seen that a value of about 0.8 gave best agreement of the effects of buoyancy in a homogeneous free shear flow: WYNGAARD [6.52] chooses an intermediate value, 0.71. What values are suggested from measurements other than in thin shear flows?

For the decay of temperature and velocity fluctuations behind a heated grid the ratio of the decay exponents of $\overline{c^2}$ and $\overline{q^2}$ should equal $R$. Nearly all the turbulence energy decay data suggest that $\overline{q^2} \propto x^{-m}$ where $m$ lies in the range 1.15–1.4. Figure 6.6, which is taken from the survey by LIN and LIN [6.60], shows that a far wider range of exponents seems to be indicated for the decay of $\overline{c^2}$. Most of these experiments, however, cover barely one decade in scalar fluctuation decay; in these circumstances one can fit a wide range of power laws through the experimental points depending on the choice of virtual origin. A re-examination[9] of the grid turbulence data cited by [6.60] has suggested that a time scale ratio of unity provides an adequate fit to most of the data, in agreement with Lumley's [6.8] conjecture (this contrasts with Hinze's [1.7] predicted value of $\frac{2}{3}$). Thus, consideration of what is still only a narrow span of flows has shown roughly a two-fold variation in $R$. Evidently this ratio is not sufficiently constant for (6.55) to serve as a general method of finding $\varepsilon_c$.

The alternative is to find $\varepsilon_c$ from its own transport equation. The exact equation may be obtained by taking the derivative of the equation for $(C+c)$ with respect to $x_l$, multiplying by $2\gamma \partial c / \partial x_l$ and ensemble averaging. The resultant equation may be written

$$\frac{D\varepsilon_c}{Dt} = \underbrace{-2\gamma \overline{\frac{\partial c}{\partial x_j} \frac{\partial u_l}{\partial x_j} \frac{\partial C}{\partial x_l}}}_{\text{(i)}} \underbrace{- 2\gamma u_l \overline{\frac{\partial c}{\partial x_j} \frac{\partial^2 C}{\partial x_l \partial x_j}}}_{\text{(ii)}}$$

$$\underbrace{- 2\gamma \overline{\frac{\partial c}{\partial x_j} \frac{\partial c}{\partial x_l} \frac{\partial U_l}{\partial x_j}}}_{\text{(iii)}} \underbrace{- 2\gamma \overline{\frac{\partial c}{\partial x_j} \frac{\partial u_l}{\partial x_j} \frac{\partial c}{\partial x_l}}}_{\text{(iv)}} \underbrace{- 2 \overline{\left( \gamma \frac{\partial^2 c}{\partial x_j \partial x_l} \right)^2}}_{\text{(v)}}$$

$$\underbrace{- \frac{\partial}{\partial x_l} \left( \overline{\varepsilon_c u_l} - \gamma \frac{\partial \varepsilon_c}{\partial x_l} \right)}_{\text{(vi)}}. \tag{6.56}$$

The equation has been discussed extensively in Ref. [4.74], p. 169 and [6.16]. Just as in the energy dissipation rate equation considered in Chapter 5, the two dominant terms in (6.56) are those of generation and destruction due to fine scale turbulence interactions, i.e., terms (iv) and

---

[9] A study in progress by Professor LUMLEY, Mr. G. NEWMAN and the writer.

(v), respectively. The direct mean-field generation terms, (i)–(iii), are negligible at high Reynolds numbers.

The problem of closing (6.56) is very much akin to that of modelling the equation for $\varepsilon$ or $\mathscr{D}$ (Section 5.3). Formal mathematical manipulation has not proved very illuminating; current attempts rest mainly on intelligent dimensional analysis. The task is rather more difficult for the $\varepsilon_c$ equation than that for $\varepsilon$, simply because the *possible* number of parameters is twice as large. For example, there are two turbulent time scales available, $\overline{q^2}/\varepsilon$ and $\overline{c^2}/\varepsilon_c$, and $P/\varepsilon$ and $P_c/\varepsilon_c$ both provide dimensionless measures of generation rates. All the published modelled forms of the $\varepsilon_c$ equation known to the writer are at best tentative. The earliest proposals [6.8, 16] contained no mechanism by which $\varepsilon_c$ could increase from its original value (though [6.16] was saved from disaster by a large diffusive "leakage" of $\varepsilon_c$ into the flow at the wall).

More recently, LUMLEY and KHAJEH-NOURI (Ref. [4.74], p. 169) proposed that the rate of creation of $\varepsilon_c$ should be proportional to $\varepsilon_c P/\overline{q^2}$. We should expect however that the production rate of $\overline{c^2}$ should be at least as influential as the production rate of kinetic energy. Suppose, for example, we create thermal grid turbulence with a transverse mean temperature gradient (experiments of this kind have been performed by WISKIND [6.61] and ALEXOPOULOS and KEFFER [6.62]). Since $P$ is zero, the proposal of LUMLEY and KHAJEH-NOURI would predict that $\varepsilon_c$ became vanishingly small as the turbulence decayed downstream, while for suitably chosen $dC/dy$, $\overline{c^2}$ would become very large—a pair of events that jointly seem implausible.

The current view of the required structure of the $\varepsilon_c$ equation may be summarized as follows:

i) The rate of dissipation of $\varepsilon_c$ is dependent on both velocity and scalar time scales. The most obvious approximation is to use two terms proportional to $\varepsilon\varepsilon_c/\overline{q^2}$ and $\varepsilon_c^2/\overline{c^2}$, respectively.

ii) The rate of creation of $\varepsilon_c$ must be sensitive to changes in both the turbulence energy and scalar yields. Here there are several terms available, e.g., $P\varepsilon_c/\overline{q^2}$, $\varepsilon_c P_c/\overline{c^2}$, perhaps terms representative of the anisotropy of the stress and heat flux yields, e.g. $(\overline{cu_i}\varepsilon)^2/q^6$. According to LUMLEY [6.11] this last group has been used by SIESS [6.63] in making computations of a number of flows.

iii) There is little point in worrying about how the diffusion of $\varepsilon_c$ can be treated until the form of the creation and destruction terms is settled.

To assist the development of a modelled form of the scalar dissipation equation there is a really urgent need for measurements of thermal turbulence in simple flows, for example where grid-generated turbulence is subject to various strain fields. At present the axisymmetric contraction study of MILLS and CORRSIN [6.64] is apparently the only experiment of this kind. This study shows that the contraction reduces the time scale ratio to about $\frac{1}{3}$ but $R$ returns rapidly to unity further downstream.

Unlike the energy dissipation rate, all the components of $\varepsilon_c$ $\left(\text{i.e., } \gamma\left[\overline{\left(\frac{\partial c}{\partial x}\right)^2} + \overline{\left(\frac{\partial c}{\partial y}\right)^2} + \overline{\left(\frac{\partial c}{\partial z}\right)^2}\right]\right)$ are fairly easily measurable[10]. Moreover, such measurements provide not just a direct measure of the dissipation rate of $\frac{1}{2}\overline{c^2}$ but also a check on the local isotropy of the flow (see, for example, [6.65] for a convincing verification of local isotropy). There is, thus, a reasonable prospect that over the next few years comprehensive measurements of the scalar field will lead to the development of a reasonably universal model form of the $\varepsilon_c$ equation.

### 6.3.3 Thermal Turbulence in Thin Shear Layers

*a) Free Shear Flows*

Pioneering studies of thermal turbulence in shear flows were undertaken in the late 1940's by CORRSIN and UBEROI [6.66] and HINZE and VAN DER HEGGE ZIJNEN [6.67] for the axisymmetric jet in stagnant surroundings and by TOWNSEND [1.22] for the plane wake. A survey of these works and of other free shear flows reported prior to 1958 is given in HINZE [1.7]. More recent studies of the jet in still air have been reported by BECKER et al. [6.68], DANCKWERTS and WILSON [6.69] and JENKINS and GOLDSCHMIDT [6.70], the former two for axisymmetric flow and the last for plane flow. JENKINS and GOLDSCHMIDT'S [6.70] survey shows substantial variations in the reported spreading rates for each of these geometries; there is satisfactory consistency, however, in the *relative* rates of spread of the velocity and scalar fields. For the plane jet, the half-width of the temperature profile is 42% greater than the velocity profile but only 25% greater in the case of the round jet. The implied average levels of the turbulent Prandtl/Schmidt number (which are inversely proportional to the square of the rates of spread) are thus 0.50 and 0.65 respectively. Estimates of $\sigma_t$ based on direct evaluation from heat-flux and temperature profiles are in reasonable agreement with the above values, confirming the difference in levels between the two flows. The recent conditionally sampled measurements of the plane jet by JENKINS and GOLDSCHMIDT [6.71] indicate that within the turbulent fluid the turbulent Prandtl number is virtually uniform across the jet with a value of only 0.41. This contrasts with the (unconditioned) variation re-

---

[10] Only a pair of resistance-thermometer wires is required.

ported by VAN DER HEGGE ZIJNEN [6.72] in which $\sigma_t$ is 0.6 on the axis, falls
to about 0.4 in the region of maximum generation, then rises to near unit
towards the edge of the jet. This pattern corresponds approximately
to the variation predicted by the simplified second-order closure of
GIBSON and LAUNDER [6.56] in which the turbulent Prandtl number becomes a function of the local ratio of turbulent energy production to
dissipation.

$$\sigma_t = 0.225(5.4 + P/\varepsilon)/(1.2 + P/\varepsilon).$$  (6.57)

For the plane jet it appears that (6.57) gives values of $\sigma_t$ that are a little
too high.

   The turbulent wake, both plane and axisymmetric, has been the subject of a number of experiments. The measurements of FREYMUTH and
UBEROI [6.65] of the wake behind a heated cylinder include determination of all terms in the $\overline{c^2}$ transport equation. Their data extend more
than 1000 diameters downstream though in most respects the flow is
self-preserving beyond 200 diameters behind the cylinder. ALEXOPOULOS
and KEFFER [6.62] also examined flow in the wake of a cylinder; in this
case, however, the cylinder was unheated and the approaching stream
had a linear variation of temperature across it, generated by passing the
flow through a heated grid some distance upstream. The wake distorts
the temperature profile anti-symmetrically about the plane of symmetry.
Consequently, the streamwise heat flux, $\overline{uc}$, falls to zero on the axis while
the transverse flux, $\overline{vc}$, remains finite; this is the opposite of the pattern
found in a shear flow with a symmetric mean temperature distribution.
A further interesting feature is that the effective thermal conductivity
shows a sharp rise near the wake axis; at $x/D = 100$, this peak value is
three times as large as that over most of the wake. No computations of
this flow seem to have been published yet; it seems likely that a number
of the measured features will provide a searching examination for
existing turbulence models. Despite the generally good internal consistency of the measurements reported in [6.62] and [6.65], in neither
case do the heat-flux profiles deduced from a mean enthalpy balance
(assuming self-preserving flow) agree particularly well with the measured
distributions of $\overline{vc}$. For [6.62] part of the discrepancy is probably due to
the neglect of mean velocity variations across the wake when evaluating
the mean enthalpy balance (the final measuring station was only 228
diameters downstream). For the Freymuth-Uberoi measurements, however, errors arising from that source should have become negligible.

   Satisfactory enthalpy balance seems to be indicated for Kovasznay
and Ali's [6.73] measurements of the thermal wake produced by a heated
flat plate. This flow develops appreciable asymmetry with passage down-

stream due apparently to buoyancy effects. The computations of this flow in [6.56] including buoyant terms in the turbulence closure account fairly well for this behavior.

The axisymmetric wake measurements of FREYMUTH and UBEROI [6.59] show much better consistency between mean temperature and lateral heat flux than their plane wake study. This is a flow where the downstream development is acutely sensitive to the precise form of the wake generator. This is possibly the reason that [6.59] displays a value of $\sqrt{c^2}/(C-C_\infty)$ on the axis of 0.55 while GIBSON et al. [6.74], for nominally the same geometry, obtained 0.38. The normalized mean temperature profile is only slightly fuller than the velocity profile in this flow indicating that the turbulent Prandtl number is close to unity; this is again in line with (6.57) since the average level of $P/\varepsilon$ is only about $\frac{1}{4}$. In the limit of very small velocity differences the axisymmetric jet in a moving stream should display the same kind of spreading laws as the axisymmetric wake. The measurements of the former flow by ANTONIA and BILGER [6.75], however, indicate a decay of centerline temperature excess proportional to $x^{-1}$ rather than the $x^{-2/3}$ decay law characteristic of asymptotic wakes. This is presumably because, at the farthest downstream station, the velocity on the axis was still 15% greater than that of the free stream. The maximum level of $\sqrt{c^2}/(C-C_\infty)$ on the axis in this study is 0.27, a value intermediate between those obtained in the round wake and that reported by DANCKWERTS and WILSON [6.69], 0.17, for a jet in stagnant surroundings.

ANTONIA and BILGER's study included measurements of the streamwise flux, $\overline{uc}$. In the region of maximum production the flux ratio $|\overline{uc}/\overline{vc}|$ is approximately 1.5. This is somewhat larger than for the homogeneous flow reported in Subsection 6.2.2. It is possible that the diffusive transport rate of $\overline{vc}$ is greater than that of $\overline{uc}$, which would produce such an effect. Mean flow data of several other co-axial jet studies, including the mixing of streams of different gases as well as of streams of the same gas at different temperatures, have been collected for the Free Shear Flows Conference held at the NASA Langley Research Center in July 1972 [2.2]. The predicted behavior of these flows obtained by thirteen groups of workers appear in a companion volume; no novel approaches to modelling heat or species transport were followed, all workers choosing uniform values of turbulent Prandtl/Schmidt number ranging from 0.5 to 1.0.

A comprehensive study of thermal turbulence in a plane mixing layer in a moving stream has been reported by WATT [6.76]. The variation of turbulent Prandtl number across this flow is similar to that noted above in the jet, with the lowest values occurring in the region of maximum $P$ (and $P_c$) and rising fairly sharply towards the edges. Unfortunately the measured level of turbulent heat flux is only about one-quarter of that needed to support the spreading rate of the mean temperature field; this undermines confidence in the correctness of the other measurements. Mean and fluctuating temperature profiles in a mixing layer in stagnant surroundings have been reported by SUNYACH and MATHIEU [6.77], the $\overline{c^2}$ profile displaying a more uniform level over the central 60% of the layer than the corresponding turbulence energy profile.

A further important mixing layer study is that of BROWN and ROSHKO [1.37] though their measurements are limited to mean field data. They studied the subsonic mixing of different velocity streams with different gases, the maximum density ratio being 7:1. The results showed that density differences had only a minor effect on the rates of spread of the shear flow, a finding which conforms with the predictions of most hydrodynamic turbulence models from the mixing length hypothesis to second-order closures. An inference to be drawn is that the marked reduction in spreading rates observed in supersonic mixing layers (see, for example, HILL and PAGE [6.78]) is a genuine Mach number effect since the accompanying density variations would not be large enough to make a significant contribution. ROSHKO presented further results from these mixing-layer experiments at the NASA Free Shear Flows Conference (Ref. [2.2], p. 629); the most startling fact was that the deduced value of turbulent Schmidt number from one of their experiments was only 0.16. This is barely one-third of the lowest value reported in any other free shear flow, and therefore needs re-confirming.

Finally, we mention an experiment in progress at the Institut de Mécanique Statistique de la Turbulence in Marseille in which a mixing layer is established with a peak in temperature within the flow (BEGUIER et al. [6.79], FULACHIER et al. [6.80]). Due to the flow asymmetry the position of maximum temperature does not coincide with the point where $\overline{vc}$ is zero. Over a portion of the flow, therefore, the rate of "generation" of $\overline{c^2}$ is negative, the loss being made good by turbulent transport.

### b) Near-Wall Flows

Most experimental data in wall-generated shear flows have been taken in conditions where departures from local equilibrium are small; there are however significant variations in structure across the flow due to the steadily diminishing influence of the wall as the distance from the surface increases. ROTTA [6.81], based on the measurements of LUDWIEG [6.82] and JOHNSON [6.33], has suggested the following formula for the variation of $\sigma_t$ across a flat plate boundary layer:

$$\sigma_t = 0.9 - 0.4(y/\delta)^2 . \tag{6.58}$$

The value of $\sigma_t$ is thus fairly uniform near the wall but falls to 0.5 at the edge of the layer. This variation is in generally good agreement with that emerging from the transpired boundary layer studies of SIMPSON et al. [6.25] and BAKER and LAUNDER [2.40], the latter with variations in density up to 50% between the inner region and the outer edge. The equilibrium formula (6.18) also suggests that $\sigma_t$ should decrease over the outer region of the layer since $\overline{uv}$ falls faster with distance from the wall than does $\overline{v^2}$. It does not seem possible to write with any certainty on the variation of the turbulent Prandtl/Schmidt number in fully developed pipe and channel flow. A number of experiments show a gradual reduction in $\sigma_t$ towards the center while others show quite the reverse (see, for example, KESTIN and RICHARDSON [6.22]).

Blom's [6.23] measurements of a thermal boundary layer developing within a thick initial turbulent boundary layer suggest that the level of turbulent Prandtl number varies with distance downstream, or, more precisely, with the relative thickness of the velocity and thermal boundary layers. For example, at $y^+ = 50$, $\sigma_t$ rises from 0.60 at 25 mm downstream from the start of heating, to 0.82 after 125 mm and to 0.92 after 425 mm. This behavior seems plausible since when the thermal boundary layer is very thin compared with the velocity layer it is likely that gradients in $\overline{cv^2}$ will be steep and hence transport significant. Possibly some of the indicated differences in the measured distributions of $\sigma_t$ in pipe flow are also attributable to transport effects, since some of the reported data have been deduced from the dispersal of samples of tracer gas in what are by no means fully developed conditions.

During the past ten years the Thermosciences Division at Stanford University has amassed a large body of carefully acquired data of thermal boundary layer development in the presence of surface mass transfer and streamwise pressure gradient. An extensive retrospective assessment of these measurements by KAYS and MOFFAT [6.83] has been published recently. Here therefore we note only the different effects of mass transfer and pressure gradients on the distribution of $\sigma_t$ for the special case of self-preserving shear layers. As remarked above, in zero pressure gradient there is no discernible effect of blowing on the turbulent Prandtl number distribution [6.25, 2.40]. BLACKWELL et al. [6.84] drew the same conclusion for flows where the free stream varied as $x^{-0.15}$. The same workers did find, however, a systematic decrease in the level of $\sigma_t$ as the adverse pressure gradient was made more severe. When $U_e$ varied as $x^{-0.20}$ the level of $\sigma_t$ over the whole boundary layer was about 0.2 below that found in zero pressure gradient. This behavior seems to be what is suggested by inspection of the equations for $U$, $C$, $\overline{uv}$ and $\overline{vc}$. The streamwise pressure gradient appears only in the first of these; thus, although through the coupling of the equations $C$, $\overline{uv}$ and $\overline{vc}$ will also feel the effects of the pressure gradient, we should expect changes in these variables to lag behind that of the mean velocity field. Now the effect of an adverse pressure gradient will be to increase $\partial U/\partial y$ since, from Bernoulli's equation, a given rise in pressure will reduce the velocity of the slow-moving fluid near the wall more than the faster-moving elements in the outer part of the boundary layer. Thus, since

$$\sigma_t \equiv \frac{\overline{uv}/(\partial U/\partial y)}{\overline{vc}/(\partial C/\partial y)}$$

we should expect the level of $\sigma_t$ to be reduced in an adverse pressure gradient. It might seem that by the same argument mass injection should also cause $\sigma_t$ to fall since this would tend to increase the mean velocity gradient. In this case, however, there is also a direct effect on the temperature profile tending to raise $\partial C/\partial y$; as a result the turbulent Prandtl number is virtually unaffected.

In practice, nearly all computations of turbulent boundary layers and pipe flows have assumed a uniform value of turbulent Prandtl number, the most popular value being 0.9 [2.72, 208, 210, 6.24, 85]. An exception is the Stanford group; for example in [6.84] they have devised a correlation of their own data in which the dimensionless local pressure gradient appears as a parameter. Baker and Launder's [2.40] computations of transpired boundary layers with and without pressure gradients adopted a modification of Rotta's proposal for $\sigma_t$ (6.58)

$$\sigma_t = 0.95 - 0.45(y/\delta)^2 \tag{6.59}$$

though, except for cases where there were steep density gradients across the flow (due to Freon injection at the wall), a uniform value of 0.90 gave just as satisfactory agreement with the data.

Finally, a few computations have been reported of thin shear flows where heat transport parallel with the surface (and normal to the main flow) is also important. LAUNDER and YING [6.2] obtained close agreement with BRUNDRETT and BURROUGHS [6.1] measured temperature profiles in a square-section duct, assuming an *isotropic* turbulent Prandtl number of 0.9; the weakness of this assumption was, however, probably disguised by the strong influence of mean-flow transport in the plane of the duct associated with the corner-induced secondary flows (Section 3.1). Certainly BERGELES et al. [6.86] have found it crucial to use distributions of $\gamma_z/\gamma_y$ similar to that indicated by Fig. 6.1 in order to predict correctly the temperature distribution downstream from a row of injection holes.

### 6.3.4 Low-Reynolds-Number Effects on Turbulent Heat Transport

When the local turbulent Reynolds number is low, molecular transport will exert important effects on the turbulent heat transport processes. CORRSIN [6.87] has shown that for isotropic turbulence the thermal microscale $\lambda_c$ is equal to $\sqrt{2/\sigma}$ times the Taylor microscale $\lambda$. It may be assumed that this result is roughly correct for strongly non-isotropic flows too. Thus for a liquid metal where $\sigma$ is typically 0.02 we should expect to find molecular effects on the thermal field at local turbulent Reynolds numbers an order of magnitude greater than those at which the velocity field becomes independent of direct viscous influence[11].

At present not much progress has been made in accounting for these influences in any basic closure of the scalar-transfer problem. Deissler's [6.88] work, perhaps the most extensive formal analysis of the problem, considered the two-point temperature-velocity transport equation specialized to the case of an infinite shear flow with linear cross-stream velocity and temperature profiles, notionally generated at time $t=0$ by passing the flow through a grid. The work extended the earlier study of isotropic turbulence of DUNN and REID [6.89]. As is permissible in the limit of very low Reynolds number, all third-order correlations were discarded from the equations (including those associated with the important pressure-strain and pressure-temperature-gradient correlations) and the equations were then transformed to wave-number space. For the case of zero strain the result is expressed as

$$\sigma_t^{-1} = 2.05\sigma/(1-\sigma)\left[1 - \left(\frac{2\sigma}{1+\sigma}\right)^{3/2}\right] \tag{6.60}$$

which displays the limiting behavior

$$\sigma_t = 0.485\sigma^{-1} \quad \text{as} \quad \sigma \to 0; \quad \sigma_t = 0.27 \quad \text{as} \quad \sigma \to \infty. \tag{6.61}$$

Except for values of the strain parameter $tdU/dy$ less than 4 an increase in mean shear brings $\sigma_t$ closer to unity for all Prandtl numbers. DEISSLER suggested that for turbulent pipe flow the effective strain parameter would be about 5% of the pipe Reynolds number based on diameter and average velocity. For such large values of $tdU/dy$ only liquid metals would have a turbulent Prandtl number significantly different from unity. It is difficult to judge how closely these results approximate those in real turbulent shear flows. Some of the neglected triple correlations are of substantial importance in the heat flux and

---

[11] The converse is only partly true. Although molecular heat or mass transport at large Prandtl/Schmidt numbers becomes insignificant at very low turbulent Reynolds numbers the viscous effects on the velocity field will still affect the scalar field.

stress equations; however, the turbulent Prandtl number contains only the *ratio* of the stresses to fluxes and hence should be relatively insensitive to such errors.

At present, no numerical results of a complete second-order closure including the low Reynolds number region seem to have been published. Mention is made of Owen's [6.16] predictions of mercury flow in a pipe, however. Although his turbulence model was devised for high Reynolds number flows only he added the effects of molecular heat transport in the mean enthalpy equation; indeed at $y^+ = 30$ (the starting point for his computations) he assumed turbulent heat transport was negligible, i.e., $\overline{uc} = \overline{vc} = 0$. Agreement with experiment was rather poor however, the predicted level of turbulent Prandtl number being virtually unaffected by the very low Prandtl number.

Lawn [6.90] has suggested that the correlation $\overline{vc}/\sqrt{\overline{v^2}}\sqrt{\overline{c^2}}$ may be relatively unaffected by Prandtl number at low turbulent Peclet numbers. His survey of the various pipe-flow measurements, spanning Prandtl numbers from 0.005 to 50, underlined the large amount of scatter in the data but at least did not disprove the suggestion. If the constancy of the correlation coefficient can be presumed over a useful range of flows, it means that Prandtl-number effects would appear only in the $\overline{c^2}$ equation—which is an easier equation for which to devise a low-Reynolds-number model.

Despite this slightly encouraging sign, it seems that significant progress in devising a reliable heat-flux closure at low Reynolds numbers will be difficult until more experimental data become available. These may be either physical data or the output of computer simulation of the (time-dependent) Navier-Stokes and energy equations using only a "sub-grid-scale" turbulence model (Section 5.5). The latter approach seems especially attractive at low Reynolds numbers where the range of eddy sizes present is smaller and the problem of accurate measurement formidable. Certainly the currently available measurements of $\overline{c^2}$ profiles for liquid flows in pipes show such wide, unsystematic variations as to be practically valueless.

Nearly all other approaches to calculating low-Reynolds-number heat transport have focussed directly on formulae for the effective diffusivity in the immediate near-wall region. An extensive review by Jayatilleke [6.91] covers most of the pertinent data (including flow over roughened surfaces) and model proposals from Prandtl's [6.5] 1910 model up to the mid 1960's. He argues that, for practical computations, the influence of the low Reynolds number can be most conveniently expressed by way of

$$C^+ = \sigma_{t,\,\infty}(U^+ + \mathscr{P}) \tag{6.62}$$

where $\sigma_{t,\infty}$, is the turbulent Prandtl number at high Reynolds numbers (assumed constant), $C^+$ is the normalized temperature $\varrho c_p(C_w - C)u_\tau/Q_w$, ($Q_w$ being the heat flux at the wall) and the so-called $P$-function is given by

$$\mathscr{P} = 9.0[(\sigma/\sigma_t)^{0.75} - 1]\,[1 + 0.28\exp(-0.007\sigma/\sigma_t)] \qquad (6.63)$$

for smooth surfaces.

REYNOLDS [6.92] has recently published a comprehensive review of models which express $\sigma_t$ as functions of $\sigma$, of position in the boundary layer and sometimes of Reynolds number (either local or, in the case of pipe flow, of bulk-flow Reynolds number). No attempt will be made here to duplicate this review. Suffice it to say that the models fall broadly into three groups:

i) *Mixing-Length Models:* according to which lumps of fluid are displaced from one region to another and, in flight, lose heat by conduction to the surrounding fluid.

ii) *Surface-Renewal Models:* developed through a series of papers by L. C. THOMAS and his colleagues[12]. Rather similar to mixing length approaches except that attention is focussed on the "residence time" of an eddy (or on its reciprocal the "mean frequency of turbulent exchange").

iii) *Lagrangian Diffusivity Models:* in which the effective diffusivities for heat or mass are expressed in terms of Lagrangian length scales and the autocorrelation coefficient for lateral velocity fluctuations.

We ought perhaps to add a fourth category: those that assume there is *no* low-Reynolds-number influence on $\sigma_t$. This assumption has been made in the computations of PATANKAR [6.95], CEBECI et al. [2.210], JONES and LAUNDER [6.24, 2.272], BAKER and LAUNDER [2.40] and many others.

Models of the first two kinds are all intuitive in approach and their physical bases sometimes seem too fragile for the mathematical structure they support. The Lagrangian diffusivity models have clear beginnings and can be developed some way with reasonable rigor; the problem, as usual in Lagrangian approaches, comes in converting the results to a form that is useful in an Eulerian framework.

As REYNOLDS [6.92] has commented, entirely different dependences of $\sigma_t$ on the molecular Prandtl/Schmidt number may be found in the literature. For example, in the limit as the turbulent Reynolds number tends to zero, $\sigma_t$ becomes proportional to $\sigma^{-1}$ according to JENKINS [6.96], WASSEL and CATTON [6.97] and others; to $\sigma^{-0.2}$ according to

---

[12] Two further articles from this group, too recent to be included in Reynolds' review, are [6.93 and 6.94].

AOKI [6.98]; and to a constant value independent of $\sigma$ according to THOMAS [6.99] and MIZUSHINA and SASANO [6.100] (as well as the "fourth category" models mentioned above). There are many more complicated formulae that defy simple categorizing; REYNOLDS [6.92] sets out most of these in full.

Such a contradictory set of proposals suggests that the experimental data likewise show inconsistencies. The task of measuring $\sigma_t$ accurately is never easy but it is many times more difficult in the sublayer region adjacent to the wall. In these circumstances the most sensible requirement is that any proposal for $\sigma_t$ should yield predictions of the measured mean temperature (or concentration) profile and surface fluxes in adequate agreement with measurements. On such a basis there seems good reason for accepting a constant value of $\sigma_t$, at least for gases and fluids with Prandtl numbers greater than unity. For example, PATANKAR [6.95] has shown that, when used in conjunction with VAN DRIEST'S [5.39] version of the mixing length hypothesis, a uniform value of 0.9 for $\sigma_t$ gives accurate predictions of pipe flow Stanton number for Prandtl numbers from 0.6 to 2000.

It may be argued that for $y^+ < 6$ the van Driest formula does not provide a particularly accurate distribution of turbulent viscosity. While of no practical consequence for flow calculations this does identify a problem in predicting concentration at very high Schmidt numbers. Significant mean gradients of species are then confined to values of $y^+$ less than about 5; consequently, mass transport rates are significantly dependent on the variation of effective diffusivity within this sublayer. For such extreme cases, therefore, one ought to concentrate on the diffusivity formula itself rather than the Schmidt number since the various turbulence models for $v_T$ display quite different asymptotic behavior as $y^+$ tends to zero.

For the other extreme, where the Prandtl number is much less than unity, LAWN [6.90] has surveyed the data for liquid-metal pipe flows and some of the proposals for turbulent Prandtl number. The measurements of SHERIFF et al. [6.101] for liquid sodium ($\sigma = 0.005$) indicate an average level of $\sigma_t$ of approximately 2.0 at a Reynolds number of $4 \times 10^4$ falling to 1.35 at $Re = 1.2 \times 10^5$; the data of FUCHS and FAESCH [6.102], BROWN et al. [6.103] also show the same trends. The appearance of turbulent Prandtl numbers greater than unity for low molecular Prandtl numbers agrees with the analyses of DEISSLER [6.88] and DUNN and REID [6.89] as well as with the majority of the simple mixing models[13]. That the turbulent Prandtl number should fall as Re is raised

---

[13] The latter predict that at low Prandtl numbers a substantial amount of heat will "leak" away from a lump of fluid in flight due to its high thermal conductivity. The turbulence interactions will thus be relatively ineffective in transporting heat.

also accords with expectations since, the higher the Reynolds number, the larger the proportion of the flow over which the heat transport is unaffected by molecular effects. As a limiting case we might cite the temperature measurements of SAKIPOV and TEMIRBAEV [6.104] in a round jet of mercury. Here there is no wall to create a low Reynolds number region and the turbulent Prandtl number (indicated by the shape of the mean temperature profile) was negligibly different from that in an air jet.

LAWN [6.90] helps put the liquid-metal pipe-flow models into perspective by plotting the implied variation with Reynolds number of the normalized turbulent heat flux at mid-radius. From this it appears that the supposition $\sigma_t = 1$, independent of Reynolds number and $\sigma$, would lead to errors in heat flux at the pipe wall not greater than 5%.

Only for air flows, where the Prandtl number is in any case close to unity, are there plentiful measurements of the distribution of $\sigma_t$ in the low Reynolds number region. The smooth-plate investigations by the Stanford group suggest that $\sigma_t$ exceeds unity in the range $10 < y^+ < 35$ (for example, most of the data of BLACKWELL et al. [6.84] indicate $\sigma_t \simeq 1.4$ at $y^+ = 10$); the same trends are also discernible in the measurements of BLOM [6.23] and BAKER and LAUNDER [2.40]. One might be tempted to attribute this rise in $\sigma_t$ as the wall is approached to the reciprocal-Prandtl-number effect predicted by DEISSLER [6.88] and DUNN and REID [6.89]. However, the recent measurements in water ($\sigma = 7.3$) by KHABAKHPASHEVA and PEREPELITSA [6.105] show virtually the same variation as the air flow measurements. Thus although the cause of the peak in $\sigma_t$ (including the possibility of systematic measuring error) cannot yet be identified, it seems unlikely that the molecular Prandtl number is a major factor.

The discussion has centered so far on situations where the shear stress across the viscosity-affected sublayer is very nearly constant; the flow structure thus corresponds closely with the "universal" form. There are, however, many important heat transfer problems where these conditions do not prevail. None of the various models for turbulent Prandtl number has been rigorously tested in such flows and there will certainly be cases where they are inadequate. Here there is space only to mention some of the more important areas and to provide an entry to the literature.

When a turbulent boundary layer is strongly accelerated it takes on some of the appearances of a laminar flow. The aerodynamic features are well documented and are discussed in Subsection 2.3.11. MORETTI and KAYS [2.226] and several subsequent Stanford studies show steep decreases in Stanton number in regions of strong acceleration. Several groups of workers [2.210, 272, 6.24] have made predictions of such flows in reasonably good agreement with measurement, taking a uniform value of turbulent Prandtl number, though in strongly accelerated flow turbulent transport is damped so rapidly that the predictions are not particularly sensitive to $\sigma_t$. A further interesting feature of accelerated flows is that while the pressure gradient limits the growth of the velocity layer there is no such mechanism of control on the thermal layer. Thus, as THIELBAHR et al. [6.106] have found, in a prolonged acceleration, the thermal boundary layer will tend to grow beyond the velocity layer, i.e., a "thermal superlayer" develops. Predictions of this phenomenon for sink-flow boundary layers are given in [6.107].

Reversion to laminar flow may be caused by several other agencies of which, perhaps, the most common is the severe heating or cooling of flow in pipes. Heating raises the viscosity of gases (thus reducing the flow Reynolds number) and causes flow acceleration too due to the reduction in density. McELIGOT and his colleagues have been providing experimental documentation of such flows over the past decade [2.264, 277, 6.108–110]. The earlier studies provided only surface temperature distributions; recently, however, PERKINS and McELIGOT [2.277] have reported mean temperature profiles in a 1-in pipe while velocity-profile and turbulence data are now being acquired [6.110]. Detailed flow measurements are greatly needed because none of the present attempts at predicting these flows has had much success.

The problem of heat transfer in fluids near the critical point has several features in common with strongly heated flows. While the temperature may vary by only 20 degrees or so across the pipe there will be large accompanying variations in fluid properties. A comprehensive review of the measured phenomena has been provided by HALL [6.111]. A complicating feature, not usually present in heated gas flows, is the strong effect of buoyant forces; as a result quite different distributions of surface temperatures have been obtained in upward and downward flow in a pipe.

In most purely natural convection flows in engineering, bulk Reynolds numbers are sufficiently modest that there is an appreciable fall in shear stress across the viscous sublayer. Here, too, it is to be expected that the distribution of mean and turbulence quantities across this region will differ from that in a constant stress layer. There have been several extensive experimental studies [4.56–58] of the vertical flat plate but the detailed near-wall structure still remains to be established.

Centrifugal force fields often produce a similar effect on the flow to that of a gravitational one. An example of this is the flow induced by a spinning disc or cone. The profile of radial velocity produced by the swirl ($\partial p/\partial r \simeq \varrho V_\theta^2/r$) is certainly similar to the velocity profile in a natural-convection boundary layer on a vertical plate. The parallel is not exact however because the steep near-wall gradients in swirl velocity generate high turbulence levels. There are many important heat-transfer problems involving spinning cones and discs especially within the gas-turbine industry. KREITH [6.112] has made an extended review of the experimental literature up to 1967. Finite-difference solutions of the swirling boundary-layer equations have been reported in [6.113] and [6.114] using a version of the mixing-length hypothesis and the low-Reynolds-number energy dissipation model, respectively; in each case $\sigma_t$ was taken as 0.9. Predicted heat-transfer rates from spinning discs and cones in air were in close agreement with measurements. The rates of diffusion of naphthalene ($\sigma = 2.4$) were, however, progressively under-predicted for spin Reynolds numbers above $3 \times 10^5$. The result suggests the discrepancy in the model is probably associated with the low Reynolds number region.

### 6.3.5 Heat Transport in Separated Flows

As remarked in Subsection 6.1.1, separated flows represent a very important class of flows in heat transfer. Especially for gases, where densities are low, it is often advantageous to promote separation, thereby raising the level of heat-transfer coefficient; the saving in capital cost (through being able to reduce the size of the heat exchanger) more than compensates for any extra pumping losses.

In general, separated flows are flows near walls where turbulent energy production and dissipation no longer dominate transport by advection and diffusion. They are thus more complex than the thin shear flows discussed above and are, physically, by no means as well understood.

One of the first contributions to shed light on the basic nature of such flows was the review by RICHARDSON [6.115]. Local measurements of heat-transfer coefficient around cylinders had been available in the literature for nearly thirty years. RICHARDSON, however, seems to have been the first to recognize that the laws governing the heat-transfer rates around the leading half of the cylinder would be quite different from those for the separated flow regime at the rear. He plotted experimental values of Nusselt number at the downstream stagnation point versus Reynolds number and discovered that the dimensionless heat-transfer rate increased as the $\frac{2}{3}$ power of Reynolds number. The measurements reported by SOGIN and BURKHARD [6.116] for heat transfer from the rear surface of a rough flat plate also displayed the same exponent dependence. Shortly afterwards a survey by SOGIN [6.117] confirmed the correctness of the $\frac{2}{3}$ power law. More recent separated flow data for a variety of configurations have mainly displayed the same power-law variation, for example the measurements of Krall and Sparrow [6.118] in the region downstream from an orifice plate in a pipe. Runchal's [6.119] measurements of flow in an abrupt pipe enlargement suggest overall an exponent of 0.55; however, if the data for Reynolds numbers below 6000 are ignored, an exponent of $\frac{2}{3}$ fits the remaining measurements satisfactorily. Finally, Jayatilleke's [6.91] survey of heat transfer from roughened surfaces showed that in the "fully rough" regime the $P$-function for many surfaces varied as the $-0.36$ power of roughness Reynolds number—which again implies that the equivalent Nusselt number varies as approximately the $\frac{2}{3}$ power of Reynolds number. As Spalding [6.120] has remarked, the dependence on Reynolds number is distinctly different from that commonly found in either a laminar or a turbulent boundary layer (where the usual exponents are 0.5 and 0.8, respectively).

A few years after Richardson's paper, Spalding [6.120] developed a one-dimensional model which provided an explanation of the main features of the observed behavior. He assumed that, except for a thin film of thickness $y_0$ immediately adjacent to the surface, the flow structure was independent of molecular transport and that the turbulence was sustained solely by diffusive transport from regions remote from the wall. The turbulent kinetic energy equation (assuming a gradient diffusion law) was thus written

$$0 = \frac{d}{dy}\left(\frac{v_T}{\sigma_q}\frac{d\frac{1}{2}\overline{q^2}}{dy}\right) - \varepsilon \tag{6.64}$$

where $\sigma_q$ denotes an "effective Prandtl number" for turbulence energy diffusion. The dissipation rate is found by assuming that the dissipation length scale increases linearly with distance from the wall. The eddy viscosity is taken as

$$v_T = c_\mu(\tfrac{1}{2}\overline{q^2})^2/\varepsilon \tag{6.65}$$

where the coefficient $c_\mu$ is taken as approximately 0.07, as required in the near-wall constant stress layer.

Within the wall film itself it was assumed that $(v_T/v)$ was a universal function of $y/y_0$ while $(\sigma_t/\sigma)$ was taken as a function of the two parameters $y/y_0$ and the molecular Prandtl/Schmidt number. The thickness of the film $y_0$ was found from the requirement that $(y_0 q/v)$ at the edge of the film be a universal constant. Thus, the higher the external energy levels the smaller the film thickness and vice versa. By integrating (6.64) and introducing the above assumptions regarding the properties of the wall film, Spalding obtained the result that the Nusselt number would vary as $Re^{0.6}$, in close agreement with experiment. Similar though somewhat more elaborate models of heat transfer processes in separated flows have been developed by RUNCHAL [6.119] and (again) by SPALDING [6.121]. WOLFSHTEIN [5.7] has provided detailed one-dimensional numerical solutions for conditions approximating those of separated flow. He devised a one-equation turbulence model for this purpose including low-Reynolds-number effects.

What, perhaps, appears a serious physical weakness in such models is the idea that the turbulence length scale is the *same* function of distance from the wall as in local equilibrium. In place of that assumption, one may solve the corresponding equation for $\varepsilon$, i.e.,

$$0 = \frac{d}{dy}\left(\frac{v_T}{\sigma_\varepsilon}\frac{d\varepsilon}{dy}\right) - c_{\varepsilon_2}\frac{\varepsilon^2}{\overline{q^2}}, \qquad (6.66)$$

in addition to (6.64). LOUGHHEAD [6.122] has done this and found the dissipation length scale still increases linearly with distance from the wall but with a slope approximately four times that found in a constant-stress wall flow. Moreover, the turbulence kinetic energy increases only as the 0.27 power of distance from the wall (compared with the four-thirds power for Spalding's analysis). As a result, with similar assumptions made for the film region as in [6.120], the Nusselt number varies approximately as $Re^{0.9}$, an exponent that is almost 40% too large!

Now the outcome of analyses such as outlined above depends critically on the assumed diffusive transport law. Loughhead's results suggest that the gradient form for the transport of $\varepsilon$ may not be adequate in separated-flow conditions. Equally it is clear that the provision of measurements of $\overline{q^2}$ and $\varepsilon$ in separated flows would greatly assist the development of better diffusive transport models.

In order to predict the behavior of practically interesting recirculating flows one must solve the full set of equations appropriate to two- (or three-) dimensional elliptic flows. The first examples of such computations (for example the pipe enlargement studied by Runchal [6.123] or Wolfshtein's [6.124] impinging jet) used relatively simple one-equation turbulence models in which the length-scale distribution through the flow was prescribed. These works mainly served to emphasize that a length-scale equation (or its equivalent) was essential to make the computational schemes suitable for design purposes. RUNCHAL [6.123] shows, for example, predictions based on four different but plausible length-scale profiles; radically different distributions of surface heat flux emerged for the four cases.

Most recent computations of heat transport in recirculating flows have been based on a two-equation eddy-viscosity model of turbulence with transport equations solved for $\overline{q^2}$ and $\varepsilon$ as described in Section 5.1; uniform values of turbulent Prandtl number have been presumed, usually 0.9. The near-wall region has been handled by matching solutions to a form of the "law of the wall" described in [6.125] in which $U^+$ is reinterpreted as $U\sqrt{0.15\overline{q^2}}/(\tau_w/\varrho)$ and $y^+$ as $y\sqrt{0.15\overline{q^2}}/\nu$. In local equilibrium, where $\tau_w/\varrho \simeq 0.15\overline{q^2}$, the above forms reduce to the usual definitions. Except where indicated the computations cited below have been based on the general two-dimensional stream function-vorticity computer program of GOSMAN et al. [2.209].

MATTHEWS and WHITELAW [6.126] provided extensive film-cooling predictions downstream from slots with thick lips where the coolant discharged over a backward-facing step; impressive agreement between the calculated and measured film-cooling effectiveness was obtained. ELGHOBASHI [6.127] has made calculations of the mixing of confined coaxial jets where the ratio of velocities of the primary and secondary streams was sufficiently large for flow separation to be provoked. Predictions are provided, *inter alia*, of concentration fluctuations associated with a tracer injected in the primary stream. Again close agreement between measurements and predictions are obtained.

DATE [6.128] has recently published computations of flow and heat transfer in tubes with "twisted-type" inserts; an ingenious polar coordinate system was used, which "twisted" with the tape in the streamwise direction. There is uncertainty, in drawing comparisons with experiment, about the "fin effectiveness" of the twisted tape. DATE [6.128] presents two sets of computed results, one in which the tape is treated as an adiabatic surface and the other where its temperature is the same as that of the pipe wall. The experimental variation of Nusselt number falls between the two predicted curves. A further important heat-exchanger geometry recently to have been the subject of numerical study is the coiled tube. PATANKAR et al. [3.34] (Subsection 3.1.2) have so far considered only the flow field behavior (using a three-dimensional "primitive-variable" method solving for $U$, $V$, $W$, and $p$ directly [6.3]). There are systematic differences between measured and predicted velocity fields; the differences are sufficiently small however to suggest that computations of the surface heat transfer rates will be close enough to the measured values for most purposes.

LE FEUVRE [6.129] has made computations of flow in in-line banks of tubes (with cross-stream axes) employing a novel scheme to reduce the false diffusion associated with upwind differencing [1.25]. A more extensive study of tube banks has recently been completed by MASSEY [6.130]. He devised an interesting system for overlapping Cartesian and cylindrical-polar grid systems, the latter being used in the neighborhood of the tubes and the former in the main part of the flow. For Reynolds

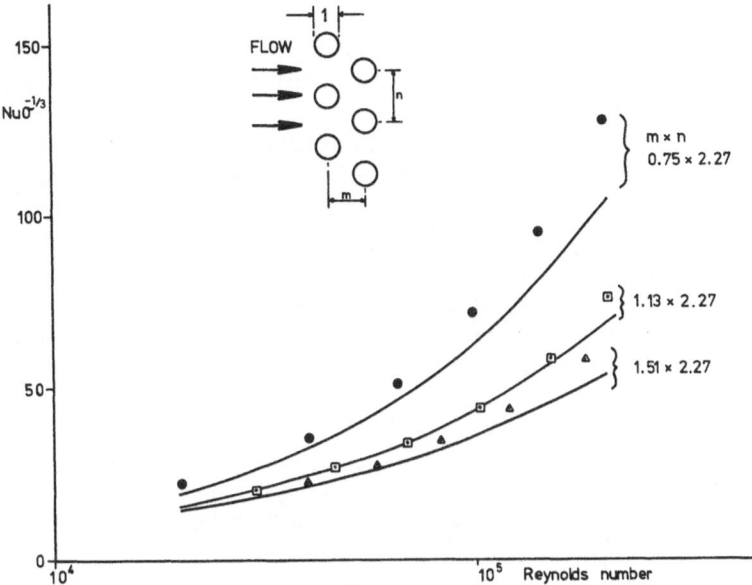

Fig. 6.7. Measurement and prediction of mean heat transfer rates in banks of staggered tubes: ●, □, △ Experiment, NORTHOVER and KELLARD [6.132]; —— Predictions, MASSEY [6.130]. Reproduced by kind permission of the Central Electricity Research Laboratories, Leatherhead, England

numbers above $2 \times 10^5$ the "standard" physical model as outlined above gave predictions of Nusselt number and pressure-loss coefficients, $C_p$, within the probable accuracy of the measurements. However, the "turbulent flow" regime in tube banks is usually considered to be well established by a Reynolds number of $2 \times 10^3$ (based on gap width between cylinders and gap velocity); for $2 \times 10^3 < Re < 2 \times 10^5$ the predicted values of $C_p$ and Nusselt number were substantially too low. The probable cause of this discrepancy emerged from PEEBLES' [6.131] measurements of the boundary layer development around a cylinder in a tube bank. His velocity profiles showed clearly that for Reynolds numbers up to $10^5$ the boundary layer over the leading half of the cylinder was laminar despite the presence of an external-stream turbulence level of approximately 25%. Limitations of computer storage made the use of the low-Reynolds-number form for the energy-dissipation model [5.16] out of the question. MASSEY therefore retained the high Reynolds number form except that for the region upstream of separation he applied the near-wall boundary conditions that $\partial \overline{q^2}/\partial r$ and $\partial \varepsilon/\partial r$ were zero and raised the additive constant in the semi-logarithmic velocity law from 5.5 to 15—a device which approximates to making the boundary layer laminar. Figure 6.7 shows the resultant predictions of mean Nusselt number for three staggered tube arrangements compared with measurements of NORTHOVER and KELLARD [6.132]. Agreement is generally very satisfactory; only for the closest pitching do differences exceed 15%.

The preceding example is perhaps an appropriate one on which to end this short survey of computations of heat and mass transport in recirculating flows. Appropriately, experiences with the tube-bank problem offer both encouragement for making further computer-based exploration of heat transfer in complex engineering flows (even with the comparatively primitive turbulence models that have so far been used) while underlining the need for continuing experimental work to help illuminate the physical processes.

*Acknowledgements.* I wish to thank S. CORRSIN, P. FREYMUTH, B. A. KOLOVANDIN, C. J. LAWN, S.-C. LIN, J. L. LUMLEY, R. J. MOFFAT, A. QUARMBY, A. J. REYNOLDS, W. RODI, and J. C. WYNGAARD who have helped me in various ways in preparing the survey.

Some of the review was prepared while I was visiting the Mechanical Engineering Department of the Pennsylvania State University. I am grateful to the Department for the resources and facilities it placed at my disposal.

# References

6.1    E. BRUNDRETT, P. R. BURROUGHS: Intern. J. Heat Mass Transfer **10**, 1133 (1967)

6.2    B. E. LAUNDER, W. M. YING: Proc. Inst. Mech. Engrs. **187**, 455 (1973)

6.3    S. V. PATANKAR, D. B. SPALDING: Intern. J. Heat Mass Transfer **15**, 1787 (1972)

6.4    S. V. PATANKAR: *Studies in Convection* **1**, 1 (Academic Press, London 1975)

6.5    L. PRANDTL: Z. Physik **11**, 1072 (1910): see also [2.10]

6.6    M. W. RUBESIN, W. C. ROSE: NASA Rept. TM X 62248 (1973)

6.7    A. S. MONIN: Atmos. Oceanic Phys. **1**, 45 (1965)

6.8    J. L. LUMLEY: Proc. Intern. Symp. on Stratified Flows, Novosibirsk (1972)

6.9    C. DU P. DONALDSON, R. D. SULLIVAN, H. ROSENBAUM: AIAA J. **10**, 162 (1972)

6.10   J. C. WYNGAARD, O. R. COTE: Boundary-Layer Meteorology **7**, 289 (1974)

6.11   B. E. LAUNDER: Rept. HTS/73/26 Mech. Engrg. Dept. Imperial College (1973)

6.12   B. E. LAUNDER: Lecture Series No. 76, Von Karman Inst., Rhode St. Genese, Belgium (1975)

6.13   S. C. CROW: J. Fluid Mech. **41**, 81 (1968)

6.14   B. E. LAUNDER: J. Fluid Mech. **67**, 569 (1975)

6.15   C. A. G. WEBSTER: J. Fluid Mech. **19**, 221 (1964)

6.16   R. G. OWEN: PhD Thesis, Mech. Engrg. Dept., Pennsylvania State Univ. (1973)

6.17   R. N. MERONEY: Project THEMIS, Rept. 28, Fluid Mechanics Program, Colorado State Univ. (1974)

6.18   J. L. LUMLEY, B. KHAJEH-NOURI: Modelling homogeneous deformation of turbulence, Dept. Aeronautics, Penn. State Univ. (1973)

6.19   B. E. LAUNDER: "On modelling the role of pressure fluctuations in homogeneous turbulent shear flow" Mech. Engrg. Dept. Rept., Imperial College (in preparation)

6.20   B. J. DALY, F. H. HARLOW: Phys. Fluids **13**, 2634 (1970)

6.21   B. J. DALY: J. Fluid Mech. **64**, 129 (1974)

6.22   J. KESTIN, P. D. RICHARDSON: Intern. J. Heat Mass Transfer **6**, 147 (1963)

6.23  J. BLOM: PhD Thesis, Dept. Physics, Tech. Univ., Eindhoven: (see also J. BLOM, Proc. IV International Heat Transfer Conference—Vol. II, Paris (1970)

6.24  W. P. JONES, B. E. LAUNDER: ASME paper 69-HT-12 (1969)

6.25  R. L. SIMPSON, D. G. WHITTEN, R. J. MOFFAT: Intern. J. Heat Mass Transfer 13, 125 (1970)

6.26  C. P. CHEN: Intern. J. Heat Mass Transfer 16, 1849 (1973)

6.27  A. QUARMBY, R. QUIRK: Intern. J. Heat Mass Transfer 15, 2309 (1972)

6.28  A. F. ORLANDO, R. J. MOFFAT, W. M. KAYS: Rept. HMT-17, Mech. Engrg. Dept. Stanford Univ. (1975)

6.29  M. M. PIMENTA, R. J. MOFFAT, W. M. KAYS: Rept. HMT-21, Mech. Engrg. Dept. Stanford Univ. (1975)

6.30  R. A. GOWEN, J. W. SMITH: Intern. J. Heat Mass Transfer 11, 1657 (1968)

6.31  T. MIZUSHINA, R. ITO, F. OGINO: Proc. 4th Intern. Heat Transfer Conference, Vol. II, Paris (1970)

6.32  A. QUARMBY, R. QUIRK: Intern. J. Heat Mass Transfer 17, 143 (1974)

6.33  D. S. JOHNSON: J. Appl. Mech. 26, 325 (1959)

6.34  J. C. WYNGAARD, O. R. COTÉ, Y. IZUMI: J. Atmos. Sci. 28, 1171 (1971)

6.35  K. BREMHORST, K. J. BULLOCK: Intern. J. Heat Mass Transfer 13, 1313 (1970); and 16, 2141 (1973)

6.36  C. J. LAWN, R. S. WHITE: Rept. RD/B/N2159, Central Electricity Generating Board (1972)

6.37  S. L. ZUBKOVSKY, L. R. TSVANG: Izv. Akad. Nauk SSSR, Fiz. Atmos. Okeana 12, 1307 (1966)

6.38  E. M. SPARROW, A. W. BLACK: Trans. ASME 89C, 258 (1967)

6.39  A. QUARMBY, R. K. ANAND: J. Fluid Mech. 38, 457 (1969)

6.40  A. QUARMBY: Intern. J. Heat Mass Transfer 15, 866 (1972)

6.41  V. P. BOBKOV, M. K. IBRAGIMOV, G. I. SABALEV: High Temperature (USSR) 6, 645 (1968)

6.42  H. RAMM, K. JOHANNSEN: Intern. J. Heat Mass Transfer 16, 1803 (1973)

6.43  P. BOURKE, D. J. PULLING: Intern. J. Heat Mass Transfer 13, 1331 (1970)

6.44  M. K. IBRAGIMOV, V. I. SUBBOTIN, G. S. TARANVO: Proc. Acad. Sci. (USSR) 183, No. 5 (1968)

6.45  S. P. S. ARYA, E. J. PLATE: J. Atmos. Sci. 26, 656 (1969)

6.46  D. B. SPALDING: Chem. Eng. Sci. 26, 95 (1971)

6.47  J. C. WYNGAARD, O. R. COTE: J. Atmos. Sci. 28, 190 (1971)

6.48  D. A. HAUGEN, J. C. KAIMAL, E. F. BRADLEY: Quart. J. Roy. Meteorol. Soc. 97, 168 (1970)

6.49  J. C. WYNGAARD, H. TENNEKES: Phys. Fluids 13, 1962 (1970)

6.50  G. MELLOR: J. Atmos. Sci. 30, 1061 (1973)

6.51  J. P. SCHON: DSc Thesis, Université Claude Bernard de Lyon (1974)

6.52  J. C. WYNGAARD: Boundary-Layer Met. 9, 44 (1975)

6.53  H. P. A. H. IRWIN: PhD Thesis, McGill University (1974)

6.54  J. LUMLEY: Phys. Fluids 18, 619 (1975)

6.55  P. BRADSHAW, D. H. FERRISS: NPL Aero Rept. 1271 (1968)

6.56  M. M. GIBSON, B. E. LAUNDER: Trans. ASME 98C, 81 (1976)

6.57  W. RODI: PhD Thesis, Imperial College, London Univ. (1972)

6.58  S. CORRSIN: J. Appl. Phys. 23, 113 (1952)

6.59  P. FREYMUTH, M. S. UBEROI: Phys. Fluids 16, 161 (1973)

6.60  S. C. LIN, S. C. LIN: Phys. Fluids 16, 1587 (1973)

6.61  H. WISKIND: J. Geophys. Res. 67, 3033 (1962)

6.62  C. C. ALEXOPOULOS, J. F. KEFFER: Phys. Fluids 14, 216 (1971): (see also Rept. 6810, Mech. Engrg. Dept., Univ. Toronto)

6.63  J. SIESS: Dr. Ing. Thesis. Université d'Aix-Marseille (1975)

6.64  R. R. MILLS, S. CORRSIN: NACA Mem 5-5-59W (1959)

6.65  P. FREYMUTH, M. S. UBEROI: Phys. Fluids 14, 2574 (1971)

6.66  S. CORRSIN, M. S. UBEROI: NACA Rept. 998 (1950)

6.67  J. O. HINZE, B. VAN DER HEGGE ZIJNEN: Appl. Sci. Res. 1 A, 435 (1949)

6.68  H. A. BECKER, H. C. HOTTEL, G. C. WILLIAMS: J. Fluid Mech. 30, 285 (1967)

6.69  P. V. DANCKWERTS, R. A. WILSON: Chem. Eng. Sci. 19, 885 (1964)

6.70  P. E. JENKINS, V. W. GOLDSCHMIDT: Trans. ASME 95 I, 581 (1973)

6.71  P. E. JENKINS, V. W. GOLDSCHMIDT: Rept. HL 74-45, Civil Engrg. Dept., Purdue University (1974)

6.72  B. G. VAN DER HEGGE ZIJNEN: Appl. Sci. Res. 7 A, 293 (1958)

6.73  L. S. G. KOVASZNAY, S. F. ALI: Paper FC 3.2 Proc. 5th Intern. Heat Transfer Conf., Tokyo (1974)

6.74  C. H. GIBSON, C. C. CHEN, S. C. LIN: AIAA J. 6, 642 (1958)

6.75  R. A. ANTONIA, R. W. BILGER: TN-F66, Mech. Engrg. Dept., Univ. Sydney (1974)

6.76  W. E. WATT: Rept. TP 6705, Dept. Mech. Engrg., Univ. Toronto (1967)

6.77  M. SUNYACH, J. MATHIEU: Intern. J. Heat Mass Transfer 12, 1672 (1969)

6.78  W. G. HILL, R. H. PAGE: Trans. ASME 91 D, 67 (1969)

6.79  C. BÉGUIER, L. FULACHIER, J. F. KEFFER, R. DUMAS: Compt. Rend. 280 B, 493 (1975)

6.80  L. FULACHIER, J. F. KEFFER, C. BÉGUIER: Compt. Rend. 280 B, 519 (1975)

6.81  J. C. ROTTA: Intern. J. Heat Mass Transfer 7, 215 (1964)

6.82  H. LUDWIEG: Z. Flugwiss. 4, 73 (1956)

6.83  W. M. KAYS, R. J. MOFFAT: Studies in Convection 1, 223 (Academic Press, London 1975)

6.84  B. F. BLACKWELL, W. M. KAYS, R. J. MOFFAT: Rept. HMT-16, Mech. Engrg. Dept., Stanford Univ. (1972)

6.85  P. STEPHENSON: Rept. RD/L/N111/74, Central Electricity Generating Board (1974)

6.86  G. BERGELES, A. D. GOSMAN, B. E. LAUNDER: "The prediction of discrete-hole cooling—II Turbulent flow", Mech. Engrg. Dept., Imperial College (1976)

6.87  S. CORRSIN: J. Appl. Phys. 22, 469 (1951)

6.88  R. G. DEISSLER: Intern. J. Heat Mass Transfer 6, 257 (1963)

6.89  D. W. DUNN, W. H. REID: NACA TM 4186 (1958)

6.90  C. J. LAWN: "Turbulent temperature fluctuations in liquid metals and other fluids". Unpubl. Rept., CEGB Berkeley Nuclear Laboratories (1973)

6.91  C. L. V. JAYATILLEKE: Prog. Heat and Mass Transfer 1, 193 (1959)

6.92  A. J. REYNOLDS: Intern. J. Heat Mass Transfer 18, 1055 (1975)

6.93  T. F. CHUNG, L. C. THOMAS: Paper FC 3.7 Proc. 5th Intern. Heat Transfer Conf., Tokyo (1974)

6.94  R. RAJAGOPAL, L. C. THOMAS: Paper FC 1.3 Proc. 5th Intern. Heat Transfer Conf., Tokyo (1974)

6.95  S. V. PATANKAR: Rept. TWF/TN/14, Mech. Engrg. Dept., Imperial College (1966)

6.96  R. JENKINS: Proc. 1951 Heat Transfer Fluid Mech. Inst., 147 (Stanford Univ., 1951)

6.97  A. T. WASSEL, I. CATTON: Intern. J. Heat Mass Transfer 16, 1647 (1973)

6.98  S. AOKI: Bull. Tokyo Inst. Technol. 54, 63 (1963)

6.99  L. C. THOMAS: Trans. ASME 92 C, 565 (1970)

6.100 T. MIZUSHINA, T. SASANO: Intern. Dev. Heat Transfer, 662 (1963)

6.101 N. SHERIFF, D. J. O'KANE, B. MATHER: TRG Rept. 2191 R, UKAEA, Risley (1973)

6.102 H. FUCHS, S. FAESCH: Proc. Intern. Heat Transfer Summer School, Trogir (1970)

6.103 H. E. BROWN, B. H. AMSTEAD, B. E. SHORT: Trans. ASME 79, 279 (1957)

6.104 Z. B. SAKIPOV, D. J. TEMIRBAEV: Teplo i Massoperenos 2, 407 (1965)

6.105 E. M. Khabakhpasheva, B. V. Perepelitsa: Paper FC 4.2, Proc. 5th Intern. Heat Transfer Conf., Tokyo (1974)

6.106 W. H. Thielbahr, W. M. Kays, R. J. Moffat: Rept. HMT-5, Mech. Engrg. Dept., Stanford Univ. (1969)

6.107 B. E. Launder, F. C. Lockwood: Trans. ASME **91** C, 229 (1969)

6.108 D. M. McEligot: PhD Thesis, Stanford Univ. (1963)

6.109 C. W. Coon, H. C. Perkins: Trans. ASME **92** C, 506 (1970)

6.110 C. J. Barney: MSE Rept., Dept. Aero/Mech. Engrg. Univ. Arizona, (in preparation)

6.111 W. B. Hall: Advances in Heat Transfer **7**, 2 (1971)

6.112 F. Kreith: Advances in Heat Transfer **5**, 129 (1968)

6.113 M. L. Koosinlin, B. E. Launder, B. I. Sharma: Trans. ASME **96** C, 204 (1974)

6.114 B. E. Launder, B. I. Sharma: Letters Heat Mass Transfer **1**, 131 (1974)

6.115 P. D. Richardson: Chem. Eng. Sci. **18**, 149 (1963)

6.116 H. H. Sogin, K. Burhard: Rept. No. 4, USAF, Aero Res. Lab. (1960)

6.117 H. H. Sogin: Trans. ASME **86** C, 200 (1964)

6.118 K. M. Krall, E. M. Sparrow: Trans. ASME **88** C, 131 (1966)

6.119 A. K. Runchal: Intern. J. Heat Mass Transfer **14**, 781 (1971)

6.120 D. B. Spalding: J. Fluid Mech. **27**, 97 (1967)

6.121 D. B. Spalding: Rept. TWF/TN/33, Mech. Engrg. Dept., Imperial College (1967)

6.122 J. N. Loughhead: Private communication

6.123 A. K. Runchal: PhD Thesis, Univ. London (1969); (also Rept. HTS/69/17, Mech. Engrg. Dept., Imperial College)

6.124 M. Wolfshtein: PhD Thesis, Univ. London (1967); (also Rept. SR/R/2, Mech. Engrg. Dept., Imperial College)

6.125 B. E. Launder, D. B. Spalding: Comput. Methods Appl. Mech. and Engrg. **3**, 269 (1974)

6.126 L. Matthews, J. H. Whitelaw: Proc. Inst. Mech. Engrs. **187**, 447 (1973)

6.127 S. Elghobashi: PhD Thesis, Univ. London (1974)

6.128 A. Date: Intern. J. Heat Mass Transfer **17**, 845 (1974)

6.129 R. F. Le Feuvre: PhD Thesis, Univ. London (1973); (also Rept. HTS/74/5, Mech. Engrg. Dept., Imperial College)

6.130 T. H. Massey: PhD Thesis, Council for National Academic Awards (1976)

6.131 W. L. Peebles: PhD Thesis, Council for National Academic Awards (1976)

6.132 E. W. Northover, P. O. Kellard: Unpubl. Rept. Central Electricity Research Laboratories, Leatherhead

# 7. Two-Phase and Non-Newtonian Flows

J. L. LUMLEY

With 9 Figures

Two-phase turbulent flow covers an enormous range of technically important situations. The two phases can be solid/fluid or fluid/fluid; if the former, the solid phase may be flexible or rigid, spherical or non-spherical. The concentration of one phase in the other may be such that individual morsels of one phase may be regarded as isolated, or such that adjacent morsels interact strongly. The characteristic dimensions of the distribution of one phase in the other may be such, relative to the dimensions characterizing the flow, that the two-phase mixture may be regarded as a continuum, or not. Most of these situations are extremely complicated, and relatively little of a scientific nature has been done, the literature consisting for the most part of the results of observation organized by dimensional reasoning and empirical models. In the space available here, we will be able to give detailed attention only to the simplest case, that of rigid, spherical particles, separated sufficiently to be considered isolated, and small enough so that the flow in their vicinity may be taken as simple. However, we will outline some of the considerations which are relevant to the other, more complicated cases.

In exactly the same way, the general subject of the turbulent flow of non-Newtonian media is extremely broad, relatively unexplored and quite beyond our scope here. Non-Newtonian media are media that depart from Newtonian behavior in any way (in a Newtonian medium, stress is proportional to strain rate). This covers a vast array of possibilities, even in laminar flow: shear thinning or thickening, visco-elasticity, plasticity and normal-stress phenomena to mention a few. We have no more than the most rudimentary ideas how turbulence behaves in most of these. In fact, a discussion of turbulence in non-Newtonian media in general would be rather out of place in this chapter. There is only one type of non-Newtonian turbulence that is at all well understood, and that is turbulence in dilute solutions of high molecular weight linear polymers, which results in a reduction of skin-friction drag, and we will limit ourselves to this type. We will find that the behavior of such solutions has several points of similarity with particulate flows, and that is the reason for its inclusion in this chapter.

## 7.1 Physical Features of Particle-Laden Flows

### 7.1.1 Particle Interaction

There is no general method for dealing with interacting particles; it is unfortunate that many of the most interesting situations involve strong interactions (blood flow, for example, or quicksand). Certain beginnings have been made. For example, there are now consistent ways to construct expansions of bulk properties about the state of infinite dilution (BATCHELOR and GREEN [7.1]). There are ways of dealing with the interactions of regular arrays of identical particles (HASIMOTO [7.2]); this is not of much help in turbulent situations, however. Certainly, if the suspension may be regarded as a continuum (Subsection 7.1.6), and if particle inertia may be neglected (in many cases of interest it may not), then the suspension may be regarded as a simple fluid (COLEMAN and NOLL [7.3]), which provides restrictions on the form of the relation between stress and deformation for the suspension. The possible forms are still far too numerous, however, and some discussion of the physics of the interaction is necessary. In the present state of the subject, the most that one can reasonably hope to do is estimate at what concentration interactions will become important.

For this purpose we may make use of the results of BATCHELOR and GREEN [7.1] and BATCHELOR [7.4]. In the first of these papers, a first-order correction to the dilute (non-interacting) expression for the viscosity of a suspension of identical spherical particles in pure strain is calculated, based on two-particle interactions. This calculation indicates that the correction will be of the order of 1% when the volume fraction is of the order 0.003. Although negligible effect on the viscosity does not necessarily mean negligible hydrodynamic interaction of individual pairs, it does suggest that the aggregate effect of the interaction is negligible. This type of flow is not the flow of most direct interest; unfortunately, certain difficulties arise in the case of a simple shear which prevent the calculation from being carried out.

For the case of non-spherical particles, the matter is not quite so clear. In BATCHELOR [7.4] a calculation is carried out for the viscosity in pure strain of a suspension of long rods or fibers, whose lateral distance from each other is small relative to their length, but large relative to their thickness. The expression found appears to join smoothly to the dilute expression (although there is a substantial range of concentration in between, where exact calculations are not presently possible). The most important result of this calculation is that the effect is dependent on the volume concentration of the smallest sphere circumscribed about the particle, and not on the actual volume concentration (which would, of course be smaller by the square of the thickness/length ratio). Using a convenient interpolation formula given in BATCHELOR [7.4], we may estimate that the correction to the apparent viscosity for interaction will be of the order of 1 % for particles of length/diameter ratio 10000 when the volume concentration of circumscribed spheres is about 0.004. This number is weakly dependent on the length/diameter ratio; for 1000, it is about 0.002, while for 100 it drops to 0.0008. Since the simplification that the interaction can be expressed in terms of the concentration of circumscribed spheres is better the larger the length/diameter ratio, it is probably fair to use as a rule of thumb for particles of arbitrary shape that a volume concentration of circumscribed spheres of roughly 0.003 corresponds to 1 % interaction. The same provision should be made with regard to the non-spherical particles as was made for the spheres: that this limit permits considerable interaction of individual pairs, but that the aggregate effect is small.

To get a feeling for these numbers, we may consider that whole human blood has a concentration of red cells by weight of about 0.4; the concentration of circumscribed spheres is probably about 1.0. The drag reduction phenomenon (Sect. 7.6) takes place in concentrations of unexpanded molecules of the order of 0.001 (there are excellent indications that it takes place at arbitrarily small concentrations, but measurements do not exist below about 0.001). When drag reduction begins, however, the molecules expand. It appears likely that the length increases at least by a factor of three, and perhaps by many orders of magnitude (again, direct measurements do not exist). This suggests that the volume concentration of circumscribed spheres increases to at least 0.01 and possibly far above unity (there is no reason why the volume concentration of circumscribed spheres should not exceed unity). Hence, again, strong interactions appear to be probable. Naturally occurring atmospheric aerosols, on the other hand, have volume concentrations nine orders of magnitude below our critical value of 0.003 [7.5]; even in a dense fog, the volume concentration is three orders of magnitude below 0.003 [7.6].

### 7.1.2 Non-Spherical Particles

Two-phase turbulent flows containing non-spherical particles have not been considered theoretically. Suspensions of non-spherical particles have been considered only in the dilute (non-interacting) limit, and only for inertia-free particles subjected to homogeneous deformation. Turbulent flow of such a suspension could thus be described by a constitutive relation (presuming that the scale of the smallest region (see Subsect. 7.1.6) of the flow is large relative to the particle dimensions). The existing work is devoted to determining the constitutive relation, and predicting the behavior in certain simple situations. In particular, we may mention the work of HINCH and LEAL [7.7, 8] and LEAL and HINCH [7.9, 10]. These authors consider ellipsoids of revolution in both the almost spherical [7.10] and strongly non-spherical [7.8] cases. They consider primarily the case of weak Brownian motion, or, equivalently, strong shear, and have considered both steady [7.7] and unsteady [7.8] motions. The behavior is generally quite complicated. Non-spherical particles weakly affected by Brownian motion rotate irregularly in a shear, spending relatively long periods aligned with the flow, and then rather abruptly flipping over (such motion being termed a Jeffery orbit). The extent to which the shear succeeds in aligning the bulk of the particles depends on the relative strength of the shear and the Brownian motion. The behavior of the suspension is characterized by two time scales, one corresponding to the period of the Jeffery orbits, and the other to the Brownian rotational diffusion.

### 7.1.3 Foams and Slurries

A foam is a liquid containing such a concentration of gas bubbles that the bubbles are virtually in contact, so that the liquid phase makes only a small contribution to the total volume. A slurry is a liquid containing such a concentration of solid particles that the particles are virtually in contact, so that the liquid phase makes a small contribution to the total volume. Thus, foams and slurries may be expected to be similar in many respects. Whipped cream is a classic example of a foam, while quicksand is a slurry. Both foams and slurries are of considerable technological importance, and occur frequently in chemical and industrial engineering practice and elsewhere. For this reason there is an extensive literature on applications and empirical laws [7.11, 12]. There is virtually no analysis, which is hardly surprising, since foams and slurries represent the case of very strong, in fact dominant, interaction. Either one may be treated as a continuum under appropriate circumstances (Subsect. 7.1.6) but the question of the appropriate constitutive relation remains. The situation is complicated because foams, at least, are plastics, that is, they display a yield stress. Slurries, too, at sufficiently low flow rates, can settle out and display similar phenomena. From the point of view of turbulent flow, the only thing that one can do is fall back on a characteristic of turbulent flow of all media. The dominant characteristics of turbulence are controlled by inertia, not by the particular constitutive relation of the medium [1.15]. Hence, when the length and time scales of the turbulence are large compared to the length and time scales characterizing the medium, we may expect the turbulence to be indistinguishable in all media. As an example of this (see Subsect. 7.6.3) the velocity profile in drag-reducing polymer flow has a log region with the classical slope, and this is also observed in pipe flows of slurries [7.12]. In a certain sense this begs the question, since the nature of the constitutive relation can have a marked effect on the bulk properties of a flow (overall drag, for instance) by changing the dynamics in the regions which are not inertially dominated (near the wall, for instance).

### 7.1.4 Particle Pickup

In a turbulent flow of liquid containing denser particles, there is often present a bed of the unsuspended particles; at the surface of the bed particles are being continually picked up and deposited by the turbulent flow. The physics of the pickup mechanism determines the concentration of particles in the flow. Familiar examples are dust or sand borne by a wind, or silt carried by a river. Less familiar, perhaps, is the pickup of coal dust by the boundary layer behind the shock wave resulting from a methane explosion in a coal-mine gallery; the amount of the dust placed in suspension is a determining factor in the resulting dust explosion. Below, we shall quote results for coal dust in air as a typical suspension[1]. Because of its considerable practical importance, there is an extensive literature (see, e.g., [7.13]). The physics of the problem, however, is extremely complicated; even in idealized cases (in which the material on the wall is loose) the pickup depends on the precise nature of the unsteady flow near the wall in a turbulent boundary layer, which is still a matter for discussion (Sect. 2.3). Real cases are complicated by the fact that the particulate matter on the wall may be compacted so that it can support a certain stress, and is subject to brittle fracture. Coal dust for example is customarily observed to go into suspension by breaking loose in large chunks, which disintegrate subsequently. Presently existing analyses are primarily scaling and dimensional analyses and correlations of data. Even the mechanism by which the boundary layer can keep heavy particles in suspension has not been analyzed in detail.

### 7.1.5 Migration

In the linear, or Stokes, regime (Subsect. 7.2.1) particles in a steady shear flow rotate in a prescribed manner (Subsect. 7.1.2) and move with the local mean velocity, experiencing no tendency to migrate across the flow. Any such tendency is associated with lift forces, which are quadratic, that is, such lift forces are associated with inertia, which is

---

[1] Typical values are: ratio of particle density to fluid density $\varrho_p/\varrho_f = 4 \times 10^3$; diameter $d = 40\,\mu\text{m}$, nominally spherical; ratio of diameter to Kolmogorov length scale, $d/\eta$, of order 0.1 in laboratory boundary layer.

excluded in the Stokes regime. Hence, we may expect migration only at non-negligible relative Reynolds numbers (based on the maximum fluid speed relative to the particle surface). Even then, if there is no velocity relative to the center of mass there can be no migration in a homogeneous shear, by symmetry; hence migration must be related to curvature of the mean profile. We might thus expect migration following an abrupt change in velocity (at a shock, say) before the particle and fluid velocities have a chance to come to equilibrium, or close to a wall where profile curvature is important. In the latter case, such migration has been analyzed [7.14], and it is predicted that particles will migrate away from a wall, in agreement with experiment. This sort of migration can be important in certain flows, since it may result in a reduced concentration of particles in the viscous sublayer. Since we are usually considering small relative Reynolds numbers, the forces producing the migration are small and the migration slow. In the turbulent part of the flow, other forces producing particle motion may be expected to be much larger, and the migration may ordinarily be neglected. Near a wall the migration must be balanced against the progressively weaker transport by the fluctuating velocities (including large eddies). Depending on the speed of the migration, it is possible for a depleted layer to develop next to the wall; such a layer will result in different properties in the sublayer and in the turbulent part of the flow, with consequent effects on the bulk properties.

### 7.1.6 The Continuum Approximation

To be considered as a continuum, we must be able to find a fine scale sufficiently small as to be smaller than the smallest scale of dynamical interest in the motion, yet large enough to contain enough particles to permit reasonable local definitions of density, velocity, etc. (see, e.g., [1.8, 7.15]). Thus, a given medium may be regarded as a continuum in one motion, and not in another. If we imagine the particles in the medium distributed in such a way that the presence of a particle in a given small region (small enough to contain at most one particle) is independent of the presence of particles in adjacent regions, then it is not difficult to show that the rms fluctuation in particle density in a larger volume is roughly equal to the inverse square root of the average number of particles in the volume. Hence, as the size of the volume increases, the rms fluctuation level will decrease as the inverse 3/2 power of a characteristic dimension. As the size of the volume increases, however, density defined over the volume will begin to average real variations in mean particle density. The error made here is approximately one-third of the square of the ratio of volume radius to a characteristic dimension of the density distribution. There is an optimum point, where the combined error from these two sources is a minimum. The error at this point is the $-2/7$ power of the number of particles in a volume having the characteristic dimension of the density distribution. This is quite a stringent requirement; for example, if we require 1 % accuracy in the local value of particle density, we must have 10 million particles in a volume of characteristic dimension. Characteristic dimension, in this case, means the smallest dynamically significant dimension. If we consider, for example, particles of radius one micron, at a non-interacting volume concentration of 0.003, there are a little less than a billion per cubic centimeter, allowing us to consider motions having scales as small as 2.4 mm on a continuum basis. If we are satisfied with 10 % accuracy in the definition of particle density locally, we need only 3000 particles in a characteristic volume, which would allow consideration of motions down to scales of 0.14 mm on a continuum basis. In turbulence, the smallest dynamically significant scale is the Kolmogorov length scale $\eta$ (Sect. 1.5). In geophysical situations, this is often of the order of a millimeter, so that probably naturally occurring aerosols and fogs in turbulent flow may safely be treated as continua. It is clear, however, that there must be very many situations of technological importance in which heterogeneous media may not be considered as continua at the smallest scales present. In such situations, it is customary and legitimate to analyze the motion into its various scales (as by Fourier transform) and to speak of the medium as being regardable as a continuum down to a certain scale. Since the larger scales in a turbulent flow are dominated by inertia, and essentially independent of the details of the mechanism of momentum transport, we can discuss the larger (continuum) scales and ignore the fact that at the smaller (non-continuum) scales the dynamics is incredibly complicated.

## 7.2 Isolated, Heavy, Rigid, Spherical Particles

Having mentioned in passing a number of the difficult aspects of two-phase flows about which, at our present stage of development, little can be done, we are ready to devote ourselves to the relatively much simpler case in which all these aspects are excluded. We choose the particles to be isolated, rigid and spherical for obvious reasons, discussed in Section 7.1. A heavy particle is one for which gravitational forces are not negligible. Our analysis (and experiment) will indicate that this includes most particles in air, a somewhat surprising result. In addition, since the response of the particles to fluctuating ambient velocity is controlled by the same parameters as the response to gravitational forces (Chapt. 4), a heavy particle is often also one for which inertia forces are not negligible. Much of the material in Subsections 7.2.2 and 7.2.3 is taken from [7.16, 17].

### 7.2.1 The Equations of Motion

Generally speaking, the equation of motion of a particle is extremely complicated. If the relative Reynolds number is not small compared to unity (permitting a Stokes approximation, [1.8]), no exact explicit equation of motion is known. The flow around the particle, and its drag, are of course determined by the Navier-Stokes equations, but the drag is not a simple function of relative velocity.

The relative Reynolds number may be determined primarily by the terminal velocity, or by the fluctuating relative velocity, depending on the particle size and density and the parameters of the turbulence in which it finds itself. For particles of density 1 g/cc in air, a relative Reynolds number of $1/2$ (about the largest at which the Stokes approximation may be safely applied) based on the terminal velocity is reached at a diameter of 63 microns. The determination of the Reynolds number based on the fluctuating relative velocity is a bit more complicated, and will be delayed for a moment. We find that in the atmosphere the restriction on relative Reynolds number based on terminal velocity is the more stringent, while in a coal-dust explosion (say) the restriction based on the fluctuating relative velocity is more stringent.

Supposing that the relative Reynolds number is less than $1/2$, exact, but rather complicated, equations can be written down [7.18]. If the particles are small relative to the smallest length in the turbulence (the Kolmogorov length scale $\eta = (v^3/\varepsilon)^{1/4} \simeq 1$ mm in the lower atmosphere) and have time constants short relative to the shortest time scales $(\tau = (v/\varepsilon)^{1/2} \simeq 0.08$ s in the atmosphere) then one can show rigorously [7.19] that the equation of motion reduces (using $v$ for the particle velocity and $u$ for the fluid velocity) to

$$dv/dt = (u - v)/a + g \tag{7.1}$$

where $g$ is the gravity vector. Here $a$ is the particle time constant, given in Stokes flow by

$$a = \varrho_p d^2 / 18 \, \mu$$

for heavy particles. If $u = 0$ and $dv/dt = 0$, then

$$v = ag . \tag{7.3}$$

We define the terminal velocity as $V_T = ag$. For Stokes particles, the length scale restriction justifying (7.1) is automatically satisfied. From the relation $V_T = ag$, the time scale restriction is satisfied for $V_T \ll 78$ cm/s in the atmosphere.

For particles having a larger relative Reynolds number, we can say only that the drag is a function of the relative velocity history. However, if the Reynolds number is low enough for the wake to be stable (say $\leq 10$) and if the time scale of velocity changes is small enough (as we will discuss in a moment) then the flow around the particle will be quasi-steady and the drag will always be aligned with the relative velocity, and will have the steady-state value at that instantaneous velocity. The time scale for change of the boundary layer on the particle is roughly $\delta^2/4v$, where $\delta$ is a typical boundary layer thickness, $d/R^{1/2}$, where $R$ is the Reynolds number. The time scale of change seen by the particle is either $\eta/V_T$ or $\tau$, whichever is smaller. The two conditions are, respectively,

$$d/\eta \ll 4, \quad (d/\eta)^2 \ll 4R. \tag{7.4}$$

Combined with the condition on the relative Reynolds number, the first condition is the more restrictive, limiting us to particles of roughly 400 μm and smaller in the atmosphere. Hence, for such particles, we may take the wake to be quasi-steady. The drag force per unit mass of particles may then be written as

$$(U/U)f(U) \tag{7.5}$$

where $U$ is the relative velocity vector, and $U$ its magnitude. If the fluctuations of $U$ are small (we will see later just how small) then we can expand (7.5) in a Taylor series in the fluctuations: $U = \bar{U} + u$,

$$(U_i/\bar{U})f(U) \simeq (\bar{U}_i/\bar{U})f(\bar{U})$$
$$+ [(f'(\bar{U}) - f(\bar{U})/\bar{U})\bar{U}_i\bar{U}_j/\bar{U}^2 + \delta_{ij}f(\bar{U})/\bar{U}]u_j \ldots \tag{7.6}$$

where $f' = df/d\bar{U}$.

If we take $f = U/a + BU^2 + \ldots$ [7.20], then the coefficient of the fluctuations in (7.6) becomes

$$\delta_{ij}/a + B\bar{U}(\delta_{ij} + \bar{U}_i\bar{U}_j/\bar{U}^2). \tag{7.7}$$

That is, the coefficients in the two directions are unequal—if $U_i = (0, 0, -V_T)$, then the coefficients in the $x_3$ and $x_1$ directions are, respectively,

$$f'(V_T) \simeq 1/a + 2BV_T + \ldots \tag{7.8}$$

$$f(V_T)/V_T \simeq 1/a + BV_T + \ldots.$$

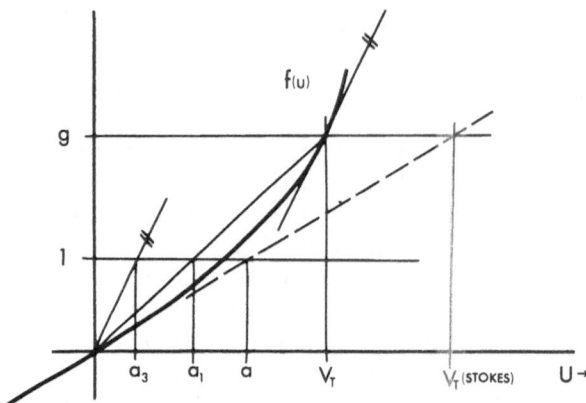

Fig. 7.1. Particle drag and time constants in the nonlinear regime. Note that $V_T < V_T$ (Stokes), and that both $a_3 < a_1 < a$ (the Stokes value). $V_T = a_1 g$ still, but $a_3 g$ is substantially smaller

Hence, evidently we can preserve an equation similar in form to (7.1), so long as we allow the coefficients in the different directions to be unequal. Of course, when the fluctuations become larger relative to $\bar{U}$, we will no longer have linearity, and cross terms will appear—that is, sideways gusts will influence the vertical drag. Using (7.5–8), we can write

$$dv/dt = g + [(u - v)/|u - v|] f(|u - v|) \tag{7.9}$$

or, in the vertical and horizontal directions, taken as $x_3$ and $x_1$, respectively, in conformity with meteorological practice,

$$dv_3/dt = f(V_T) + (u_3 - v_3 - V_T)/a_3 - g = (u_3 - v_3)/a_3 - V_T/a_3$$

$$dv_1/dt = (u_1 - v_1)/a_1 \tag{7.10}$$

where $1/a_3$ and $1/a_1$ are given, respectively, by the terms in (7.8). The relationship between the various terms is shown in Fig. 7.1. Note that, although $a_1 g = V_T$ (determined by $f(V_T) = g$), $V_T/a_3$ is substantially larger than $g$.

The condition for the validity of the expansion (7.6) can be shown to be (with a little algebra, and using the exact expression for the drag given in [7.20]) that the Reynolds number based on the rms fluctuating relative velocity should be small compared to unity. We will have to wait until the next section to calculate this condition explicitly, but it is clear from (7.10) that if $a_1$ and $a_3$ are small enough relative to some time scale (to be found), then we can make the fluctuations in relative velocity as small as we please.

### 7.2.2 The Inertia Effect and the Relative Reynolds Number

We are going to obtain estimates for the relative size of the time derivatives in (7.10). We will find that, if $a_1$ and $a_3$ are made small enough, the time derivative and hence the relative fluctuating velocity can be made as small as we please. In the limit, the relative fluctuating velocity will be zero in the horizontal and equal to the terminal velocity in the vertical. Hence, also the value of the fluctuating relative Reynolds number will be determined by the values of $a_1$ and $a_3$.

To estimate the relative size of the time derivative, we must have an estimate for the statistical behavior of $u$. This is not straightforward, since $u$ is the velocity at the particle location. If $u(x, t)$ is the Eulerian velocity at a fixed laboratory point $x$ at $t$, then what the particle experiences is $u(X(t), t)$ where $X$ is the instantaneous particle position. The statistics of $u(X(t), t)$ are difficult to determine, since the $u(x, t)$ field determines $X(t)$. This is similar to the following situation: if a climber enters a mountain range determined never to exceed a certain gradient, the altitude statistics he will experience will be quite different from those he would experience if he were determined to follow a fixed compass heading.

There are two limiting cases which are a little easier to handle, however. The first is $V_T \to 0$, in which case $u$ is the Lagrangian velocity seen by the wandering fluid point. The second is $V_T \to \infty$, in which case the falling particle cuts rapidly through the frozen turbulence, so that $u$ is the Eulerian velocity sampled along a vertical line. In either case, using high Reynolds number Kolmogorov scaling (Sect. 1.5), approximate forms for the spectra of $u$ can be obtained. These are shown in Fig. 7.2. These spectra are determined for the case of homogeneous turbulence, an approximation that we will use nearly everywhere. For a derivation of these spectral forms, see [1.15].

Now, since $a_3 < a_1$, we will examine the horizontal case; if the acceleration term is small in the horizontal, it will be smaller in the vertical. Using standard techniques of Fourier analysis, it is easy to show that

$$\overline{v_1^2} = \int_{-\infty}^{+\infty} E(\omega)/(1 + a_1^2 \omega^2) d\omega$$

$$\overline{(u_1 - v_1)^2} = \int_{-\infty}^{+\infty} E(\omega) a_1^2 \omega^2/(1 + a_1^2 \omega^2) d\omega \qquad (7.11)$$

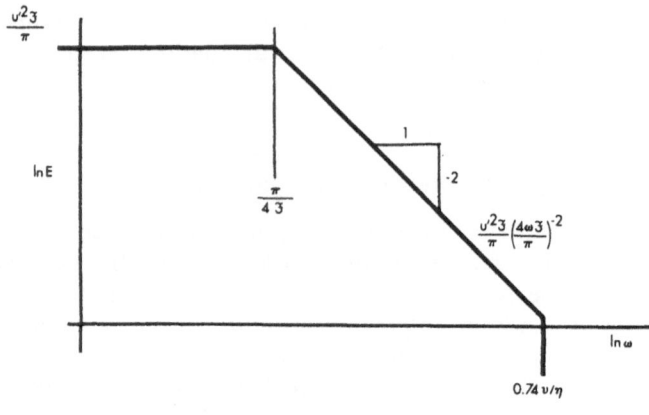

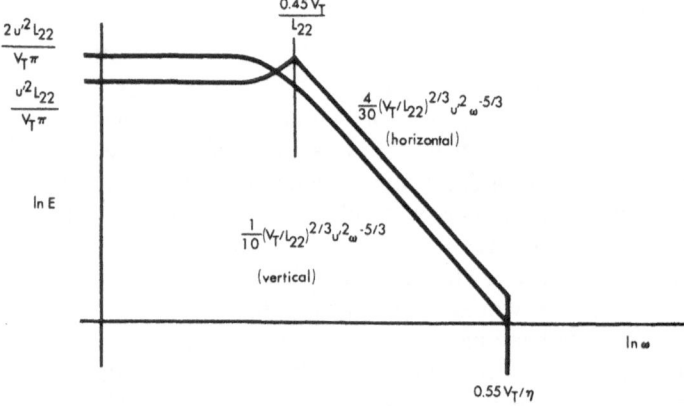

Fig. 7.2. Limiting cases of velocity spectra. (a) Lagrangian spectrum for either horizontal or vertical fluctuating velocity. $\mathcal{T} = 4L_{22}/3u'^2$. (b) Spectra for vertical and horizontal fluctuating velocities in the case $V_T \to \infty$. $L_{22}$ is the transverse integral scale

where $E(\omega)$ is the frequency spectrum of the fluid velocity seen by the particle, analogous to the Eulerian spectra discussed in Section 1.5, and is either of the spectra in Fig. 7.2 in the two limiting cases. A simplified integration of the Lagrangian case gives (where we have assumed that $1/a_1$ is larger than either cutoff frequency)

$$\overline{(u_1 - v_1)^2} \simeq 0.87 \varepsilon a_1^2 (\varepsilon/v)^{1/2} \tag{7.12}$$

where $\varepsilon$ is the dissipation of energy per unit mass. The same sort of simplified integration gives for the Eulerian case

$$\overline{(u_1 - v_1)^2} \simeq 0.23 V_T^2 \varepsilon a_1^2 / v . \tag{7.13}$$

Using the Stokes form for $a_1$, we can obtain expressions for the relative Reynolds number [based on the square root of (7.12) or (7.13)] and the relative velocity; for the Lagrangian case we have (writing $u_1 - v_1 = \Delta u$)

$$R \simeq (\varrho_p/\varrho_f)(d/\eta)^3/19 \qquad (7.14)$$

$$(\overline{\Delta u^2})^{1/2}/u' \simeq (\varrho_p/\varrho_f)(d/\eta)^2/19 R_l^{1/4} \qquad (7.15)$$

where $u' = (\overline{u^2})^{1/2}$, and $R_l = u'l/\nu$, $l = u'^3/\varepsilon$. The Eulerian case gives

$$R \simeq (\varrho_p/\varrho_f)^2(d/\eta)^5(lg/\overline{u^2})/676 R_l^{1/4}, \qquad (7.16)$$

$$(\overline{\Delta u^2})^{1/2}/u' \simeq (\varrho_p/\varrho_f)^2(d/\eta)^4(lg/\overline{u^2})/676 R_l^{1/2}. \qquad (7.17)$$

If we take $\varrho_p/\varrho_f = 4 \times 10^3$ (coal dust in air) then, considering the Lagrangian case first, if we require $R \leq 0.5$ we have from (7.14)

$$d/\eta \leq 0.13 \qquad (7.18)$$

and (7.15) gives, using (7.18),

$$(\overline{\Delta u^2})^{1/2}/u' \leq 3.56/R_l^{1/4}. \qquad (7.19)$$

In the atmosphere, $R_l$ may frequently be as large as $10^6$, while in the laboratory it is usually between $10^2$ and $10^3$. Hence, in the atmosphere, on the Lagrangian basis, the time derivative may be neglected, while in the laboratory it usually cannot.

The evaluation of the Eulerian case is a little more complicated, both because a term in $R_l$ is present in both (7.16) and (7.17), and because of the term in $lg/u^2$. In a thermally driven situation $u^2/lg \simeq \theta/T$ (where $\theta$ is the rms temperature fluctuation, and $T$ the absolute temperature), while in a mechanically driven situation it may have any value in principle, but is virtually always much less than unity. Taking the case of the atmosphere, with $R_l = 10^6$ and $u^2/lg = 2 \times 10^{-2}$, for $\varrho_p/\varrho_f = 4 \times 10^3$, we find for $R \leq 0.5$ from (7.16)

$$d/\eta \leq 0.11 \qquad (7.20)$$

which gives (from (7.17))

$$(\overline{\Delta u^2})^{1/2}/u' \leq 0.15. \qquad (7.21)$$

Hence, the time derivative may probably be neglected in the atmosphere on the Eulerian basis also, and the Eulerian Reynolds number restriction is more severe.

In the general case, we may take the ratio of either (7.15)/(7.14) or (7.17)/(7.16) to give

$$\overline{(\Delta u^2)}^{1/2}/u' \simeq R/(d/\eta)R_l^{1/4} . \tag{7.22}$$

If we require $R \leq 0.5$, then the time derivative will be negligible ($\leq 0.1$)) if

$$(d/\eta)R_l^{1/4} \leq 5 . \tag{7.23}$$

We can also take the ratio of the $d/\eta$ ratio required to give $R \leq 0.5$ in the Lagrangian and Eulerian cases, to give

$$(d/\eta)_{\mathrm{L}}/(d/\eta)_{\mathrm{E}} \simeq 0.67(\varrho_{\mathrm{p}}/\varrho_{\mathrm{f}})^{1/15}(\overline{u^2}/lg)^{-1/5}R_l^{-1/20} . \tag{7.24}$$

This has a value of 1.28 for coal dust in the atmosphere; reducing $R_l$ to $10^3$ changes it only to 1.81. Few particles have a density below 1 gm/cc or above 8 gm/cc, producing a total change in the ratio (7.24) of only 15 %; the quantity $u^2/lg$ even in mechanical turbulence is rarely greater than unity, which could reduce the ratio (7.24) by about a factor of two.

We may conclude that the two specifications in (7.24) seldom differ by as much as a factor of two, and that (7.20) is therefore typical. Hence, from (7.23) we may conclude that the time derivative may probably be neglected in the atmosphere, but not elsewhere.

If we take $\eta = 1$ mm in the atmosphere, then (7.20) restricts us to particles of roughly 110 μm diameter to satisfy the restrictions on fluctuating relative Reynolds number.

We have not mentioned the requirement that the flow in the vicinity of the particle appear to be at most a simple shear. For this it is roughly sufficient that the particle diameter be at most 1/6 of the shortest wavelength present in the turbulence; this is approximately $\eta/2\pi$, so that this requirement is satisfied approximately by $d \leq \eta$. Hence, of the three requirements [this one, (7.4) and that on the fluctuating relative Reynolds number], the requirement on the fluctuating relative Reynolds number is the most restrictive.

Let us consider the horizontal particle diffusivity (we could treat the vertical equally well, by removing $V_{\mathrm{T}}$), considering dispersion relative to a point moving downward with velocity $V_{\mathrm{T}}$. The mean square horizontal displacement may always be written

$$\overline{X_1^2} = \overline{2v_1^2}t \int_0^t (1 - \tau/t)P_{11}(\tau)d\tau \tag{7.25}$$

where

$$\overline{v_1^2} = \overline{v_1^2(t)}; \quad \overline{v_1^2}P_{11}(\tau) = \overline{v_1(t)v_1(t+\tau)} \tag{7.26}$$

which is valid only in a homogeneous flow. We will obtain the diffusivity in a homogeneous flow, and use it as a first approximation in an inhomogeneous one. For large times, we have

$$\overline{X_1^2} \simeq 2\overline{v_1^2} t \mathcal{T} \tag{7.27}$$

where

$$\mathcal{T} = \int_0^\infty P_{11}(\tau) d\tau, \quad \text{the particle integral scale.} \tag{7.28}$$

If we begin from

$$dv_1/dt + v_1/a_1 = u_1/a_1 \tag{7.29}$$

we can write

$$v_1 = (1/a_1) \int_0^\infty e^{-x/a_1} u_1(t - x) dx \tag{7.30}$$

where $u_1$ is the fluid velocity seen by the particle. With a little manipulation, we can form

$$\overline{v_1(t)v_1(t+\tau)} = (\overline{u_1^2}/a_1) \int_0^\infty e^{-x/a_1} [g_{11}(\tau + x) + g_{11}(\tau - x)]/2dx \tag{7.31}$$

where $\overline{u_1^2}$ is the mean square fluid velocity, and

$$\overline{u_1^2} g_{11}(\tau) = \overline{u_1(t)u_1(t+\tau)}. \tag{7.32}$$

If the particle is falling like a stone, $\overline{u_1^2}$ is the mean square fluid velocity seen in laboratory coordinates; if the particle moves like a Lagrangian point, $\overline{u_1^2}$ is also the same as the value in laboratory coordinates [7.21]. It is not unreasonable to suppose that it is the same for all $a_1$.

Now, form

$$2\overline{v_1^2} \mathcal{T} = \int_{-\infty}^{+\infty} \overline{v_1(t)v_1(t+\tau)} d\tau \tag{7.33}$$

the diffusion coefficient. We find

$$2\overline{v_1^2} \mathcal{T} = \overline{u_1^2} \int_{-\infty}^{+\infty} g_{11}(\tau) d\tau \tag{7.34}$$

so that the time constant does not directly affect the asymptotic diffusion coefficient [7.17]. It certainly affects $\overline{v_1^2}$ (because particle transport is due to the largest, slowest eddies):

$$\overline{v_1^2} = (\overline{u_1^2}/a_1) \int_0^\infty e^{-x/a_1} g_{11}(x) dx \leq \overline{u_1^2}. \tag{7.35}$$

Evidently, $\mathscr{T}$ is larger than that for a fluid point, while $\overline{v_1^2}$ is smaller, and the two compensate.

## 7.2.3 The Crossing-Trajectories Effect

We know that our particle is drifting downward with velocity $V_T$, and hence does not remain with the parcel of fluid with which it began. For this reason, the fluid velocity seen by the particle loses correlation, the rate depending on $V_T$. We can make a rough estimate of this effect, following [7.16].

Consider the correlation of vertical velocity fluctuations. For very large Reynolds numbers, we may consider the large-scale structure of the velocity field to be determined entirely by the velocity scale $u'$ and the length scale $L_{11}$. $u'$ is the rms turbulent velocity in laboratory coordinates, in principle the vertical component; however, we will suppose that the turbulence is not too far from isotropy, so that $u'$ is characteristic of any component. $L_{11}$ is the longitudinal integral scale (Sect. 1.3). The correlation will evidently also be a function of $V_T$ and $\tau$, the time lag. Thus, we can write

$$u'^2 g_{33}(\tau) = \overline{u_3(t+\tau)u_3(t)} = u'^2 f(u'\tau/L_{11}, V_T\tau/L_{11}) \qquad (7.36)$$

and $f(u'\tau/L_{11}, 0)$ corresponds to the Lagrangian case and $f(0, V_T\tau/L_{11})$ to the Eulerian case in which the particle falls like a stone. Now, there are no data to speak of, but there is good reason to believe on theoretical grounds that the Lagrangian and Eulerian longitudinal correlations have similar shapes, both positive and monotone decreasing. With this assumption, it seems natural to set

$$f(u'\tau/L_{11}, V_T\tau/L_{11}) = h_L[(u'^2/\beta^2 + V_T^2)^{1/2}\tau/L_{11}] \qquad (7.37)$$

letting the iso-correlation contours be ellipses. $\beta$ is a constant yet to be determined, and the shape of $h_L$ is fixed by the Eulerian longitudinal correlation. We have

$$\int_0^\infty h_L(V_T/L_{11})d\tau = L_{11}/V_T; \qquad \int_0^\infty h_L(x)dx = 1. \qquad (7.38)$$

At the other limit, we must obtain the Lagrangian integral scale

$$\mathscr{T}_L = \beta L_{11}/u' \qquad (7.39)$$

and from theoretical considerations we know that $\beta = 2/3$, [1.15]. We can now write

$$2\overline{v_3^2}\mathcal{T} = 2u'^2 \int_0^\infty h_L[(u'^2/\beta^2 + V_T)^{1/2}\tau/L_{11}]d\tau \tag{7.40}$$

$$= 2u'^2 L_{11}/(u'^2/\beta^2 + V_T^2)^{1/2} = 2u'^2 \mathcal{T}_L/(1 + 4V_T^2/9u'^2)^{1/2} \tag{7.41}$$

where $\mathcal{T}_L$ is the Lagrangian fluid integral scale [1.15]; the numerator is the dispersion of a fluid point. Hence, the crossing-trajectories effect reduces the dispersion by the factor in the denominator.

If we attempt to apply the same reasoning to the horizontal dispersion we run into trouble because, while the Lagrangian correlation never goes negative (presumably), the transverse Eulerian correlation must; by continuity, there must be enough back-flow across a plane to make the net mass-flux vanish. In an isotropic flow this means that $\int_0^\infty x h_T(x)dx = 0$ if $h_T$ is the transverse velocity correlation. On the other hand, measurements of SNYDER [7.22] indicate that the particle velocity autocorrelations are similar to the Eulerian transverse correlation for all $V_T$ except the smallest, $V_T = 1.26$ cm/s in a flow where $u' = 13$ cm/s. Hence, we can probably make an assumption like (7.37) even though it will not be valid right down to $V_T = 0$ (see Fig. 7.3).

With such an assumption, we obtain

$$2\overline{v_1^2}\mathcal{T} = 2u'^2 \mathcal{T}_L/(1 + 16V_T^2/9u'^2)^{1/2} \tag{7.42}$$

where the coefficient is about 16 (rather than 4) because the transverse integral scale based on $\overline{u_1(x)u_1(x+r_2)}$ is half the longitudinal integral scale based on $\overline{u_1(x)u_1(x+r_1)}$ in isotropic turbulence [1.15], and fairly close to this value in most practical cases.

### 7.2.4 Estimates for the Variances

We may obtain rough estimates for the velocity variances if we make a crude approximation

$$h_L = e^{-x}. \tag{7.43}$$

Then we have [with $\gamma = (u'^2/\beta^2 + V_T^2)^{1/2}/L_{11}$]

$$\overline{v_3^2} = (u'^2/a_3) \int_0^\infty e^{-x/a_3}e^{-\gamma x}dx = (u'^2/a_3) \int_0^\infty e^{-x(1/a_3+\gamma)}dx$$

$$= u'^2/(1 + a_3\gamma) = u'^2/\{1 + (a_3/\mathcal{T}_L)[1 + (4/9)V_T^2/u'^2]^{1/2}\} \tag{7.44}$$

where $\mathcal{T}_L = 4L_{22}/3u' = 2L_{11}/3u'$.

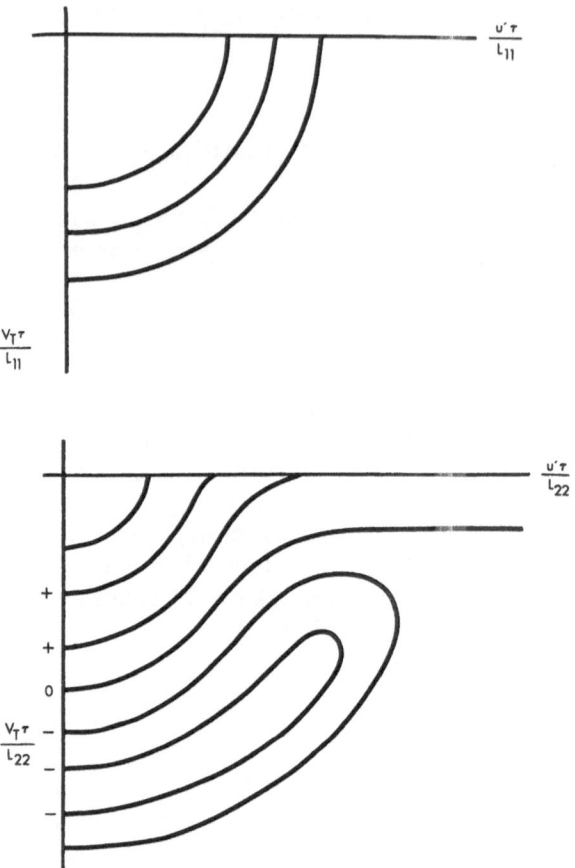

Fig. 7.3. Isocorrelation contours for longitudinal and lateral correlation

In the horizontal direction it is a little more complicated, because we have to pick a form for $h_T$ that satisfies $\int_0^\infty x h_T(x)dx = 0$ as well as $\int_0^\infty h_T(x)dx = 1$ (and $h_T(0) = 1$). If we pick $h_T = (1-x/4)e^{-x/2}$, we get

$$\overline{v_1^2} = u'^2[1+(a_1/4\mathcal{T}_L)(1+16V_T^2/9u'^2)^{1/2}]$$
$$/[1+(a_1/2\mathcal{T}_L)(1+16V_T^2/9u'^2)^{1/2}]^2 \qquad (7.45)$$

and, of course, the integral scales have the inverse factors.

If we consider 40 µm coal particles in a boundary layer with $U = 15$ m/s and $u' = 0.5$ m/s, then $V_T/u' = 1/5$. This results in a 1% decrease in vertical diffusivity and a 3% decrease in horizontal diffusivity.

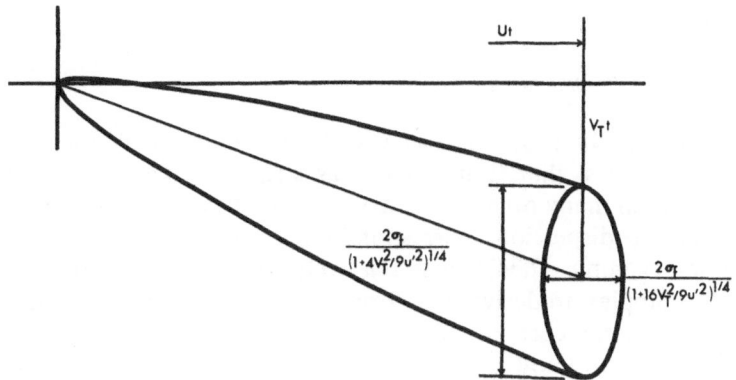

Fig. 7.4. Schematic of single particle dispersion in the asymptotic limit

It is evident that the variances will depend primarily on the ratios $a_1/\mathcal{T}_L$ and $a_3/\mathcal{T}_L$. We may write

$$a/\mathcal{T}_L = (\varrho_p/\varrho_f)(d/\eta)^2/6R_l^{1/2}\ . \tag{7.46}$$

Comparing this with (7.14) and (7.16), it is evident that, if $R_l$ is not too large, a non-negligible value of (7.46) is possible. If we take coal dust, $\varrho_p/\varrho_f = 4 \times 10^3$, $d/\eta = 0.1$, $R_l = 167$ (typical of a laboratory boundary layer), we have $a/\mathcal{T}_L = 0.52$. Hence, ignoring differences between $a$, $a_1$ and $a_3$, $v_3^2$ would be reduced to about 66% of the Lagrangian value, while $v_1^2$ would be reduced to about 71% of the Lagrangian value.

In applying these results, it is evident that, qualitatively speaking, if we take the example of heavy particle dispersion downwind of a stack in the atmosphere, the concentration maximum will be closer to the stack, and higher, than for gaseous diffusion. Unfortunately, if we wish a quantitative answer, the matter is more difficult. We cannot apply the asymptotic formulae obtained above, because the asymptotic state is never reached. That is, if effluent is released from a stack of height $H$, roughly speaking, the process stops when the effluent reaches the ground, or when

$$V_T t + (2u'^2 t \mathcal{T})^{1/2} = H\ . \tag{7.47}$$

Crudely, if $t$ reaches $\mathcal{T}$ before the left side of (7.47) reaches $H$, we have reached the asymptotic state. That is, the condition for the asymptotic state is roughly

$$(V_T + u'2^{1/2})\mathcal{T} < H\ . \tag{7.48}$$

We can write this as

$$(V_T/u' + 2^{1/2})\,(4/3)/(1 + 4V_T^2/9u'^2)^{1/2} \leqq H/L_{22}\,. \tag{7.49}$$

In the atmosphere, the right-hand side is of order unity, while the left-hand side is never less than 2 and can be as large as 3. Thus, in a real atmosphere, and in many other boundary layer applications, one should use forms for the dispersions corresponding to finite times [thus invalidating, for example, (7.34)]. It is relatively straightforward, though algebraically complex, to derive such forms, using the crude approximations like (7.43) for the correlation functions.

If the variance is known, the concentration distributions may be predicted using the fact that, in a homogeneous situation, they are observed to be Gaussian both before, and in, the asymptotic state. Although there is no theoretical support for this fact [7.23], it suggests the use of a diffusion equation [1.18].

The analysis described here is often applied to much larger particles (up to several hundred μm diameter) (7.17) but there is essentially no information on how good an approximation this is likely to be.

## 7.3 Effect of Particles on Shear Flows

The effect of even a dilute suspension of non-interacting particles on a turbulent shear flow is, in the general case, much too complicated to be treated rigorously. Experimentally [7.24] many phenomena are observed in various parameter ranges. In the transport of sediment (that is, with *heavy* particles in a boundary layer *above* a horizontal surface), the profile is always steeper, and the turbulence somewhat suppressed; the drag is usually increased, but may be reduced. In horizontal pipe flow of similar particles, the results are similar. In pipe flow of neutrally buoyant particles, the drag is frequently reduced. The picture is somewhat complicated by the fact that most experimental data refer to particles which are much larger than the Kolmogorov length scale, at concentrations high enough for strong interaction to occur. We will limit ourselves here to sketching physical explanations for a few of these phenomena under very restrictive assumptions.

Specifically, we will deal with two simplified situations: the first, in which the inertia effect is negligible, but the crossing trajectories effect is not, corresponding roughly to dust particles in the atmosphere; and the second, in which the reverse is true, corresponding to coal dust in the boundary layer following a methane explosion. These will serve to explain several of the phenomena mentioned above; drag reduction by neutrally buoyant particles will only be touched upon qualitatively. We will assume throughout this section that the concentration is high enough so that the suspension may be regarded as a continuum to some approximation, but low enough to ignore interaction, and we will presume that we are in the Stokes regime.

### 7.3.1 Heavy, Inertia-Free Particles

From Subsection 7.2.2 we know that there are circumstances corresponding roughly to dust particles in the atmosphere, under which the inertia term may be neglected, and the equations of motion become

$$v_1 = u_1, \qquad v_2 = u_2$$
$$v_3 = -V_T + u_3. \tag{7.50}$$

It is instructive to consider averages of these velocities, having in mind specifically a boundary layer above a horizontal surface, in which the particles are suspended. Then the vertical component of particle mean velocity, $v_3$, must be essentially zero (disregarding the slow growth of the boundary layer) since the particle may not penetrate the surface or leave the top of the layer. Hence $\overline{u_3} = V_T$. At first glance this result seems surprising, since $\overline{u_3} = 0$ in laboratory coordinates. This $u_3$, however, is the $u_3$ seen by the particle. Evidently the particle must seek out (on the average) rising currents, to compensate this tendency to fall out. No measurements have been made but in all likelihood the particles which fall to the bottom are swept into windrows between the longitudinal big eddies and are lifted and ejected into the upper part of the layer by the updraft (and associated "bursting" phenomenon) between them; see Section 2.3 and [2.11]. This sort of behavior may be seen in dry snow-flakes on a hard-surface road, or in sand blowing across a beach.

We must now switch to laboratory coordinates. The $v_i$ at a laboratory point is the $v_i$ of the particles in a small region near that point. These particles have necessarily the same velocity [from (7.50)] in a region in which $u_i$ is essentially uniform. Now if we average, we have that $\overline{u_3} = 0$, so that the average velocity of two particles at any point is $-V_T$. Now we must consider, however, the fluctuations in particle volume concentration $C$. The net upward particle mass flux must vanish (again ignoring boundary layer growth), since we presume that the concentration distribution is not changing. This gives

$$\varrho_p \overline{CV_3} = -\varrho_p \bar{C} V_T + \varrho_p \overline{Cu_3} = 0 \tag{7.51}$$

or

$$\overline{Cu_3} = \bar{C} V_T; \tag{7.52}$$

that is, there is an upward turbulent particle flux to balance the fallout. This is the equivalent statement in laboratory coordinates of our previous conclusion in particle coordinates. We may obtain a crude concentration distribution if we make a mixing length assumption, which may not be bad here (see [1.15])

$$\overline{Cu_3} = -\bar{C}_{,3}\overline{u_3^2}\mathcal{T} \qquad (7.53)$$

where $\bar{C}_{,3} = \partial\bar{C}/\partial x_3$ and $\mathcal{T} \cong q^2/3\varepsilon$. In a constant stress layer (Section 1.8) $q^2$ and $\overline{u_3^2}$ are nominally constant, and $\mathcal{T} \propto x_3$, and we may easily integrate (7.52) with (7.53) to obtain

$$\bar{C} \propto x_3^{-\alpha} \qquad (7.54)$$

where

$$\alpha = (V_T/u_\tau)(3u_\tau^4/\overline{u_3^2}\kappa q^2) \cong 1.70 V_T/u_\tau . \qquad (7.55)$$

We have used values corresponding to a constant-stress layer without particles; hence, if the concentration is large enough to influence the structure of the layer (the only interesting case) then (7.54) and (7.55) would have to be modified. Qualitatively, however, the conclusion would be unaltered: $\bar{C}$ decreases upward, the dimensionless rate of decrease being $\alpha$, and the inhomogeneity is monotone increasing in $V_T/u_\tau$.

Now, the particles exert a force on the fluid, and vice versa, the force being dependent on the relative velocity. Since the relative velocity is only nonzero in the vertical, this is the only nonzero component of the force. From (7.50) and (7.10), the net vertical force on the fluid per particle is simply $-mg$ and the net force per unit volume is $-\varrho_p Cg$. We can write equations of motion for the fluid and particle components, giving (if $C \ll 1$, as it must be to avoid interaction)

$$\dot{u}_i + u_{i,j}u_j = -P_i/\varrho_f - g\delta_{3i}(\varrho_p C/\varrho_f) + \nu u_{i,jj} \quad \text{and} \quad u_{i,i} = 0 \qquad (7.56\text{a, b})$$
$$\dot{C} - V_T C_{,3} + (Cu_i)_{,i} = 0 \qquad (7.57)$$

where the tensor conventions of Chapter 1 apply, a dot denotes differentiation with respect to time, and $u_i$ is the instantaneous fluid velocity. The gravitational term remains as $C \to 0$ if we fix $\varrho_p C/\varrho_f$ (called the loading, $= \mathcal{L}$, say) corresponding to denser and denser particles. This is physically reasonable; for dust in air we may have $\varrho_p/\varrho_f \sim 10^3$, while for non-interaction we must have $C \sim 10^{-3}$, so that the product may indeed be of order unity, while $C \ll 1$. Equation (7.56) has precisely the form of the Boussinesq approximation for temperature fluctuations in the atmosphere (Subsection 4.1.1 and [7.15,30]). There the buoyancy term has the form $g\theta/\Theta_0$ where $\theta$ is the temperature difference from an adiabatic atmosphere and $\Theta_0$ the mean potential temperature. Thus we may conclude that the primary dynamic effect of the particles in our boundary layer is 1) to modify the mean vertical pressure gradient corresponding to (7.54); and 2) if the fluctuating velocity causes a parcel of fluid, having a particle concentration equal to the local value, to change levels, it will feel a spurious buoyancy (actually an anomaly in drag force) tending to return it to its original level. Our boundary layer then corresponds to a stable, stratified atmospheric boundary layer, with downward heat flux. Equation (7.57), of course, is not quite the usual temperature equation, because of the term in $V_T$, so that the concentration and temperature distribution will differ, but the principle is the same.

There is an extensive literature on stratified boundary layers (Chapter 4). We will mention here only a few of the simplest results. If we separate $U_i$ into mean and fluctuating quantities $\overline{U}_i + u_i$, and form the equation for $u_i$ from (7.56), we may then form the equation for $q^2$ [1.15]. From this we find that the total production of turbulent energy may be written, in $x$, $y$, $z$ coordinates with $z$ vertical and $\overline{U}' = \partial\overline{U}/\partial z$, as

$$-\overline{U'uw}(1 + \mathscr{L}V_T g/\overline{U}'\overline{uw}) \tag{7.58}$$

where use has been made of (7.52) and of the definition of the loading $\mathscr{L}$ in the last paragraph. This permits us to define a flux Richardson number [1.15], analogous to the buoyant-flow definition following (4.13), as $-$(extra production)/(shear production), or

$$R_f = -\mathscr{L}V_T g/\overline{U}'\overline{uw} . \tag{7.59}$$

From the equation for the mean velocity, it is straightforward to determine that a constant-stress layer exists in this flow as in a particle-free flow [1.15], something which is borne out by experiment [7.25]. If we make use of the relation (1.48) for a particle-free logarithmic layer, $\overline{U}'\overline{uw} = -u_\tau^3/\kappa z$, we may define a Monin-Obukhov length as

$$L = u_\tau^3/\mathscr{L}V_T g\kappa . \tag{7.60}$$

In the constant stress layer of a density stratified flow this would be constant; here it is slowly varying due to (7.54) and (7.55). Using (7.60), we have

$$R_f = z/L . \tag{7.61}$$

This is an exact result, within our assumptions, unlike the corresponding result for buoyancy. There is a highly developed similarity theory for stratified boundary layers (Chapter 4 and [1.18]) based on the idea that the only relevant length is $L$, and that everything must be a function of $z/L \equiv \zeta$. In particular, one writes

$$\partial \bar{U}/\partial z = (u_\tau/\kappa z)\phi_m(\zeta) \tag{7.62}$$

and the function $\phi_m(\zeta)$ has been determined from theory and experiment [1.18]; $\phi_m(0) = 1$, and $\phi_m$ is monotone increasing; for $\zeta \to -\infty$, $\phi_m \propto |\zeta|^{-1/3}$, while for $\zeta \to +\infty$, $\phi_m \propto \zeta$. Thus, for $\zeta > 0$, which corresponds to our case, the mean velocity gradient (and hence the mean velocity) is everywhere larger than the value without particles, for the same shear velocity.

The drag is, however, not determined simply by the shear velocity. The wall shear is given by

$$\tau_{13} = -\varrho_f \overline{(1-C)u_1 u_3} - \varrho_p \overline{C v_1 v_3} \tag{7.63}$$

and the leading term can be written as (making various semi-empirical assumptions regarding third moments)

$$\tau_{13} = \varrho_f u_\tau^2 (1 + \mathscr{L}) + O(V_T^2/u_\tau^2). \tag{7.64}$$

Thus, the loading substantially increases the mean density, and hence, the shear stress (at the same shear velocity). We can thus say that the mean velocity profile corresponds to drag reduction relative to a non-stratified flow of the *same density*, but whether this corresponds to drag reduction relative to the same fluid *without particles* is another question, dependent on whether $\tau_{13}/\varrho_f \bar{U}^2$ increases or decreases. Other things being equal, $\bar{U}$ increases roughly linearly with the loading, and so does $\tau_{13}$; thus, for small loading, both $\tau_{13}$ and $\bar{U}^2$ increase linearly, and there will be drag reduction if the coefficient of the loading in $\bar{U}^2$ is greater than unity. This coefficient is roughly

$$2\beta z V_T g/\bar{U} u_\tau^2 \tag{7.65}$$

for $\zeta$ not too large, where $\beta \sim 6$ [1.18]. The coefficient depends on a number of parameters describing the boundary layer and the particles, but probably the most influential is $V_T/u_\tau$; the remainder is often of the order of 10. Hence, for sufficiently small values of $V_T/u_\tau$ drag is increased over the clear fluid, while for values probably of order unity drag should be reduced. In the first case, the stratification effects are small, and the fluid appears simply as an unstratified, but higher-density fluid.

Most cases of sediment transport correspond to relatively small values of $V_T/u_\tau$, since they take place in liquids so that the particle/fluid density difference is small, and considerable agitation (relatively speaking) is necessary to suspend the sediment. Hence, drag is usually increased.

### 7.3.2 Light Particles with Inertia

We turn now to the other limiting case: particles for which inertia cannot be neglected, but for which fallout can. This corresponds roughly to a high-speed, relatively low Reynolds number flow. In such a flow, we may use Lagrangian estimates. Due to the particle inertia, there is a fluctuating relative velocity between the fluid and the particles. In Subsection 7.3.1 also, we had a relative velocity between the fluid and the particles, which was independent of scale; that is, every eddy, no matter what size, experienced a downward force due to the particles falling through it. Here, however, large slow motions can take the particles along, and no slip is experienced between the phases; only the small, high-frequency motions cause a relative velocity. This is easily seen from the spectrum of relative velocity [e.g., (7.11)] which rises as $\omega^2$ at low frequencies.

We cannot use (7.12) to estimate the rate at which work is done by these fluctuating relative velocities, because of the restriction placed on (7.12) that $1/a_1$ is larger than the cutoff frequency. It turns out that most of the interesting phenomena happen when $1/a_1$ is smaller than the cutoff frequency. A simple modification of our crude integration (p. 300) for this case gives for this additional energy dissipation per unit total mass (including the particles)

$$1.18\varepsilon(2 - 1/0.74sa)\mathscr{L}/(1 + \mathscr{L}), \quad 0.74sa > 1 \tag{7.66}$$

where we have written $s$ for the cutoff frequency which we wrote before as $(\varepsilon/\nu)^{1/2}$.

Now, the presence of an additional mechanism for energy dissipation at high frequencies does not imply that the over-all dissipation of energy is increased. As pointed out in Section 1.7, it is an experimentally observed characteristic of turbulent flows that, so long as the dissipation is confined to the small scale motions, the larger scales are dominated by inertia, and are unaffected by the nature and efficiency of the dissipating mechanism. Hence, if the efficiency of the dissipative mechanism is increased (always presuming that it is confined to the small scales), the scale at which it is effective is increased to keep the total amount dissipated the same. If the nature of the mechanism is changed, the detailed structure of the spectrum at high frequencies will change, but the low-frequency part of the spectrum will remain unchanged.

In expression (7.66), the $\varepsilon$ which appears is the rate at which energy is being cascaded down the spectrum, which will not be changed by the presence of the particles. The cutoff frequency $s$, on the other hand, will no longer be given by $(\varepsilon/v)^{1/2}$, but will be reduced by the presence of the particles. In the same way the viscous dissipation will be given by $vs^2$, where $v$ is the value for the fluid *with* particles. As an aside, we should mention that the *dynamic* viscosity is essentially unchanged by the presence of the particles; the ratio (with particles)/(without particles) $\cong 1 + 5C/2$ [7.1], and with $C \cong 10^{-3}$ we can ignore this effect.

Since the viscous dissipation plus the dissipation associated with the particles must dissipate all of $\varepsilon$, we may write

$$vs^2 + 1.18\varepsilon(2 - 1/0.74sa)\mathscr{L}/(1 + \mathscr{L}) = \varepsilon \qquad (7.67)$$

which provides an equation to determine $s$. The most convenient form in which to place (7.67) involves the ratio of $s$ to its value when $\mathscr{L} = 0$, say $\hat{s} = s(v_f/\varepsilon)^{1/2}$, where $v_f$ is the kinematic viscosity for the particle-free fluid. We find another parameter, say $\gamma = 0.74a(\varepsilon/v_f)^{1/2}$, the product of the particle time scale (7.2) and the cutoff frequency in the particle-free fluid, with the factor 0.74 in (7.67) included here to simplify the final result. We can estimate

$$a(\varepsilon/v_f)^{1/2} = (d/\eta)^2 \varrho_p/18\varrho_f \qquad (7.68)$$

where, of course, $\eta$ is determined from $v_f$. If $d/\eta \sim 0.1$, $\varrho_p/\varrho_f \sim 10^3$, then (7.68) gives a value of 0.56; for denser particles it may rise to the order of 2.

With these definitions, (7.67) becomes

$$\mathscr{L} = (1 - \hat{s}^2)/(1.36 - 1.18/\gamma\hat{s}) . \qquad (7.69)$$

Of more relevance in a shear flow is the cutoff viscous wave number, $k_d$. This is proportional to $(s/v)^{1/2}$. If we take its ratio to the value at $\mathscr{L} = 0$, which is proportional to $(\varepsilon/v_f^3)^{1/4}$, and designate the ratio by $\hat{k}_d$, we can write

$$\hat{k}_d = (\hat{s}(1 + \mathscr{L}))^{1/2} . \qquad (7.70)$$

Now, there are two conflicting phenomena here again: the increased density due to the presence of the particles reduces the effective $v$, and tends to increase $k_d$, while the increased dissipation tends to reduce it.

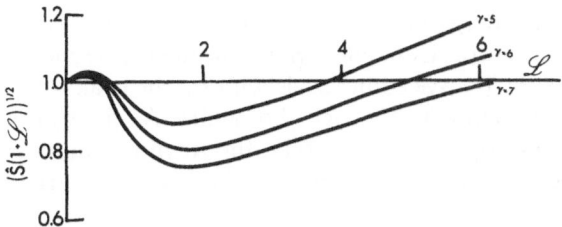

Fig. 7.5. Variation of dimensionless cutoff wavenumber with loading, for various values of the particle/fluid time scale ratio

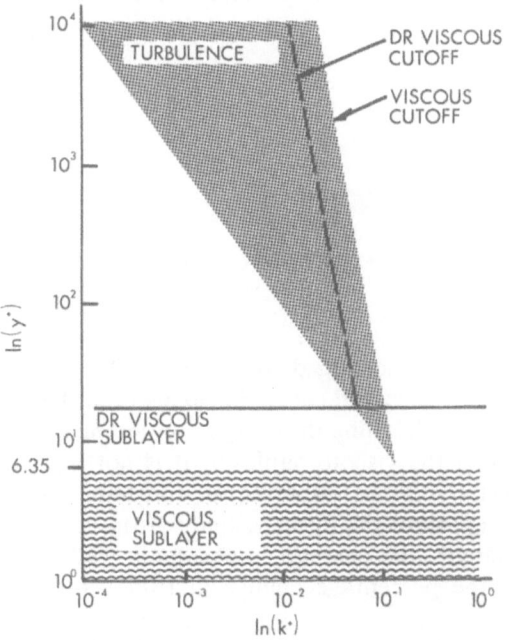

Fig. 7.6. Scaling relations in the viscous and inertial sublayers with and without polymers, from [7.25]. *DR* refers to drag reduction; the *DR* viscous cutoff and sublayer indicated correspond to an increase in viscosity in the turbulent part of the fluid, without a corresponding increase in the viscous sublayer

From (7.69) and (7.70), we find that there is a relatively narrow range of values of $\mathscr{L}$ in which, for sufficiently large values of $\gamma$, modest reductions in $k_d$ occur (see Fig. 7.5).

In the wall region of a particle-free turbulent shear flow (Section 2.3), the occurrence of turbulence may be plotted as in Fig. 7.6. The abscissa is wave number, scaled with $u_\tau$ and $v$ and the ordinate is distance from the wall, scaled in the same way. The turbulence is contained in the shaded region. That is, the energy-containing eddies scale with distance from the wall; the peak of the spectrum corresponds to the left boundary of the shaded region. The dissipative eddies scale with $k_d$; the peak of the dissipation spectrum corresponds to the right boundary. As the wall is approached, the two scales approach each other, the Reynolds number dropping accordingly, and in the viscous sublayer the two scales are of the same order.

Now, the presence of particles evidently can (in the right parameter range) reduce the value of $k_d$. Away from the wall, a reduction in the value of $k_d$ will have no effect on the energy containing eddies, and hence will leave the momentum transport mechanism, and consequently, the

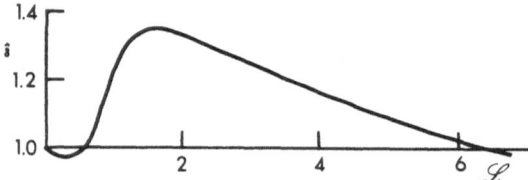

Fig. 7.7. Variation of dimensionless viscous sublayer thickness with loading, for a rather large value of the particle/fluid time scale ratio

slope of the mean velocity profile, unchanged. Just outside the viscous sublayer, however, this reduction in $k_d$ will cause the apex of the shaded region to move farther from the wall, killing the energy-containing eddies there, and effectively thickening the viscous sublayer. It is not hard to show that a thickening of the viscous sublayer at constant $u$ corresponds to a reduction in drag, since the mean velocity at the sublayer edge increases. For small changes, the percentage change in wall shear stress is about the same as the percentage change in sublayer thickness.

Again, we must be very careful to consider the effect of the change in density. If the sublayer is thicker than for a clear fluid of the same density, drag has not necessarily been reduced; this will happen only if the sublayer has been thickened relative to the thickness in the *suspending fluid* (without particles).

If we define the sublayer thickness by the intersection of the right and left edges of the diagram for the flow with particles corresponding to Fig. 7.6, we can write

$$\hat{\delta} = [\hat{s}(1 + \mathscr{L})]^{-2/3} \tag{7.71}$$

where $\hat{\delta}$ is the ratio of sublayer thickness to that in the clear fluid. $\hat{\delta}$ is essentially equal to the inverse of the drag ratio. This is plotted in

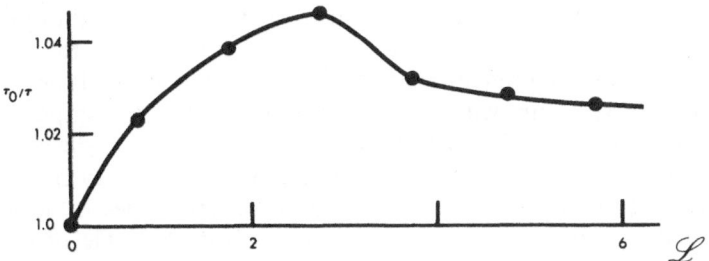

Fig. 7.8. Effect of solid particles on pressure drop (expressed as a ratio of wall stresses) as a function of loading from [7.29]. 35 micron MgO in air in a smooth 5 in. dia cylindrical tube. $U = 140$ fps, Reynolds number based on air $2.9 \times 10^5$

Fig. 7.7 for a value of $\gamma = 7$, $\gamma$ being evaluated at the distance from the wall corresponding to the sublayer thickness (as defined above) in the clear fluid. Relations corresponding to the clear fluid (for $\varepsilon$, etc.) were used.

For a qualitative comparison, Fig. 7.8 shows a measured reduction in drag due to particles. The particle size is too large, and the concentration too great for the analysis to be applicable but the general shape and range of $\mathscr{L}$ is qualitatively similar.

### 7.3.3 Neutrally Buoyant Particles

If a particle is neutrally buoyant, it can neither fall out, producing an artificial buoyancy, nor dissipate extra energy through inertia. Yet flows of neutrally buoyant particles are observed to display drag reduction (cf. [7.26] and the references therein).

Certainly if the particles are spherical and small relative to $\eta$, we would expect no influence, and experiment bears this out. Anomalous effects are observed only in flows in which at least one dimension of the particle exceeds the Kolmogorov length scale near the wall. We can understand this in the following way: if the particles are much smaller than $\eta$, then the velocity field in their vicinity can be regarded as a homogeneous deformation. With fallout and inertia negligible, only the effect on viscosity is important, and if the velocity field is a homogeneous deformation, the classical formulas of [7.1.4] can be used. These indicate that at the concentrations used moderate changes in viscosity should occur. In any event since these changes in viscosity would affect equally the turbulence and the viscous sublayer, they would leave a diagram such as Fig. 7.6 unchanged, and hence could not result in a reduction in drag.

The observation that the particles are larger than $\eta$ in some part of the flow, but cause a relatively small change in viscosity, suggests the explanation for the observed phenomena. If the change in viscosity is negligible, the sublayer structure will remain essentially unchanged. In particular, the normalization of Fig. 7.6, if based on the viscosity in the sublayer, will remain unchanged. On the other hand, particles may be expected to exert a strong damping effect on turbulent eddies of scale smaller than the particles. As the wall is approached, the scale of the eddies is progressively reduced. At some point the particle size is reached, a fixed size being represented by a vertical line in Fig. 7.6. Again, the momentum transport and hence the velocity profile should be unaffected until the *energy-containing eddies* reach the particle size, that is, until the vertical line crosses the left boundary of the shaded region in Fig. 7.6. We may expect that eddies below that point will be strongly attenuated, and the sublayer effectively thickened (in the absence of turbulent momentum transport in this region, the mean velocity profile will break farther from the wall), resulting in drag reduction.

This explanation predicts that the slope of the mean velocity profile in the logarithmic region will remain unchanged (since the turbulence there is dominated by inertia), whereas observations indicate that it is usually increased. To a certain extent this result may be due to experimental technique, since the measurements usually cover only the defect region of the profile (Section 2.3 and 1.15), or it may be a result of small buoyant effects (which we have seen produce an increase in slope). It may also be an effect of interaction; the concentrations are usually large enough so that interaction cannot be neglected.

With the larger particles there is also the possibility of migration out of the wall region, which would have a beneficial effect to the extent that the viscosity in the wall region would not be increased, while that in the turbulent part would.

## 7.4 Drag Reduction by Polymer Additives

A relatively complete summary of the characteristics of this phenomenon can be found in [7.25]. Briefly, the phenomenon takes place in solutions, as dilute as a few parts per million by weight, of linear flexible polymers having molecular weights above $10^5$. Reductions of drag by as much as 80% can be achieved. The shape of the mean velocity profile in the logarithmic region remains unchanged, but the sublayer appears to be thickened. There is a well-defined onset; that is, there is in a given experiment a well-defined speed below which there is no observable difference from the flow of the solvent.

A careful examination of the evidence [7.25] suggests that probably the phenomenon is not dependent on interaction of the molecules (i.e., can take place in arbitrarily dilute solutions; however, there may be some question even about this from recent unpublished measurements of BERMAN), but that most observations are of flows in which interaction is not negligible. That this should be so at such low concentrations is at first surprising; however, when a molecule is placed in solution, it expands considerably, forming a tangled ball (mostly filled with solvent). The *volume* fraction occupied by these balls may be close to unity at *weight* fractions of a few tens of parts per million if the molecular weight is high enough ($\sim 10^6$). As we will see, flow extends the molecules so that the volume fraction of circumscribed spheres may be much larger.

The influence of the suspended molecules on the viscosity in shear is very small at these concentrations, usually causing an increase of no more than a percent or so.

The molecules are several orders of magnitude below the Kolmogorov length scale in size in the unexpanded state; even when expanded they are smaller than $\eta$.

We will not attempt here to examine all aspects of this interesting phenomenon, but will limit ourselves to those aspects which have certain similarities to flows with particles.

### 7.4.1 Behavior of Macromolecules in Homogeneous Deformation

Due to the negligibility of inertia, and the negligible density differences, we may presume that isolated molecules always move with the fluid in which they are imbedded (ignoring thermal agitation). Thus, we expect a molecule to rotate with the local vorticity. If the fluid motion in which the molecule is imbedded is not a rigid motion, however, the molecule (being an elastic structure) will in general not be able to follow, and there will be relative flow over various parts of the molecule, producing drag forces on the molecule. For example, consider a molecule in equilibrium in a pure plane strain. The center of drag (presumably the same as the center of mass) will see no relative motion, but there will be relative motion of the fluid past other parts of the molecule, proportional to their distance from the center of drag. The molecule will then experience a force tending to elongate it in the direction of the positive strain rate. There is a restoring force (called the entropy restoring force) arising from thermal agitation, which acts like a spring force, attempting to restore the molecule to spherical symmetry. If the strain rate is large enough, however, the drag forces exceed the restoring force, and the molecule expands. Just how far it expands is a difficult question, since the restoring force is certainly not linear as the molecule becomes untangled.

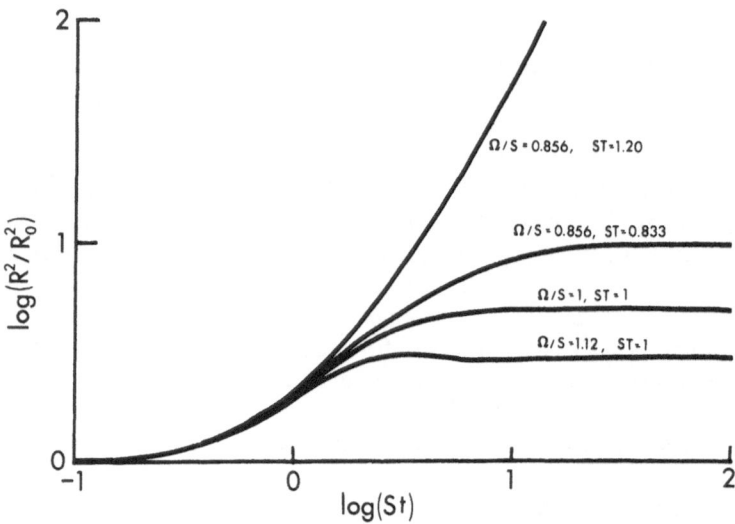

Fig. 7.9. Mean-square molecular radius as a function of time in a two-dimensional flow, for various values of vorticity and strain rate (relative to the relaxation time $T$) from [7.25]

There is evidence, however, that the expansion can become nearly complete—that is, the molecule may become almost totally elongated.

The restoring force (for small departures from spherical symmetry) is measured by the terminal relaxation time $T$, the time constant for the (exponential) return to spherical symmetry. If $S$ is the strain rate, a typical value of $e_{il}$ defined by (1.7), the expansion takes place when $2ST$ first exceeds unity.

The drag force due to the relative velocity over different parts of the molecule represents a dissipation of energy. The greater the distance from the center of drag of the most remote part of the molecule, the greater the drag, and the greater the dissipation. Thus, there is an increase in viscosity when the molecules expand, and since the expansion can be very great, the increase in viscosity can be also. The viscosity is measured by the relative increase in viscosity per unit concentration, the *intrinsic viscosity*. In a pure strain, the intrinsic viscosity increases by a factor of roughly the number of submolecular units in the chain, which may be of the order of $10^4$ or more.

In a mixed flow field, containing both strain and rotation, we find another phenomenon. Since the molecule rotates with the average angular velocity of the fluid, it is in general rotated past the direction of maximum positive strain rate. Since any given direction on the molecule spends less time aligned with the strain rate, the strain rate must be more intense to produce the same extension. The criterion for large expansion becomes in two dimensions $2T(S^2 - \Omega^2)^{1/2} \geqq 1$. In three dimensions there are so many possible orientations of vorticity $2\Omega$ relative to the principal axes of strain rate that it is not possible to give a simple criterion, though the same principles apply. Note that, in a simple shear, $\Omega = -S$, and consequently the criterion for full expansion can never be attained.

Finally, the conclusions above are somewhat modified by unsteadiness. The expansion, even supposing that the criterion is satisfied, takes significant time, rising exponentially with a time constant of the order of $S$. In an unsteady flow full expansion is never achieved, though there is a marked difference between the expansions achievable below and above the expansion criterion. Figure 7.9 shows mean-square molecular radius as a function of time in a two-dimensional flow. It is evident that, when the expansion criterion is not exceeded, the total change in radius is of the order of a factor of three at most, while when the criterion is exceeded, the growth is exponential.

### 7.4.2 Macromolecules in Turbulent Flows

When a macromolecule is in the fully turbulent part of a turbulent flow it is subjected to a random combination of vorticity and strain rate, with both the intensity of each and the orientation of the axes of one relative to the other fluctuating widely. The mean square level of strain rate is equal to the mean square level of vorticity in a high Reynolds number turbulent flow [1.15], and both scale with $\varepsilon/\nu$; in a high Reynolds number flow, the two are uncorrelated, and both have distributions with longer tails (higher kurtosis) than the Gaussian. Hence, the probability of finding regions of relatively rotation-free straining is good. We may expect that a wandering molecule will be expanded if some mean square measure of the excess of instantaneous strain rate over instantaneous vorticity exceeds a threshold. We would also expect that the criterion would contain a measure of the persistence of the relatively rotation-free straining fields, since if the fields were very impermanent it would have an effect similar to a large vorticity. That is, the molecule would not be significantly stretched before the field reversed.

This complex problem is analyzed in [7.27]. There it is found that the criterion for expansion is $2T\overline{|\lambda|^2}\mathscr{T} \geqq 1$, where $\lambda$ is an eigenvalue (possibly complex) of $u_{i,j}$. In two dimensions $\overline{|\lambda|^2} = |S^2 - \Omega^2|$, but in three dimensions no such simple expression is possible. $\mathscr{T}$ is the Lagrangian integral scale of persistence of these strain fields. It is clear that $\overline{|\lambda|^2}$ must scale with $\varepsilon/\nu$. The value of $\mathscr{T}$ is less clear. There is a number of possible time scales, ranging from $l/u'$, the Lagrangian integral time scale for the energy containing motion, to $\lambda_g/u'$, the time scale of the fluctuating strain rate (where $\lambda_g$ is the Taylor microscale), to $\lambda_g/u'(R_\lambda/30)^{1/2}$ (where $R_\lambda = \lambda_g u'/v$), the Lagrangian time microscale [7.27]. All of these may be expressed as $l/u'$ times some power of $R_l$; for example, the Lagrangian time microscale (the most likely possibility) becomes $(l/u')R_l^{-1/4}$. Hence, the criterion becomes, in a boundary layer flow, $u_\tau^2 T/v \geqq b$, where $b$ is a number weakly dependent on Reynolds number, perhaps $\propto R_l^{1/4}$.

Thus, in a boundary layer flow, we expect the molecules to become extended where $u_\tau^2 T/v$ exceeds some critical value (weakly dependent on Reynolds number), being unextended below that size. Experiment [7.28] indicates that the onset of the drag reduction phenomenon does occur at a critical value of $u_\tau^2 T/v$ which scales properly when $v$ is varied, and which increases slowly with Reynolds number. Unfortunately there is also a dependence on concentration (and probably on interaction) which clouds the issue, as well as dependence on polydispersity. That is, samples of polymers do not consist of a single molecular weight, but of a spectrum of weights, and presumably only part of the distribution is active at onset of drag reduction.

It is clear that there are still a number of aspects that we do not understand; nevertheless, the overall picture is consistent with expansion of the macromolecules, with consequent increase of intrinsic (and actual) viscosity at onset.

### 7.4.3 Effect of Macromolecules in Shear Flows

Just as in the case of light particles with inertia (Subsection 7.3.2) we have a mechanism which can significantly increase the viscosity in the turbulent part of the flow, while leaving it virtually unchanged in the viscous sublayer. That is, since the sublayer is a simple shear, in which vorticity $\cong$ strain rate, we do not expect that the molecules will be expanded there; diffusion is relatively slow, so that expanded molecules from the turbulent part of the flow will contract again by the time they have been transported down into the sublayer.

As we have seen (above) an increased viscosity in the turbulent part of the flow, and not in the sublayer, causes an increase in the minimum scale of the turbulence; this has no effect on the energy-containing eddies except where they are smallest, just outside the sublayer. Hence, the slope in the log region will be unchanged (the turbulence here is still inertia dominated). The suppression of the eddies just outside the sublayer reduces the Reynolds stress there, moving the point at which the velocity profile slope changes farther from the wall, effectively thickening the sublayer and reducing the drag.

The expansion of the molecules permits an interesting feedback mechanism: the increased viscosity due to the expanded molecules reduces the strain rate. Since a certain level of strain rate is necessary to expand the molecules, too great a reduction will cause their collapse; thus, the strain rate is held quite close to the critical value, and an increase in concentration results in the molecules collapsing just enough to hold the viscosity (and hence the strain rate and drag reduction) about the same. Of course, at low concentrations, the molecules are fully expanded, since the increased viscosity is not enough to reduce the strain rate enough to permit collapse. As a consequence of this, the drag reduction at first increases with concentration, but then saturates, and does not increase further [7.25].

The structure of the wall region remains essentially the same. The large eddies which are there, whose scale is determined by the sublayer thickness, grow with the sublayer. The velocity fluctuations which those eddies produce from the mean velocity profile actually increase (as opposed to the smaller scale fluctuations which bear the Reynolds stress, which are suppressed), since the net mean velocity difference over the height of an eddy is increased [7.25]. It seems very likely that a similar phenomenon takes place in drag-reducing flows with particles, although no observations exist.

*Acknowledgements*

Much of the material in this chapter derives from work which was supported in part by: the U.S. Office of Naval Research, Fluid Dynamics Branch; the U.S. National Science Foundation. Atmospheric Sciences Section; the U.S. Environmental Protection Agency through its Select Research Group in Meteorology; and the Naval Sea Systems Command through the Garfield Thomas Water Tunnel.

# References

7.1   G.K.BATCHELOR, J.J.GREEN: J. Fluid Mech. **56**, 401 (1972)
7.2   H.HASIMOTO: J. Fluid Mech. **5**, 317 (1959)
7.3   B.D.COLEMAN, W.NOLL: Ann. NY Acad. Sci. **89**, 672 (1961)
7.4   G.K.BATCHELOR: J. Fluid Mech. **46**, 813 (1971)
7.5   D.H.SLADE (ed.): *Meteorology and Atomic Energy* 1968 (USAEC Division of Technical Information Extension, Oak Ridge, Tennessee 1968)
7.6   L.T.MATVEEV: *Fundamentals of General Meteorology: Physics of the Atmosphere* (Israel Program for Scientific Translations, Jerusalem 1967)
7.7   E.J.HINCH, L.G.LEAL: J. Fluid Mech. **52**, 683 (1972)
7.8   E.J.HINCH, L.G.LEAL: J. Fluid Mech. **57**, 753 (1973)
7.9   L.G.LEAL, E.J.HINCH: J. Fluid Mech. **46**, 685 (1971)
7.10  L.G.LEAL, E.J.HINCH: J. Fluid Mech. **55**, 745 (1972)

7.11  J. J. BIKERMAN, J. M. PERRI, R. B. BOOTH, C. C. CURRIE: *Foams: Theory and Industrial Applications* (Reinhold Publishing Corp., New York 1953)

7.12  W. C. MIH, C. K. CHEN, J. F. ORSBORN: *Bibliography of Solid-Liquid Transport in Pipelines* (Albrook Hydraulic Laboratory, College of Engineering Research Division, Washington State University, Pullman, Washington 1971)

7.13  S. YALIN: *Mechanics of Sediment Transport* (Pergamon Press, New York 1972)

7.14  P. G. SAFFMAN: J. Fluid Mech. **22**, 385 (1965)

7.15  J. L. LUMLEY, H. A. PANOFSKY: *The Structure of Atmospheric Turbulence* (Interscience Publishers, New York 1964)

7.16  M. I. YUDINE: Advances in Geophysics **6**, 185 (1959)

7.17  G. T. CSANADY: J. Atmospheric Sci. **2**, 201 (1963)

7.18  S. CORRSIN, J. L. LUMLEY: Appl. Sci. Res. A **6**, 114 (1965)

7.19  J. L. LUMLEY: PhD Thesis (Aeronautics), The Johns Hopkins University, Baltimore (1957)

7.20  M. VAN DYKE: *Perturbation Methods in Fluid Mechanics* (Academic Press, New York 1964)

7.21  J. L. LUMLEY: In: A. FAVRE (Ed.): *The Mechanics of Turbulence* (Gordon and Breach, New York 1964)

7.22  W. H. SNYDER, J. L. LUMLEY: J. Fluid Mech. **48**, 41 (1971)

7.23  J. L. LUMLEY: *Lecture Notes in Physics*, Vol. 12: *Statistical Models and Turbulence* (Springer, Berlin, Heidelberg, New York 1972)

7.24  W. C. MIH: *Sedimentation* (H. W. SHEN, P. O. Box 606, Fort Collins, Colorado 1972)

7.25  J. L. LUMLEY: J. Polymer Sci.: Macromolecular Reviews **7**, 263 (1973)

7.26  R. C. VASELESKI, A. B. METZNER: AIChE J. **20**, 301 (1974)

7.27  L. LUMLEY: Symposia Mathematica **9**, 315 (1972)

7.28  N. S. BERMAN, W. GEORGE: Phys. Fluids **17**, 250 (1974)

7.29  S. L. SOO: *Fluid Dynamics of Multiphase Systems* (Blaisdell, Waltham, Mass. 1967)

7.30  G. GOLITSYN: J. Atmospheric Sci. **31**, 1917 (1974)

# Additional References with Titles

(Updated to early 1978)

Numbers in square brackets are relevant page numbers of main text.

Anon: Flow separation. AGARD Conf. Proc. 168 (1975) [68, 144]

Anon: Three dimensional and unsteady separation at high Reynolds numbers. AGARD Lecture Series No. 94 (1978) [90]

B. U. Achia, D. W. Thompson: Structure of the turbulent boundary in drag-reducing pipe flow. J. Fluid Mech. 81, 439 (1977) [321]

N. Afzal, R. Narasimha: Axisymmetric turbulent boundary layer along a circular cylinder at constant pressure. J. Fluid Mech. 74, 113 (1976) [76]

Q. A. Ahmad, R. E. Luxton, R. A. Antonia: Characteristics of a turbulent boundary layer with an external turbulent uniform shear flow. J. Fluid Mech. 77, 369 (1976) [65]

S. F. Ali: Structure of the turbulence in the plane wake behind a heated flat plate. Ph.D. Thesis, The Johns Hopkins University, Baltimore (1975) [268]

P. S. Andersen, W. M. Kays, R. J. Moffat: Experimental results for the transpired turbulent boundary layer in an adverse pressure gradient. J. Fluid Mech. 69, 353 (1975) [55]

J. C. André: Une approche statistique de la turbulence inhomogene: Thése, Docteur ès Sciences, Université Pierre et Marie Curie, Paris (1976) [259]

J. C. André, G. de Moor, P. Lacarrere, R. du Vachet: Turbulence approximation for inhomogeneous flows. Part I: The clipping approximation. Part II: The numerical simulation of a penetrative convection experiment. J. Atmos. Sci. 33, 476, 482 (1976) [259]

R. A. Antonia, A. Prabhu, S. E. Stephenson: Conditionally sampled measurements in a heat turbulent jet. J. Fluid Mech. 72, 455 (1975) [56, 267]

R. A. Antonia, H. Q. Danh, A. Prabhu: Response of a turbulent boundary layer to a step change in surface heat flux. J. Fluid Mech. 80, 153 (1976) [271]

R. A. Antonia, C. W. van Atta: Statistical characteristics of Reynolds stresses in a turbulent boundary layer. AIAA J. 15, 71—75 (1977)

R. A. Antonia, C. W. van Atta: Structure functions of temperature fluctuations in turbulent shear flows. J. Fluid Mech. 84, 561—580 (1978) [23]

S. P. S. Arya: Buoyancy effects in a horizontal flat-plate boundary layer. J. Fluid Mech. 68, 321 (1975) [257]

M. A. Badri Narayanan, S. Rajagopalan, R. Narasimha: Experiments on the fine structure of turbulence. J. Fluid Mech. 80, 237 (1977) [13]

R. G. Batt: Some measurements on the effects of tripping the two-dimensional shear layer. AIAA J. 13, 245 (1975) [43]

P. M. Bevilaqua, P. S. Lykoudis: Some observations on the mechanism of entrainment. AIAA J. 15, 1194—1196 (1977) [57]

A. J. Bilanin, M. E. Teske, C. du P. Donaldson, R. S. Snedeker: Viscous effects in aircraft trailing vortices. NASA Symposium on Wake Vortex Minimization, Washington (1976) p. 55 [46]

R. F. Blackwelder, R. E. Kaplan: On the wall structure of the turbulent boundary layer. J. Fluid Mech. 76, 89—112 (1976) [56]

P. Bradshaw: A skin-friction law for compressible turbulent boundary layers based on the full Van Driest transformation. AIAA J. 15, 212 (1977) [97]

P. Bradshaw: Compressible turbulent shear layers. Ann. Rev. Fluid Mech. 9, 33 (1977) [90]

G. L. Brown, A. S. W. Thomas: Large structure in a turbulent boundary layer. Phys. Fluids 20, S 243 (1977)

D. M. Bushnell, A. M. Cary, Jr., B. B. Holley: Mixing length in low Reynolds number compressible turbulent boundary layers. AIAA J. 13, 1119 (1975) [60, 94, 97]

T. Cebeci, K. Kaups, J. A. Ramsey: A general method for calculating three-dimensional compressible laminar and turbulent boundary layers on arbitrary wings. NASA CR-2777 (1977) [84, 220]

T. Cebeci, A. Khattab: Prediction of turbulent-free-convective heat transfer from a vertical flat plate. Trans. ASME 97 C, 469 (1975) [189]

C. P. Cerasoli: Experiments on buoyant-parcel motion and the generation of internal gravity waves. J. Fluid Mech. 86, 247 (1978) [175]

T. L. Chambers, D. C. Wilcox: Critical examination of two-equation turbulence closure models for boundary layers. AIAA J. 15, 821 (1977) [197]

F. H. Champagne: The fine-scale structure of the turbulent velocity field. J. Fluid Mech. 86, 67 (1978) [24]

F. H. Champagne, C. A. Friehe, J. C. LaRue, J. C. Wyngaard: Flux measurements, flux estimation techniques, and fine-scale turbulence measurements in the unstable surface layer over land. J. Atmos. Sci. 34, 515—530 (1977) [173]

N. V. Chandrasekhara Swamy, P. Bandyopadhyay: Mean and turbulence characteristics of three-dimensional wall jets. J. Fluid Mech. 71, 541—562 (1975) [85]

C. Chandrsuda, R. D. Mehta, A. D. Weir, P. Bradshaw: Effect of free-stream turbulence on large structure in turbulent mixing layers. J. Fluid Mech. 85, 693 (1978) [43]

C. J. Cheu, W. Rodi: A mathematical model for stratified turbulent flows and its application tion to buoyant jets. Proc. 16th. Congress Int. Assoc. for Hydraulic Res., São Paulo (1975) [250, 278]

H. W. Coleman, R. J. Moffat, W. M. Kays: The accelerated fully rough turbulent boundary layer. J. Fluid Mech. 82, 507—528 (1977) [72]

P. Cooper, E. Reshotko: Turbulent flow between a rotating disk and a parallel wall. AIAA J. 13, 573 (1975) [119]

G. M. Corcos, F. S. Sherman: Vorticity concentration and the dynamics of unstable freeshear layers. J. Fluid Mech. 73, 241 (1976) [43]

J. Counihan: Adiabatic atmospheric boundary layers: a review and analysis of data from the period 1880–1972. Atmos. Environment 9, 871 (1975) [173]

J. Cousteix, R. Michel: Theoretical analysis and prediction of three-dimensional turbulent boundary layers. ONERA T.P. No. 1975-43 (1975) [90]

M. D. Crawford, W. M. Kays, R. J. Moffat: Heat transfer to a full-coverage film-cooled surface with 30-deg slant-hole injection. NASA CR-2786 (1976) [271]

D. E. Daney: Turbulent natural convection of liquid deuterium, hydrogen and nitrogen within enclosed vessels. Intern. J. Heat Mass Transf. 19, 431 (1976) [189]

A. E. Davies, J. F. Keffer, W. D. Baines: Spread of a heated plane turbulent jet. Phys. Fluids 18, 770 (1975) [267]

W. R. Davis: Three-dimensional boundary layer computation on the stationary end-walls of centrifugal turbomachinery. Trans. ASME 98 I, 431 (1976) [161]

I. N. Dorokhov, V. V. Kafarov, R. I. Nigmatulin: Methods of continuous medium mechanics for defining polyphase multicomponent mixtures with chemical reactions and heat- and mass-transfer. J. Appl. Math. Mech. (PMM) 39, 461 (1975) [290]

J. P. Dussauge, J. Gaviglio: Comportement d'un ecoulement turbulent de proche sillage, a vitesse supersonique. (Behaviour of a near-wake turbulent flow at supersonic speed). Rech. Aerospatiale No. 1975-3, 145 (1975) [97]

A. ELSENAAR, B. VAN DEN BERG, J. P. F. LINDHOUT: Three-dimensional separation of an incompressible turbulent boundary layer on an infinite swept wing. AGARD Conf. Proc. 168, 34 (1975) [88]

T. K. FANNELOP, P. A. KROGSTAD: Three-dimensional turbulent boundary layers in external flows. A report on Euromech 60. J. Fluid Mech. 71, 815 (1975) [85]

J. H. FERZIGER: Large eddy numerical simulations of turbulent flows. AIAA J. 15, 1261—1267 (1977) [204]

P. J. FINLEY, KHOO CHONG PHOE, CHIN JECK POH: Velocity measurements in a thin turbulent water layer. Houille Blanche 21, 713 (1966) [136]

D. E. FITZGARRALD: An experimental study of turbulent convection in air. J. Fluid Mech. 73, 693 (1976) [186]

L. S. FLETCHER, D. G. BRIGGS, R. H. PAGE: A review of heat transfer in separated and reattached flows. Israel J. Tech. 12, 236 (1974) [103] (Ref. 2.222)

P. FREYMUTH: Search for final period of decay of the axisymmetric turbulent wake. J. Fluid Mech. 68, 813 (1975) [269]

L. FULACHIER: Contribution a l'etude des analogies des champs dynamique et thermique dans une couche limite turbulente. Thése Docteur ès Sciences, Université de Provence (1972) [270]

R. A. McD. GALBRAITH, M. R. HEAD: Eddy viscosity and mixing length from measured boundary layer developments. Aeronaut. Quart. 26, 133 (1975) [55]

M. D. GIBSON, B. E. LAUNDER: On the calculation of horizontal turbulent shear flows under gravitational influence. ASME J. Heat Transf. 98, 81 (1976) [250]

P. S. GRANVILLE: A modified law of the wake for turbulent shear flows. Trans. ASME 98 I, 578 (1976)

D. E. GUITTON, B. G. NEWMAN: Self-preserving wall jets over convex surfaces. J. Fluid Mech. 81, 155 (1977) [113]

E. GUTMARK, I. WYGNANSKI: The planar turbulent jet. J. Fluid Mech. 73, 465 (1976) [46]

K. HANJALIC, B. E. LAUNDER: Contribution towards a Reynolds stress closure for low-Reynolds-number turbulence. J. Fluid Mech. 74, 593 (1976) [197]

V. G. HARRIS, J. A. H. GRAHAM, S. CORRSIN: Further experiments in nearly homogeneous turbulent shear flow. J. Fluid Mech. 81, 657 (1977) [199]

M. R. HEAD: Equilibrium and near-equilibrium turbulent boundary layers. J. Fluid Mech. 73, 1 (1976) [65]

R. A. HERRINGE, M. R. DAVIS: Structural development of gas-liquid mixture flows. J. Fluid Mech. 73, 97 (1976) [292]

G. A. HERZINGER: Note on the settling of small particles in a recirculating flow. AIAA J. 13, 837 (1975) [293]

C. C. HORSTMAN, G. S. SETTLES, I. E. VAS, S. M. BOGDONOFF, C. M. HUNG: Reynolds number effects on shock-wave turbulent boundary-layer interactions. AIAA J. 15, 1152—1158 (1977) [94]

M. S. HOSSAIN, W. RODI: Influence of buoyancy on the turbulence intensities in horizontal and vertical jets. Proc. 1976 ICHMT Seminar on Turbulent Buoyant Convection, Dubrovnik, Yugoslavia (1976) [250, 278]

W. HUMPHRIES, J. H. VINCENT: An experimental investigation of the detention of airborne smoke in the wake bubble behind a disk. J. Fluid Mech. 73, 453 (1976) [130]

A. K. M. F. HUSSAIN, W. C. REYNOLDS: Measurements in fully developed turbulent channel flow. J. Fluids Eng. Trans. ASME 97, 568—580 (1975) [133]

R. G. JACKSON: Sedimentological and fluid-dynamic implications of the turbulent bursting phenomenon in geophysical flows. J. Fluid Mech. 77, 531—560 (1976) [54]

B. A. KADER, A. M. YAGLOM: Turbulent heat and mass transfer from a wall with parallel roughness ridges. Intern. J. Heat Mass Trans. **20**, 345—357 (1977)  [72]

J. C. KAIMAL, J. C. WYNGAARD, D. A. HAUGEN, O. R. COTE, Y. IZUMI, S. J. CAUGHEY, C. J. READINGS: Turbulence structure in the convective boundary layer. J. Atmos. Sci. **33**, 2152—2169 (1976)  [173]

L. H. KANTHA: Turbulent entrainment at the density interface of a two-layer stably-stratified fluid system. John Hopkins Geophys. Fluid Dyn. Lab. TR 75-1, AD-A 016 607/49 A (1975)  [187]

C. W. KITCHENS, R. SEDNEY, N. GERBER: The role of the zone of dependence concept in three-dimensional boundary layer calculations. Ballistic Research Lab. Rept. 1821, AD-A 016 896/3 GA (1975)  [87]

B. A. KOLOVANDIN: Transfer of scalar substance in turbulent shear flows. Proc. 1968 Summer School and Heat and Mass Transfer in Turbulent Boundary Layers: **1**, 359 (1972)  [259]

B. A. KOLOVANDIN, I. A. VATUTIN: Statistical transfer theory in non-homogeneous turbulence. Intern. J. Heat. Mass Transf. **15**, 2371 (1972)  [259]

J. P. KRESKOWSKY, S. J. SHARMROTH, H. MCDONALD: Application of a general boundary layer analysis to turbulent boundary layers subjected to strong favourable pressure gradients. Trans. ASME **97** I, 217 (1975)  [68]

A. J. LADERMAN, A. DEMETRIADES. Turbulent fluctuations in the hypersonic boundary layer over an adiabatic slender cone: Phys. Fluids **19**, 359 (1976)  [96]

R. G. LAMB, W. H. CHEN, J. H. SEINFELD: Numerico-empirical analyses of atmospheric diffusion theories. J. Atmos. Sci. **32**, 1794 (1975)  [185]

L. S. LANGSTON, M. L. NICE, R. M. HOOPER: Three-dimensional flow within a turbine cascade passage. Trans. ASME **99** A, 21 (1977)  [161]

J. C. LARUE, P. A. LIBBY: Statistical properties of the interface in the turbulent wake of a heated cylinder. Phys. Fluids **19**, 1864—1875 (1976)  [58]

B. E. LAUNDER, C. H. PRIDDIN, B. I. SHARMA: The calculation of turbulent boundary layers on curved and spinning surfaces. J. Fluids Eng. Trans. ASME **99**, 237 (1977)  [113]

O. LAWACZECK, K. A. BUTEFISH, H. J. HEINEMANN: Vortex streets in the wakes of subsonic and transonic turbine cascades. Revue Francaise de Mechanique, Supplement, 9 (also IAA abstract item A 77-20154)  [165]

R. H. C. LEE, P. M. CHUNG: Buoyancy effects on a turbulent shear flow. AIAA J. **13**, 1592 (1975)  [267]

M. A. LEMONE: Modulation of turbulence energy by longitudinal rolls in an unstable planetary boundary layer. J. Atmos. Sci. **33**, 1308—1320 (1976)  [178]

H. R. LESLIE (ed.): Reviews in viscous flow – Proceedings of the Lockheed-Georgia Viscous Flows Symposium. Lockheed-Georgia Co., Marietta, Ga., Rept. LG77ER0044 (1977)  [28]

R. P. LOHMANN: The response of a developed turbulent boundary layer to local transverse surface motion. Trans. ASME **98** I, 354 (1976)  [88]

R. R. LONG: A theory of mixing in a stably stratified fluid. J. Fluid Mech. **84**, 113—124 (1978)  [187]

R. R. LONG: The influence of shear on mixing across density interface. J. Fluid Mech. **70**, 305 (1975)  [187]

J. L. LUMLEY, O. ZEMAN, J. SIESS: The influence of buoyancy on turbulent transport. J. Fluid Mech. **84**, 581—597 (1978)  [238]

A. D. MCEWAN: Angular momentum diffusion and the initiation of cyclones. Nature **260**, 127 (1976)  [153]

L. M. MACK: A numerical method for the prediction of high-speed boundary layer transition using linear theory. NASA SP 347 (1975)  [80]

A. MARTELLUCI, A. L. LAGANELLI: Hypersonic viscous flow over a slender cone. I: Mean flow measurements. AIAA paper 74-533 (1974) [96]

H. U. MEIER: Investigation of the heat transfer mechanism in supersonic turbulent boundary layers. Wärme- u. Stoff. **8**, 159 (1975) [78, 96]

A. MELLING, J. H. WHITELAW: Turbulent flow in a rectangular duct. F. Fluid Mech. **78**, 289 (1976) [84]

G. L. MELLOR, H. J. HERRING: Simple eddy viscosity relations for three-dimensional turbulent boundary layers. AIAA J. **15**, 886—887 (1977) [220]

V. MIKULLA, C. C. HORSTMAN: Turbulence stress measurements in a non-adiabatic hypersonic boundary layer. AIAA J. **13**, 1606 (1975) [94]

V. MIKULLA, C. C. HORSTMAN: Turbulence measurements in hypersonic shock-wave boundary-layer interaction flows. AIAA J. **14**, 568—575 (1976) [94]

H. NAKAGAWA, I. NEZU: Prediction of the contributions to the Reynolds stress from bursting events in open-channel flows. J. Fluid Mech. **90**, 99 (1977) [50]

D. NAOT: Two-dimensional unidirectional turbulent flow in a local equilibrium. Phys. Fluids **18**, 1813 (1975) [126]

R. NARASIMHA, P. R. VISWANATH: Reverse transition at an expansion corner in supersonic flow. AIAA J. **13**, 693 (1975) [82]

J. C. J. NIHOUL, F. C. RONDAY: Coherent structures and negative viscosity in marine turbulence. J. de Mécanique **15**, 119 (1976) [179]

K. NOTO, R. MATSUMOTO: Turbulent heat transfer by natural convection along an isothermal vertical flat surface. Trans. ASME **97**C, 621 (1975) [188]

G. R. OFFEN, S. J. KLINE: A proposed model of the bursting process in turbulent boundary layers. J. Fluid Mech. **70**, 209 (1975) [54]

D. D. PAPAILIOU, P. S. LYKOUDIS: Turbulent free convection flow. Intern. J. Heat Mass Transf. **17**, 161 (1974) [186]

K. PAPAILIOU, R. FLOT, J. MATHIEU: Secondary flows in compressor bladings. J. Eng. Power Trans. ASME **99**, 211—224 (1977) [161]

A. E. PERRY, C. J. ABELL: Asymptotic similarity of turbulence structures in smooth- and rough-walled pipes. J. Fluid Mech. **79**, 785 (1977) [133]

A. E. PERRY, B. D. FAIRLIE: A study of turbulent boundary layer separation and reattachment. J. Fluid Mech. **69**, 657 (1975) [68]

F. J. PIERCE, S. H. DUERSON, JR.: Reynolds stress tensors in an end-wall three-dimensional channel boundary layer. Trans. ASME **97**I, 618 (1975) [88, 161]

L. M. PISMEN, A. NIR: On the motion of suspended particles in stationary homogeneous turbulence. J. Fluid Mech. **84**, 193—206 (1978) [290]

B. R. RAMAPRIAN, B. G. SHIVA PRASAD: An experimental study of the effect of mild longitudinal curvature on the turbulent boundary layer. AIAA J. **15**, 189 (1977) [77]

A. K. RASTOGI, W. RODI: Three-dimensional calculation of heat and mass dispersion in open channel flows. Paper HMT 17-75. Proc. 3rd National Heat Transfer Conf., Indian Inst. Tech. Bombay (1975) [272]

K. REHME: Turbulence measurements in smooth concentric annuli with small radius ratios. J. Fluid Mech. **72**, 189 (1975) [137]

J. K. REICHERT, R. S. AZAD: Nonasymptotic behaviour of developing turbulent pipe flow. Canad. J. Phys. **54**, 268 (1976) [126]

A. J. REYNOLDS: The prediction of turbulent Prandtl and Schmidt numbers. Intern. J. Heat Mass Trans. **18**, 1055 (1975) [245]

H. K. RICHARDS, J. B. MORTON: Experimental investigation of turbulent shear flow with quadratic mean-velocity profiles. J. Fluid Mech. **73**, 165 (1976) [47]

W. RODI: A new algebraic relation for calculating the Reynolds stresses. ZAMM **56**, T 219 (1976) [261]

H. A. Rose: Eddy diffusivity, eddy noise and subgrid-scale modelling. J. Fluid Mech. **81**, 719 (1977) [204]

J. C. Rotta: A family of turbulence models for three-dimensional thin shear layers. DFVLR IB-251-76 A25. (Paper presented at the Symp. on Turbulent Shear Flows, Penn State Univ., 1977) [202]

B. Roux, P. Bontoux: Supersonic turbulent obundary layer in the symmetry plane of a cone at incidence. AIAA J. **13**, 705 (1975) [87]

M. W. Rubesin et al.: A critique of some recent second-order turbulence closure models for compressible boundary layers. AIAA Paper 77—128 (1977) [202]

J. Sabot, G. Comte-Bellot: Intermittency of coherent structures in the core region of fully developed turbulent pipe flow. J. Fluid Mech. **74**, 767—796 (1976) [128]

D. J. Schlien, S. Corrsin: Dispersion measurements in a turbulent boundary layer. Intern. J. Heat Mass Transf. **19**, 285 (1976) [185]

W. D. Schofield: Measurements in adverse-pressure gradient turbulent boundary layers with a step change in surface roughness. J. Fluid Mech. **70**, 573 (1975) [75]

H. C. Seetharam, W. H. Wentz, Jr.: Experimental investigation of subsonic turbulent separated boundary layers and wake on an airfoil. J. Aircraft **14**, 51—55 (1977) [68]

N. Seki, S. Fukusako, T. Hirata: Turbulent fluctuations and heat transfer for separated flow associated with a double step at entrance to an enlarged flat duct. J. Heat Transf., Trans. ASME **98**, 588—593 (1976) [152]

R. L. Simpson, D. B. Wallace: Laminarescent turbulent boundary layers: Experiments on sink flow. Project Squid TR SMU-1-PU (1975) [82]

R. L. Simpson: Characteristics of a separating incompressible turbulent boundary layer. AGARD Conf. Proc. 168, 14.1 (1975) [68, 140]

R. L. Simpson: Interpretating laser and hot-film anemometer signals in a separating boundary layer. AIAA J. **14**, 124, (1976) [68, 140]

R. L. Simpson, J. H. Strickland, P. W. Barr: Features of a separating turbulent boundary layer in the vicinity of separation. J. Fluid Mech. **79**, 553—594 (1977) [68]

J. A. Smith, J. F. Driscoll: The electron-beam fluorescence technique for measurements in hypersonic turbulent flows. J. Fluid Mech. **72**, 695 (1975) [93]

R. N. Smith, R. Grief: Turbulent transport to a rotating cylinder for large Prandtl or Schmidt numbers. Trans. ASME **97**C, 594 (1975) [118]

R. M. C. So: A turbulence velocity scale for curved shear flows. J. Fluid Mech. **70**, 37 (1975) [76]

K. R. Sreenivasan, R. Narasimha: Rapid distortion of axisymmetric turbulence. J. Fluid Mech. **84**, 497—516 (1978) [26]

J. R. Usher, A. D. D. Craik: Non-linear wave interactions in shear flows. J. Fluid Mech. **70**, 437 (1975) [80]

P. J. Sullivan: Dispersion of a line source in grid turbulence. Phys. Fluids **19**, 159 (1976) [185]

N. H. Thomas, P. E. Hancock: Grid turbulence near a moving wall. J. Fluid Mech. **82**, 481 (1977) [65]

S. M. Thompson, J. S. Turner: Mixing across an interface due to turbulence generated by an oscillating grid. J. Fluid Mech. **67**, 349—368 (1975) [187]

Y. Tsuji, A. Y. Morikawa: Turbulent boundary layer with pressure gradient alternating in sign. Aero. Quart. **27**, 15 (1975) [68]

G. C. Vliet, D. C. Ross: Turbulent natural convection on upward and downward facing inclined constant heat flux surfaces. Trans. ASME **97**C, 549 (1975) [188]

J. M. Wallace, R. S. Brodkey: Reynolds stress and joint probability density distributions in the u-v plane of a turbulent channel flow. Phys. Fluids **20**, 351—355 (1977) [41, 54]

R. D. Watson: Wall cooling effects on hypersonic transitional/turbulent boundary layers at high Reynolds numbers. AIAA Paper 75-834 (1975) [81, 97]

F. M. WHITE, R. C. LESSMANN, G. H. CHRISTOPH: A three-dimensional integral method for calculating incompressible turbulent skin friction. Trans. ASME 97I, 550 (1975) [212]

W. W. WILLMARTH: Pressure fluctuations beneath turbulent boundary layers. Ann. Rev. Fluid Mech. 7 (1975)   [7, 39]

W. W. WILLMARTH: Structure of turbulence in boundary layers. Adv. Appl. Mech. 15, 159 (1975)   [58]

K. G. WINTER: An outline of the techniques available for the measurements of skin friction in turbulent boundary layers. R.A.E. TM Aero. 1656 (1975)   [55, 96]

D. H. WOOD, R. A. ANTONIA: Measurements in a turbulent boundary layer over a d-type surface roughness. ASME J. Appl. Mech. 42, 591 (1976)

J. M. WU, C. H. CHEN, T. H. MOULDEN: Experimental study of compressible turbulent boundary layer separation as influenced by upstream disturbances. AIAA Paper 75-830 (1975)   [97]

S. T. B. YOUNG: Some turbulence measurements of a density stratified shear flow; a preliminary report. I.C. Aero. TN 73-102 (1975)   [242]

O. ZEMAN: The dynamics of entrainment in the planetary boundary layer; a study in turbulence modelling and parametrization. PhD Thesis, The Pennsylvania State Univ. (1975)   [259]

# Subject Index

Main references are in **bold** type; "def." means definition of subject, "diag." means diagram. Only introductory references are given for oft-mentioned subjects.

# Applied Physics

*A monthly journal*

| | |
|---|---|
| Board of Editors | **S. Amelinckx,** Mol. · **V. P. Chebotayev,** Novosibirsk<br>**R. Gomer,** Chicago, Ill. · **H. Ibach,** Jülich<br>**V. S. Letokhov,** Moskau · **H. K. V. Lotsch,** Heidelberg<br>**H. J. Queisser,** Stuttgart · **F. P. Schäfer,** Göttingen<br>**A. Seeger,** Stuttgart · **K. Shimoda,** Tokyo<br>**T. Tamir,** Brooklyn, N.Y. · **W. T. Welford,** London<br>**H. P. J. Wijn,** Eindhoven |
| Coverage | application-oriented experimental and theoretical physics:<br>*Solid-State Physics*     *Quantum Electronics*<br>*Surface Sciences*     *Laser Spectroscopy*<br>*Solar Energy Physics*     *Photophysical Chemistry*<br>*Microwave Acoustics*     *Optical Physics*<br>*Electrophysics*     *Integrated Optics* |
| Special Features | **rapid** publication (3–4 months)<br>**no** page charge for **concise** reports<br>prepublication of titles and abstracts<br>**microfiche** edition available as well |
| Languages | Mostly English |
| Articles | original reports, and short communications<br>review and/or tutorial papers |
| Manuscripts | to Springer-Verlag (Attn. H. Lotsch), P.O. Box 105280<br>D-69 Heidelberg 1, F.R. Germany |

Place North-American orders with:
Springer-Verlag New York Inc., 175 Fifth Avenue, New York. N.Y. 10010, USA

# Springer-Verlag
# Berlin Heidelberg New York

S. Haken
# Synergetics
An Introduction

Nonequilibrium Phase Transitions and Self-Organization in Physics, Chemistry and Biology

2nd enlarged edition 1978.
153 figures. Approx. 360 pages
ISBN 3-540-08866-0

*Contents:*
Goal. – Probability. – Information. – Chance. – Necessity. – Chance and Necessity. – Self-Organization. – Physical Systems. – Chemical and Biochemical Systems. – Applications to Biology. – Sociology: A Stochastic Model for the Formation of Public Opinion. – Chaos. – Some Historical Remarks and Outlook.

M. Holt
# Numerical Methods in Fluid Dynamics
1977. 107 figures, 2 tables.
VIII, 253 pages
(Springer Series in Computational Physics)
ISBN 3-540-07907-6

*Contents:*
General Introduction. Brief Review of Concepts of Numerical Analysis. – The Godunov Schemes. – The BVLR Method. – The Method of Characteristics for Three-Dimensional Problems in Gas Dynamics. – The Method of Integral Relations. – Telenin's Method and the Method of Lines.

Springer-Verlag
Berlin
Heidelberg
New York

# Nonlinear Problems in the Physical Sciences and Biology

Proceedings of a Battelle Summer Institute, Seattle, July 3–28, 1972

Editors: I. Stakgold, D. D. Joseph, D. H. Sattinger

1973. 63 figures. VIII, 357 pages
(Lecture Notes in Mathematics, Volume 322)
ISBN 3-540-06251-3

*Contents:*
Lyapunov methods and equations of parabolic type. – Multiple solutions of nonlinear partial differential equations. – Fading memory and functional-differential equations. – Singular perturbation by a quasilinear operator. – Remarks on branching from multiple eigenvalues. – Asymptotic analysis of a class of nonlinear integral equations. – Remarks about bifurcation and stability of quasi-periodic solutions which bifurcate from periodic solutions of the Navier-Stokes equations. – Bifurcation of periodic solutions into invariant tori: the work of Ruelle and Takens. – Ergodic theory and statistical mechanics of non-equilibrium processes. – Mathematical problems in theoretical biology. – On predator-prey equations simulating an immune response. – Bifurcation theory for gradient systems. – Six lectures on the transition to instability. – Groundwater flow as a singular perturbation problem and remarks about numerical methods. – Nonlinear problems in nuclear reactor analysis. – Some nonlinear problems in statistical mechanics and biology. – Stability properties and period behavior of controlled biochemical systems.